T0202217

Islands and Snakes

Islands and Snakes

Volume II: Diversity and Conservation

Edited by

Harvey B. Lillywhite and Marcio Martins

OXFORD
UNIVERSITY PRESS

OXFORD
UNIVERSITY PRESS

Oxford University Press is a department of the University of Oxford. It furthers
the University's objective of excellence in research, scholarship, and education
by publishing worldwide. Oxford is a registered trade mark of Oxford University
Press in the UK and certain other countries.

Published in the United States of America by Oxford University Press
198 Madison Avenue, New York, NY 10016, United States of America.

© Oxford University Press 2023

CIP data is on file at the Library of Congress

ISBN 978–0–19–764152–1

DOI: 10.1093/oso/9780197641521.001.0001

Printed by Integrated Books International, United States of America

We dedicate this work to numerous persons who were influential in our careers and inspired a fascination for many aspects of biodiversity, creativity, and the importance of place, including islands. In particular, we owe a debt of gratitude to Rodolfo Ruibal, George Bartholomew, Harold Heatwole, Henry Fitch, Brian McNab, Ivan Sazima, and Harry Greene. We also express gratitude to contemporary colleagues, too numerous to name, and the many students who have been part of our journey of scientific exploration. We also express sincere and ongoing love and gratitude for our respective families.

—Harvey B. Lillywhite
—Marcio Martins

Contents

Preface

In 2019, we published the first volume *Islands and Snakes: Isolation and Adaptive Evolution*. Our overall aim in collecting and publishing contributions to that volume was to illustrate how geographic isolation has enabled rapid, meaningful, and often dramatic evolutionary changes in the vertebrate inhabitants of the world's islands, with a focus on snakes that successfully inhabit many of these places. Snakes have been very successful colonizers of islands where, quite often, they have replaced endothermic vertebrates as top carnivores and are key elements in food webs.

In this second volume of *Islands and Snakes*, we have shifted focus somewhat to emphasize the diversity of insular snakes and to point out the need for conservation of the numerous and remarkable insular species. Aspects of insular ecology and evolution necessarily continue as a thread or patchwork of discussion in this second volume, as in the first. However, the new emphasis on diversity and conservation includes some important topics that were not addressed in the first volume, for example, reproductive biology; biogeography, historical taxonomy, and discoveries of insular species; snake diversity on large islands; threats and collapse of insular systems; and invasive species impacting islands. This second volume also includes descriptions of natural history and diversity of snakes on important islands not previously given much popular attention.

Snakes are present or often dominant on islands because of speciation *in situ*, invasion from the sea, and various immigration events involving dispersal or human introductions. Insular snakes are important and interesting for several reasons beyond the significance of islands and insular fauna as field sites for ecological analysis. First, the abundance of insular snakes can be extraordinary and offers opportunities to observe snakes and their behaviors up close, often with the added benefit of "insular tameness." Second, various snake species on islands exhibit novel or unusual adaptations of form and function (*e.g.*, rattle-less rattlesnakes, gigantic and melanistic tiger snakes, and pair-bonding of insular cottonmouths that also express other attributes of unusual social behavior). Third, insular snakes have associated natural histories that provide the world with examples of spectacular natural phenomena related to displays of aggregations, movements, feeding, or other interactions that might be observable to island visitors.

Following the first volume, we are committed to providing a discussion of insular snakes that heightens the awareness of conservation issues and hopefully promotes interest and mitigation actions to preserve these wonderful creatures and the islands on which they live. As with all biodiversity, insular snakes are important in several well-known contexts including evolutionary history, scientific value, ecosystem services, and enhancing the richness of life's experiences here on Earth. We further emphasize that conservation of insular snakes is important for aesthetic and cultural values as well as the scientific reasons discussed in this and the previous volume.

Acknowledgments

We are grateful to the many persons who have made this book possible, especially our families, the various authors who have contributed thoughtful and stimulating chapters, and the numerous colleagues, reviewers, and others who have encouraged our adventures in science, including numerous visits to exciting islands and ideas for topics for inclusion and coverage in this volume. We also thank the editors and production staff at Oxford University Press for their professional guidance and assistance throughout this project. In the beginning, Jeremy Lewis encouraged us to produce a second volume that would follow and expand on the first volume that we published in 2019. We appreciate his ongoing encouragement during this project. Marcio Martins thanks Fundação de Amparo à Pesquisa do Estado de São Paulo for a research grant (# 2020/12658-4) that helped to support the production of this book. We hope that readers will find the enjoyment and satisfaction of reading this book that we have intended for them to discover.

Contributors

Selma M. Almeida-Santos
Laboratório de Ecologia e Evolução
Instituto Butantan
São Paulo, Brazil

Ligia G. S. Amorim
Laboratório de Ecologia e Evolução
Instituto Butantan
São Paulo, Brazil

Gustavo Arnaud
Centro de Investigaciones Biológicas del
Noroeste
La Paz, Baja California Sur, México

Dragan Arsovski
Macedonian Ecological Society
Skopje
North Macedonia

Fausto E. Barbo
Laboratório de Coleções Zoológicas
Instituto Butantan
São Paulo, Brazil

Xavier Bonnet
Centre d'Etudes Biologiques de Chizé
Centre National de la Recherche
Scientifique—La Rochelle Université
Villiers en Bois, France

Henrique B. Braz
Laboratório de Ecologia e Evolução
Instituto Butantan
São Paulo, Brazil

François Brischoux
Centre d'Etudes Biologiques de Chizé
Centre National de la Recherche
Scientifique—La Rochelle Université
Villiers en Bois, France

Gordon M. Burghardt
Departments of Psychology and Ecology &
Evolutionary Biology
University of Tennessee
Knoxville, Tennessee, USA

S. R. Chandramouli
Department of Ecology and Environmental
Sciences
School of Life Sciences
Pondicherry University
Puducherry, India

Diego F. Cisneros-Heredia
Instituto de Biodiversidad Tropical
IBIOTROP & Galapagos Science Center
Colegio de Ciencias Biológicas y
Ambientales
Universidad San Francisco de Quito USFQ
Quito, Ecuador

Keara L. Clancy
Department of Wildlife Ecology and
Conservation
University of Florida
Gainesville, Florida, USA

Natalie M. Claunch
Department of Wildlife Ecology and
Conservation
University of Florida
Gainesville, Florida, USA

Indraneil Das
Institute of Biodiversity and Environmental
Conservation
Universiti Malaysia Sarawak
Sarawak, Malaysia

Pedro Galán
Grupo de Investigación en Bioloxía
Evolutiva (GIBE)
Departamento de Bioloxía, Facultad de
Ciencias
Universidad de A Coruña
Coruña, Spain

James C. Gillingham
Department of Biology
Central Michigan University
Mt. Pleasant, Michigan, USA

Ana Golubović
Institute of Zoology, Faculty of Biology
University of Belgrade
Belgrade, Serbia

Felipe G. Grazziotin
Laboratório de Coleções Zoológicas
Instituto Butantan
São Paulo, Brazil

Madison E. A. Harman
Department of Wildlife Ecology and
Conservation
University of Florida
Gainesville, Florida, USA

Daniel Haro
Department of Wildlife Ecology and
Conservation
University of Florida
Gainesville, Florida, USA

Arik Hartmann
Department of Biology
University of Florida
Gainesville, Florida, USA

Masami Hasegawa
Department of Biology
Toho University
Chiba, Japan

Kodiak C. Hengstebeck
School of Natural Resources and
Environment
University of Florida
Gainesville, Florida, USA

Robert C. Jadin
Department of Biology and Museum of
Natural History,
University of Wisconsin Stevens Point
Stevens Point, Wisconsin, USA

Michael J. Jowers
Departamento de Zoología
Facultad de Ciencias
Universidad de Granada
Granada, Spain

Diego Juárez-Sánchez
Department of Wildlife Ecology and
Conservation
University of Florida
Gainesville, Florida, USA

Karina N. Kasperoviczus
Laboratório de Ecologia e Evolução
Instituto Butantan
São Paulo, Brazil

Mark A. Krause
Department of Psychology
Southern Oregon University
Ashland, Oregon, USA

Harvey B. Lillywhite
Department of Biology
University of Florida
Gainesville, Florida, USA

Simon Maddock
School of Natural and Environmental
Sciences
Newcastle University
Newcastle, United Kingdom

Otavio A. V. Marques
Laboratório de Ecologia e Evolução
Instituto Butantan
São Paulo, Brazil

Marcio Martins
Department of Ecology
University of São Paulo
São Paulo, Brazil

Rebecca K. McKee
Department of Wildlife Ecology and
Conservation
University of Florida
Gainesville, Florida, USA

Akira Mori
Department of Zoology
Graduate School of Science
Kyoto University
Kyoto, Japan

John C. Murphy
Science & Education Field Museum
Chicago, Illinois, USA

Mark O'Shea
Faculty of Science and Engineering
University of Wolverhampton
Wolverhampton, United Kingdom

John S. Placyk, Jr.
Science Division
Trinity Valley Community College
Athens, Texas, USA

Carolina Reyes-Puig
Instituto de Biodiversidad Tropical (IBIOTROP)
Colegio de Ciencias Biológicas y Ambientales
Universidad San Francisco de Quito
Quito, Ecuador

Christina M. Romagosa
Department of Wildlife Ecology and
Conservation
University of Florida
Gainesville, Florida, USA

Mark R. Sandfoss
US Geological Survey
Fort Collins Science Center, Everglades
National Park
Homestead, Florida, USA

Ricardo J. Sawaya
Centro de Ciências Naturais e Humanas
Universidade Federal do ABC
São Bernardo do Campo, Brazil

Amber Sutton
Department of Wildlife Ecology and
Conservation
University of Florida
Gainesville, Florida, USA

Ljiljana Tomović
Institute of Zoology, Faculty of Biology
University of Belgrade
Belgrade, Serbia

Mariaguadalupe Vilchez
Department of Wildlife Ecology and
Conservation
University of Florida
Gainesville, Florida, USA

John C. Weber
Department of Geology
Grand Valley State University
Allendale, Michigan, USA

1

Snakes of a Continental Island

History and Patterns of Discovery of the Snake Fauna of Borneo to the Start of the Anthropocene

Indraneil Das

Introduction

Its biological diversity "hidden in plain sight" (Poe 1844), the gigantic tropical island of Borneo (Figure 1.1) was known to ancient seafarers as "Land Below the Wind" (Keith 1940), yet not worthy of conquest or exploration (this was before the realization of the value of timber or petroleum). The English traveler Earl (1837) wrote that the north coast was "scarcely known even to the native trader," and it is thus unremarkable that scientific research and explorations of the island were to commence much later compared to the adjacent and smaller islands of Sumatra and Java.

Croizat (1958) described Borneo as a geological composite. As a part of Sundaland, the island is situated on the eastern rim of the Sunda Shelf, a Laurasian continental plate. Pleistocene glaciations saw sea levels drop 120–200 m below current levels (Wang and Wang 1990), both connecting it to the Asian mainland and joining the other islands of the Sundas (Morley and Flenley 1987). Reconstructions of the region's archipelago systems during the period are in Heaney (1991) and Voris (2000). Stretching between 04°S and 07°N and from 109° to 119°E, Borneo is the second largest tropical island in the world (after New Guinea), and it covers a land area of approximately 743,380 km². A major part of the island falls in the Indonesian portion referred to as Kalimantan (area: 539,460 km²), most of the balance within the east Malaysian states of Sarawak (124,450 km²) and Sabah (73,710 km²). Nearly enclosed by Sarawak State is the Sultanate of Negara Brunei Darussalam (5,760 km²). Based on the hosted biota, Ali (2018) classified it as a shelf island, also being geologically contiguous with the Asian mainland, and with a shallow intervening seabed.

This chapter describes four phases in the history of description of the snakes of Borneo, classified according to a temporal timeline that correlated with regional and international sociopolitical events and highlighting notable personalities, species, and localities.

Indraneil Das, *Snakes of a Continental Island* In: *Islands and Snakes*. Edited by: Harvey B. Lillywhite and Marcio Martins, Oxford University Press. © Oxford University Press 2023. DOI: 10.1093/oso/9780197641521.003.0001

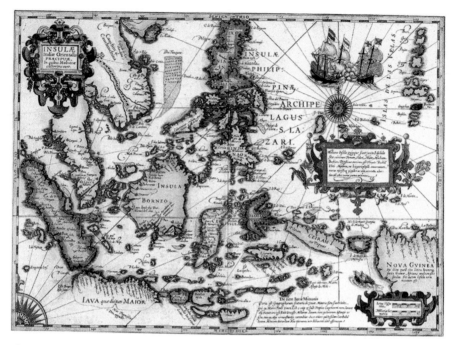

Figure 1.1 An early 17th-century map of Borneo and adjacent regions of southeast Asia by Jodocus Hondius (1563–1612), as reproduced in Gerard Mercator's (1816) atlas.

Phase One: The Age of Linnaeus and Cabinets of Curiosities

> Plants and animals from the extra-European world became what the science studies scholar Bruno Latour has referred to as "Immutable and combinable mobiles," objects that became portable and stable could be compared and combined, allowing for simultaneous study on a global scale.
>
> **—Parsons and Murphy (2012)**

The first snakes known from Borneo do not bear precise localities (*e.g., Coluber pelias*, equivalent to *Chrysopelea pelias* and *Coluber buccatus*, equivalent to *Homalopsis buccatus*, both mentioned as from "*in* Indiis," referring to either the East or West Indian Archipelagos), and are contained in Linnaeus's (1758) *Systema Naturae* (10th edition). The former (holotype of *C. pelias*) originate from Museum De Geer, the private collection of the Dutch industrialist and amateur entomologist, Baron Charles de Geer (1720–1779). The latter (holotype of *C. buccatus*) was donated to Uppsala University by the Councilor of Commerce, Jonas Alstromer, and was formerly in the Museum Adolphi Fredrici, the personal collection of Adolf Fredrik (1710–1771),

King of Sweden (Das 2012). The provenance of the collection remains unclear, and at least some of the ichthyological types were thought to have been obtained during the voyages of ships of the Swedish East India Company along the India–China route, according to Ng and Kottelat (2008).

Two other snakes known from Borneo and with Linnaean names exist: *Coluber laticaudatus* Linnaeus 1758 (currently *Laticauda laticaudata*) and *Anguis platura* Linnaeus 1766 (currently *Hydrophis platurus*); the provenance of the first is given as "*in* Indiis," that of the latter is unspecified in this subsequent work which formed the 12th edition. The first mentioned originate from the Museum Adolphi Federici, the latter from "Mus. Fr. Ziervogel Pharmac," the private collection of Friedrich Ziervogel (1727–1782), Apothecary Royal and in charge of packing the royal collections, who accompanied Linnaeus on his travels (Sandermann Olsen 1997).

For the 23 names of Asian taxa dealt with by Linnaeus, which represent 21 biological species, Das (2012) showed that three localities could have served as the combined geographical source. Apart from the extralimital nature of the first (Sri Lanka), the Malay Peninsula, the island of Java, and their surrounding seas are suspected to have been the source of all four species of snakes now known from Borneo.

A major source of material early descriptions were private collections of rich aristocrats and merchants, sometimes referred to as "cabinets of curiosities," the most famous ones being those accumulated by the Dutch apothecary Albertus Seba (1665–1736). His two-volume (1734–1735) *Locupletissimi Rerum Naturalium Thesauri Accurata et Descriptio, et Iconibus Artificiosissimus Expressio, per Universam Physices Historiam*, published by Janssonio-Waesbergios, J. Wetstenium, and Gul. Smith in Amsterdam, was indeed a source of several Linnean (and other early) names.

The late 18th century thus marked the first period of discovery for Borneo (as for the rest of the world) and resulted in the naming of specimens in private collections using poorly associated geographical data.

Phase Two: More Cabinets and the Rise of European Museums

> [Private natural history collections or cabinets of curiosities were] the centre of literary gatherings, where a fresh and continuous dialogue was established between the expert and the simply curious. This dialogue took place between experts and the wealthy aristocratic owner of the collections.
>
> —Valdecasas *et al.* (2006)

The next phase coincides with the work of early zoologists on cabinets of the rich and famous as well as the efforts of museum curators based in Europe, primarily the German philologist and zoologist Johann Gottlob Theaenus Schneider (1750–1822) and, secondarily, English botanist and zoologist George (sometimes written as "Georgius") Kearsley Shaw (1751–1813) between the end of the 18th and the first decade of the 19th centuries. Global in scope (see Bauer and Lavilla 2022), their work described numerous species, and, relevant for Borneo were the several familiar ones, in the case of Schneider.

His two-volume work, *Historiae Amphibiorum naturalis et literariae*, contained the description of five (three in volume one, 1799; two in volume two, 1801) snake species that occur on Borneo. These were based on Schneider's study of material in private holdings as well as literature.

Species on Borneo in the said works and their type localities are listed below, although there is no evidence that any were actually collected from the island. The original orthography is retained here.

1. *Hydrus Colubrinus* Schneider 1799; currently *Laticauda colubrina* (Hydrophiidae). The type locality of this nominal species was not specified in the original description, although the data associated with the holotype (ZMB 9078) indicate that it is from "ostindisches Meer" east Indian seas), according to Bauer (1998). It was originally examined at the Lampe collection (see below).

2. *Hydrus Granulatus* Schneider 1799; currently *Acrochordus granulatus* (Acrochordidae). The holotype was originally in the Lampe Collection and is at present untraced and presumably lost (*fide* McDowell 1979). The type locality was given as "Indici" (from India) and was restricted by David and Vogel (1996: 44) to "Madras, India" (Chennai, in southeastern India). However, the presumed area of activity of its collector, Christoph Samuel John (1747–1813), a Danish missionary based at the Protestant mission who sent material to a fellow member of the Gesselschaft für Naturforschender Freunde zu Berlin, Marcus (Markus) Elieser (Elisar) Bloch (1723–1799), was Tranquebar (currently, Tharangambadi, a town in Mayiladuthurai District, Tamil Nadu State, more than 250 km south of Chennai), on the Coromandel coast (see Das 2004a).

3. *Hydrus Enhydris* Schneider 1799, currently *Enhydris enhydris* (Homalopsidae). This was based on the description and color plates in Russell (1796: 35; Pl. XXX) of a specimen from "Ankapilly Lake" (Anakapalle, East Godavari District, Andhra Pradesh, southeastern India). The type locality is mentioned as "Indiae orientalis" in the original description. Patrick Russell (1726–1805), surgeon and polymath-naturalist (Das 2015), collected snakes and prepared two folio volumes on the fauna of the southeast coast of India, employing vernacular names to refer to species.

4. *Boa Reticulata* Schneider 1801, currently *Malayopython reticulatus* (Pythonidae). The species was based on specimens in the Göttingen Museum (at present not extant). Thus, the species is not just based on color plates in Seba (1734: Pl. lxii; 1735: Pl. lxxix), as generally mentioned (*e.g.*, David and Vogel 1996: 41; McDiarmid *et al.* 1999: 179). The type locality is mentioned as "Orient"

(Seba 1734: Pl. 62), "mountains of Japan" (Seba 1735; Pl. 79; in error), and "Nova Hispania" (Seba 1735; Pl. 80; in error) and emended to "Java" (in the Greater Sundas, Indonesia) by Brongersma (1972).

5. *Pseudoboa Fasciata* Schneider 1801, currently *Bungarus fasciatus* (Elapidae); ZMB 2771–72 (syntypes; fide Bauer 1998). The species is based on additional material in the collection of Linck (formerly at Leipzig; see A. M. Bauer and Wahlgren 2013) as well as on the description and color plates in Russell (1796), whose specimen was from "Mansoor Cottah, Bengal" (at present Gopalpur-on-Sea, c. 24 km south of Ganjam, Odisha State, southeastern India). The type locality is not specified in the original description, although the ZMB catalog indicates that the provenance of their syntypes is "Indien" (India): these were originally in Musei Blochiani.

A few words on three of the most important private collections may be pertinent here. The first, Musei Blochiani, was the private collection of Marcus Elieser Bloch (1723–1799), a German medical doctor, zoologist, and publisher based in Berlin. Between 1782 and 1795, he published *Allgemeine Naturgeschichte der Fische*, a 12-volume work on fishes that is considered encyclopedic in scope (Hirschberg 1913). His natural history cabinet was enriched through donations from his correspondents in Asia. Of the famous Lampe Collection, rather little is known of Johann Bodo Lampe (1738–1802) except for his appointment as Leibchirurgus (personal surgeon of the monarch; see Goldmann *et al.* 2021) in Hannover. His collections included zoological (Bauer and Lavilla 2022) and human specimens that were presumably of relevance for anatomical studies particularly related to surgery (Goldmann *et al.* 2021). Finally, the Linck Collection was amassed by Heinrich Linck (1638–1717) and his grandson, Johann Heinrich Linck the Younger (1734–1807), the latter an acquaintance of Schneider (Bauer and Lavilla 2022); see Bauer and Wahlgren (2013) for a history of the collection.

In contrast, all Bornean species named by George Shaw (1751–1813) in his 1802 work are members of the genus *Hydrophis* (Hydrophiidae) and based on vouchered specimens that are extant at the Natural History Museum, London. Those occurring on Borneo are enumerated below (although, as in Schneider's case, there is no evidence of a Bornean provenance). Shaw was a preeminent biologist of his time, served as Keeper of the Natural History Department of the British Museum for 22 long years, and co-founded the Linnean Society of London (Adler 1989). While Shaw was neither a collector nor traveler, the Museum at the time was the recipient of material from the numerous voyages, scientific as well as associated with administration of Britain's colonies worldwide. These provided much material for his *General Zoology or Systematic Natural History*, a series of 16 volumes, of which Shaw prepared the first eight.

1. *Hydrus Gracilis* Shaw 1802, BMNH 1946.1.17.37 (ex-BMNH III.4.1.a; holotype; fide McCarthy 1993: 235), type locality not specified in the original description.

2. *Hydrus Caerulescens* Shaw 1802, BMNH 1946.1.3.90 (ex-BMNH III.6.13.a; holotype; fide McCarthy 1993: 232), "an East-Indian species." Restricted to Vizagapatnam (Vishakhapatnam, Andhra Pradesh, southeastern India) by Cogger *et al.* (1983: 250).

3. *Hydrus Curtus* Shaw 1802, BMNH 1946.1.17.59 (ex-BMNH III.2.1.a; holotype; fide McCarthy 1993: 244), reportedly "An East-Indian species."

4. *Hydrus Spiralis* Shaw 1802, BMNH 1946.1.6.94 (ex-BMNH III.6.10.c; holotype; fide McCarthy 1993: 241), type locality "Indian Ocean" and unspecified in original description.

This second phase of snake discoveries was doubtless important in the naming of species that are familiar today. Nonetheless, locality knowledge continued to remain poor, and specimens are suspected to have been brought in by commercial activities, such as via sale by sailors and apothecaries. The challenges of transporting specimens over water during voyages lasting months and perhaps years must have been formidable (see Parsons and Murphy 2012), and specimens preserved in brandy, wine, or rum were particularly vulnerable to consumption by sailors "unconcerned with the protein content" (Stearns 1952).

Phase Three: The First Explorations

> ... a strong motive for the donors of specimens to the Society's Museum must have been that they were supporting the body which would publish the papers they submitted, or which would gain them an audience for their studies being read at a meeting.... Another category of presents of specimens which the Society would have had difficult in refusing when offered were those from distinguished patrons or men of science, and collections which were known to be important.
>
> —Wheeler (1995)

The third phase, and arguably the golden age of discovery of Borneo's snake fauna, can be seen from the 1820s to the end of that century. The period coincided with numerous sociopolitical events in Europe, from the dawn of the Victorian Era and progress in science and the Industrial Revolution, to the establishment of colonial rule over the region and entrenchment of bureaucracy, the realization of the intrinsic and scholarly value of knowledge, and, importantly, the travel of enlightened naturalists to those far-flung corners of the world. In the case of Borneo, it included, among many others, the traveling naturalists Alfred Russel Wallace (1823–1913), Marquis Giacomo Doria of Genoa (1840–1913), Odoardo Beccari (1843–1920), and John Whitehead (1860–1899), whose purpose were to provide botanical, zoological, and/ or geological specimens for European museums. Another source of collections or observations at the time can be attributed to civil servants of the administration, particularly that of Sarawak State, including Hugh Low (1824–1905), Alfred Hart Everett (1849–1898), Eric Mjöberg (1882–1938), Charles Hose (1863–1929), and Edward Bartlett (1836–1908). Specimens were more carefully acquired, particularly in the case of employees of the Sarawak Museum, and had at least the most basic locality information. Often, enough details of their work stations and travel itineraries remain in archives to permit a resolution of collection localities. Such materials (with some exceptions: Das and Leh 2005 mentioned colonial collections that were retained/returned to the Sarawak Museum) were sent to European collections (chiefly the British Museum of Natural History in London and, in some cases, as by Whitehead, to the Paris Museum). This often intense attention to the local biodiversity and accession of remote regions during the period (Gunung Kinabalu, for instance) resulted in several new species of snakes (including *Paraxenodermus borneensis*, initially allocated to *Stoliczkia*, an Indian genus of the Xenodermatidae). Figure 1.2 illustrates some of these "local" collectors and describers of species, their vitae sometimes

Figure 1.2 Portraits of significant Bornean collectors and authors of early snake descriptions. Top row, left, Odoardo Beccari (1843–1920); center, Hugh Low (1824–1905); right, Eric Mjöberg (1882–1938). Bottom row, left Robert Shelford (1872–1912); center, Alfred Russel Wallace (1823–1913); right, John Whitehead (1860–1899).

unknown outside of their "domiciled" or "adopted" countries (see Das 2004b). These greatly enriched European collections, particularly those in London and Paris, and also those in Berlin and Turin.

Phase Four: Research in the Anthropocene

> Species are going extinct rapidly, while taxonomic catalogues are still incomplete for even the best-known taxa. Intensive fieldwork is finding species so rare and threatened that some become extinct within years of discovery.
>
> **—Lees and Pimm (2015)**

The two world wars had marked influence on species descriptions, with the relevant decades showing no accretion to the fauna. The next six decades, in fact, show little activity in terms of new species discoveries, which picked up only around the end of the 20th century. It would appear that the newly independent nations (in this case, Malaysia, Brunei, and Indonesia) had other priorities and/or less local expertise or visitations by foreign researchers with an interest in snakes. This final phase of snake species discovery is worthy of comment, with the description of a number of species that are endemic to the island. Nearly all of these are the result of either careful morphological analyses (possible through the accumulation of often rare rainforest species in numbers) or the utilization of quantitative methods and/or molecular (particularly DNA) techniques of recognition of cryptic species (or species morphologically similar but genetically unique; see Bickford *et al.* 2007).

Table 1.1 presents the current checklist of the snake fauna of Borneo. Among the changes from the previously published list is the recognition of the family Pseudaspididae for members of the genus *Psammodynastes* (the so-called mock vipers) in southeast Asia and with other representatives in the Ethiopian region (Zaher *et al.* 2019). The lineage was formerly allocated to Lamprophiidae.

Generic changes include recognition of a novel or revived names: *Paraxenodermus* to accommodate *Stoliczkia borneensis* Boulenger 1899; *Malayotyphlops* for *Typhlops koekkoeki* Brongersma 1934; *Malayopython* for *Python reticulatus* Schneider 1801; *Craspedocephalus* for *Trimeresurus borneensis* Peters 1872; *Miralia* for *Enhydris alternans* Reuss 1834; *Phytolopsis* for *Enhydris punctatus* Gray 1849; *Gonyosoma* for *Gonyophis margaritatum* Peters 1871; and *Hebius* for all Bornean species formerly allocated to *Amphiesma* of the Natricidae. Furthermore, *Macropisthodon* was synonymized under the medically important genus *Rhabdophis*, within the family Natricidae, including two species from Borneo, *flaviceps* (Duméril *et al.* 1854) and *rhodomelas* (Boie 1827).

Faunal revisions have been associated with species accretions in general. For instance, the review of the genus *Calamaria* by Inger and Marx (1965) produced eight

Table 1.1 Checklist of the snakes of Borneo (current: November 20, 2022)

Acrochordidae—Wart snakes
Acrochordus granulatus (Schneider 1799)
Acrochordus javanicus Hornstedt 1787

Anomochilidae—Giant blind snakes
Anomochilus leonardi Smith 1940
*Anomochilus monticola Das *et al.* 2008
Anomochilus weberi (van Lidth de Jeude 1890)

Cylindrophiidae—Pipe snakes
*Cylindrophis engkariensis Stuebing 1994
*Cylindrophis lineatus Blanford 1881
Cylindrophis ruffus (Laurenti 1768)

Pythonidae—Pythons
Malayopython reticulatus (Schneider 1801)
*Python breitensteini Steindachner 1881

Xenopeltidae—Sunbeam snakes
Xenopeltis unicolor Reinwardt 1827

Colubridae—Typical snakes and reed snakes
Colubrinae- Typical snakes
Ahaetulla fasciolata (Fischer 1885)
Ahaetulla prasina (Boie 1827)
Boiga cynodon (Boie 1827)
Boiga drapiezii (Boie 1827)
Boiga jaspidea (Duméril *et al.* 1854)
Boiga melanota (Boulenger 1896)
Boiga nigriceps (Günther 1863)
Chrysopelea paradisi Boie 1827
Chrysopelea pelias (Linnaeus 1758)
Coelognathus erythrurus (Duméril *et al.* 1854)
Coelognathus flavolineatus (Schlegel 1837)
Coelognathus radiatus (Boie 1827)
Dendrelaphis caudolineatus (Gray 1834)
Dendrelaphis formosus (Boie 1827)
Dendrelaphis haasi van Rooijen and Vogel 2008
Dendrelaphis kopsteini Vogel and van Rooijen 2007
Dendrelaphis pictus (Gmelin 1789)
Dendrelaphis striatus (Cohn 1905)
Dryophiops rubescens (Gray 1835)

(continued)

Table 1.1 Continued

Elapoidis fusca Boie 1826
Gongylosoma baliodeirum Boie 1827
Gongylosoma longicauda (Peters 1871)
Gonyosoma margaritatum Peters 1871
Gonyosoma oxycephalum (Boie 1827)
Liopeltis tricolor (Schlegel 1837)
Lycodon albofuscus (Duméril *et al.* 1854)
Lycodon capucinus (Boie 1827)
Lycodon effraenis Cantor 1847
Lycodon subannulatus (Duméril *et al.* 1854)
Lycodon tristrigatus (Günther 1858)
Oligodon annulifer (Boulenger 1893)
*Oligodon everetti Boulenger 1893
Oligodon octolineatus (Schneider 1801)
Oligodon purpurascens (Schlegel 1837)
Oligodon signatus (Günther 1864)
*Oligodon vertebralis Günther 1865
Oreocalamus hanitschi Boulenger 1899
Orthriophis taeniurus (Cope 1861)
Ptyas carinata (Günther 1858)
Ptyas fusca (Günther 1858)
Ptyas korros (Schlegel 1837)
Sibynophis geminatus (Boie 1826)
Sibynophis melanocephalus (Gray 1834)
*Stegonotus borneensis Inger 1967
*Stegonotus caligocephalus Kaiser *et al.* 2020
Xenelaphis ellipsifer Boulenger 1900
Xenelaphis hexagonotus (Cantor 1847)

Calamariinae–Reed snakes
*Calamaria battersbyi Inger and Marx 1965
Calamaria bicolor Duméril *et al.* 1854
*Calamaria borneensis Bleeker 1860
*Calamaria everetti Boulenger 1893
*Calamaria grabowskyi Fischer 1885
*Calamaria gracillima (Gunther 1872)
*Calamaria griswoldi Loveridge 1938
*Calamaria hilleniusi Inger and Marx 1965
*Calamaria lateralis Mocquard 1890
Calamaria leucogaster Bleeker 1860
Calamaria lovii Boulenger 1887

Table 1.1 Continued

Calamaria lumbricoidea Boie 1827
*Calamaria lumholtzi Andersson 1923
Calamaria melanota Jan 1862
Calamaria modesta Duméril *et al.* 1854
*Calamaria prakkei van Lidth de Jeude 1893
*Calamaria rebentischi Bleeker 1860
Calamaria schlegeli Duméril *et al.* 1854
*Calamaria schmidti Marx and Inger 1955
Calamaria suluensis Taylor 1922
Calamaria virgulata Boie 1827
*Pseudorabdion albonuchalis (Günther 1896)
*Pseudorabdion collaris (Mocquard 1892)
Pseudorabdion longiceps (Cantor 1847)
*Pseudorabdion saravacense (Shelford 1901)

Natricidae—Water snakes
*Hebius arquus (David and Vogel 2010)
*Hebius flavifrons (Boulenger 1887)
*Hebius frenatus (Dunn 1923)
Hebius petersii (Boulenger 1893)
Hebius saravacensis (Günther 1872)
*Hydrablabes periops (Günther 1872)
*Hydrablabes praefrontalis (Mocquard 1890)
*Opisthotropis typica (Mocquard 1890)
Rhabdophis chrysargos (Schlegel 1837)
Rhabdophis conspicillatus (Günther 1872)
Rhabdophis flaviceps (Duméril *et al.* 1854)
*Rhabdophis murudensis (Smith 1925)
Rhabdophis rhodomelas (Boie 1827)
Xenochrophis maculatus (Edeling 1864)
Xenochrophis trianguligerus (Boie 1827)

Pseudaspididae—Mock vipers
Psammodynastes pictus Günther 1858
Psammodynastes pulverulentus (Boie 1827)

Pseudoxenodontidae—False cobras
*Pseudoxenodon baramensis (Smith 1921)

Elapidae-Cobras and kraits, coral and sea snakes
Elapinae–Elapid snakes
Bungarus fasciatus (Schneider 1801)

(*continued*)

Table 1.1 Continued

Bungarus flaviceps Reinhardt 1843
Calliophis bivirgatus (Boie 1827)
Calliophis intestinalis (Laurenti 1768)
Calliophis nigrotaeniatus (Peters 1863)
Naja sumatrana Müller 1890
Ophiophagus hannah (Cantor 1836)

Hydrophiinae–True sea snakes
Aipysurus eydouxii (Gray 1849)
Hydrophis annandalei (Laidlaw 1901)
Hydrophis anomalus (Schmidt 1852)
Hydrophis atriceps Günther 1864
Hydrophis brooki Günther 1872
Hydrophis caerulescens (Shaw 1802)
Hydrophis curtus (Shaw 1802)
Hydrophis cyanocinctus Daudin 1803
Hydrophis gracilis (Shaw 1802)
Hydrophis jerdonii (Gray 1849)
Hydrophis klossi Boulenger 1912
Hydrophis melanosoma Günther 1864
Hydrophis ornatus (Gray 1842)
Hydrophis platurus (Linnaeus 1766)
Hydrophis schistosus Daudin 1803
*Hydrophis sibauensis Rasmussen *et al.* 2001
Hydrophis spiralis (Shaw 1802)
Hydrophis torquatus Günther 1864
Hydrophis viperinus (Schmidt 1852)

Laticaudinae–Sea kraits
Laticauda colubrina (Schneider 1799)
Laticauda laticaudata (Linnaeus 1758)

Homalopsidae—Puff-faced water snakes
Cerberus schneiderii (Schlegel 1837)
Enhydris enhydris (Schneider 1799)
Fordonia leucobalia (Schlegel 1837)
Gerarda prevostiana (Eydoux and Gervais 1822)
Homalopsis buccata (Linnaeus 1758)
*Homalophis doriae Peters 1871
*Homalophis gyii (Murphy *et al.* 2005)
Hypsiscopus plumbea (Boie 1827)

Table 1.1 Continued

Miralia alternans (Reuss 1834)
Phytolopsis punctata Gray 1849

Pareidae—Slug-eating snakes
Aplopeltura boa (Boie 1828)
*Asthenodipsas borneensis Quah *et al.* 2020
*Asthenodipsas ingeri Quah *et al.* 2021
Asthenodipsas laevis (Boie 1827)
*Asthenodipsas jamilinaisi Quah *et al.* 2019
*Asthenodipsas stuebingi Quah *et al.* 2019
Asthenodipsas vertebralis (Boulenger 1900)
Pareas carinatus Wagler 1830
*Pareas nuchalis (Boulenger 1900)

Viperidae—Vipers and pitvipers
*Craspedocephalus borneensis (Peters 1872)
*Garthius chaseni (Smith 1931)
*Trimeresurus malcolmi Loveridge 1938
Trimeresurus sumatranus (Raffles 1822)
Trimeresurus sabahi Regenass and Kramer 1981
Tropidolaemus subannulatus (Gray 1842)

Xenodermatidae—Strange-skinned snakes
*Paraxenodermus borneensis (Boulenger 1899)
Xenodermus javanicus Reinhardt 1836

Xenophidiidae—Spine-jawed snakes
*Xenophidion acanthognathus Günther and Manthey 1995

Typhlopidae—Blind snakes
Argyrophis muelleri (Schlegel 1839)
Indotyphlops braminus (Daudin 1803)
*Malayotyphlops koekkoeki (Brongersma 1934)
Ramphotyphlops lineatus (Schlegel 1839)
*Ramphotyphlops lorenzi (Werner 1909)
Ramphotyphlops olivaceus (Gray 1845)

Endemic and near-endemic species are marked with an asterisk; parentheses indicate species placement in a genus different from the original allocation (following Article 51.3 of the ICZN).

new species within its range, two of which are from Borneo. Similarly, work on another typically montane lineage, the Pareidae ("Slug-eating snakes") by Quah *et al.*
(2019, 2020) has led to a doubling of the Bornean component of the fauna.

One does not fail to notice the paradox, though, of enhanced knowledge of global
environmental issues, from climate change to rainforest loss and the loss of technical
knowledge associated with species descriptions (*i.e.*, taxonomic procedures) and the
relegation of associated disciplines (such as systematics and evolutionary biology) in
general to second-class science. There seems to be no shortcut solutions to the issue
(see Drew 2011).

Conclusion

Figure 1.3 shows the nature of snake discoveries on Borneo. The distinct temporal
phases of research peaks are clearly visible, both in terms of overall species descriptions and those of species endemic to Borneo.

Endemic lineages on Borneo are worthy of note. These may be classified into those
strictly restricted to the island (true endemics) and a few others (near-endemics) that
co-occur on adjacent land masses (all smaller offshore islands and considered part
of the main island during certain glacial phases, such as Palawan in the Philippines
(*Calamaria everetti*), Pulau Bunju (*Malayotyphlops koekkoeki*), Pulau Miang Besar

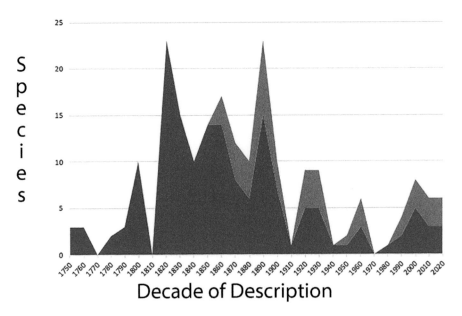

Figure 1.3 Descriptions of snake species known from Borneo, by decade (1760s to
2020s). Dark areas refer to all nominal species, pale areas refer to Bornean endemic and
near-endemic species.

Table 1.2 Snake species of the major islands of the western Sundas and their land areas (in km²)

	Borneo	Sumatra	Java	Bali	Natuna	Palawan
Species	163	160	109	44	20	37
Land area	743,330	473,481	124,413	5,780	2,001	14,650
Source	Table 1.1	David and Vogel 1996	de Lang 2017	Somaweera 2020	Iskandar and Colijn 2001	Griffin 1909

Source for the faunal richness indicated, updated where relevant from Uetz *et al.* (2022).

(*Ramphotyphlops lorenzi*), and the Natunas (*Lycodon tristrigatus*), the last three localities in Indonesia. Of the 163 currently recognized Bornean species, just 46 (or 28%) are restricted to the island system. Borneo's geologically recent (ca. 20,000 years before present, during the Last Glacial Maximum, or the last phase of the Pleistocene; Hanebuth *et al.* 2009) connection to the Asian mainland appears to be a factor influencing the low endemicity in its biota. Indeed, an examination of endemics demonstrate that a majority are restricted to mountain tops (such as all species of *Hydrablabes* and *Opisthotropis*, a majority of species of the genus *Calamaria* and of the Pareidae). Other endemic snake species are residents of special habitats, including inland waters (*Hydrophis sibauensis* and at least two species each of *Hebius* and *Homalopsis*), karst and limestone regions (*Cylindrophis lineatus*), beach forests (*Pseudorabdion sarawacense*), or small, remote islands (*Malayotyphlops koekkoeki*).

As reflected in the title of this chapter, the island of Borneo's low endemicity and similarity to the Malay Peninsula and other islands of Sundaland is somewhat unsurprising in the light of discoveries in botanical research. Employing species distribution models, Raes *et al.* (2009) demonstrated richness and endemicity within specific regions within the island that are characterized by a relatively small range in annual temperature but with seasonality in temperatures within that range least affected by El Niño Southern Oscillation drought events and a number of other, local factors.

Borneo, the largest island of Sundaland, predictably has the greatest snake species richness (Table 1.2; Figure 1.4), as can be expected from the species-area power model (Conor and McCoy 2013). Nonetheless, Sumatra comes a close second, especially with the addition of the faunas of the adjacent Mentawai Islands which shows significant relictual elements in its vertebrate fauna (Wilting *et al.* 2012). In O'Shea and Maddock's (Chapter 2, this volume) comparison of the snake faunas of New Guinea, another large tropical island (785,753 km²) and Borneo, the former has fewer families (nine) compared to Borneo (18, or 16 using the taxonomy followed in this work), although the total species counts of non-marine species are about the same (138 for New Guinea, 140 for Borneo).

A few remarks on the ecological distribution of the snake fauna of Borneo may be relevant. Apart from the marine/coastal obligates, including members of the families

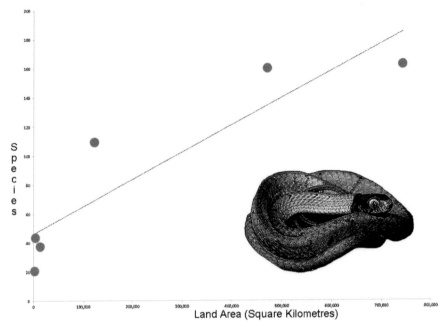

Figure 1.4 Species-area (in km²) relationships of snake faunas of the major islands (and associated satellite archipelagos) of the western Sundas (R² = 0.8259).
Photograph of *Pareas nuchalis* by Indraneil Das.

Hydrophiidae (excluding *H. sibauensis* that reaches in central Borneo, hundreds of kilometers upriver) and representatives of the Acrochordidae (*Acrochordus javanicus*) and the Homalopsidae (*Cerberus schneiderii, Fordonia leucobalia* and *Gerarda prevostiana*), totaling 24 species, the Bornean snake fauna is largely linked to specific vegetational zones that correspond to elevations.

The greatest diversity (105 species, or 80.2% of the non-marine/coastal fauna) of snakes are linked to lowland rainforests (at elevations up to 1,600 m asl), and a smaller subset (26 species, or 19.8%) are restricted to montane forests (>1,600 m asl). Included in these figures are species with limited ecological data, such as *Hebius arquus*, described in 2010, and *Hydrablabes praefrontalis*, described in 1890, both based on specimens collected over a century ago and assumed to be a lowland and highland species, respectively, based on respective collector itineraries.

A few species defy the broad generalizations followed, such as the swamp-sluggish river–inhabiting *Acrochordus javanicus*, and, furthermore, a few species have wide distributions across elevations, ranging from lowland species to montane limits: *Calamaria schlegelii, Gongylosoma baliodeirum, Python breitensteini*, and *Rhabdophis chrysargos* (and were thus left out of the calculations). For a number of species pairs, a clear ecological replacement is evident: the familiar *Trimeresurus sumatranus* of the lowlands appears replaced at higher elevations by *T. malcolmi*;

Stegonotus borneensis, also a lowland-dwelling taxon, is replaced by the newly described *S. caligocephalus*; and, finally, the widespread lowland *Calliophis intestinalis* by the montane *C. nigrotaeniatus*. A majority of Bornean endemics are unsurprisingly restricted to mountain peaks and ridges, such as a number of species of *Calamaria*, *Anomochilus monticola*, *Rhabdophis murudensis*, *Pseudoxenodon baramensis*, and *Paraxenodermus borneensis*.

Explorations of snake diversity in terms of species descriptions have had a long history in southeast Asia, particularly Sumatra and Java, and Borneo was initially given relatively less importance. The early collectors in Sarawak and Sabah were colonial administrators and European explorers who collected for museums in Europe, chiefly London and Paris. Far less important were Dutch collectors, although the activities of the latter were significant for both Java and Sumatra. The resulting body of knowledge was thus widely distributed in scholarly journals (and not always readily available).

Haile (1958) presented the first comprehensive pan-Bornean snake checklist, in a paper remarkable for also including a key (albeit not dichotomous) to all recognized species. The next important attempts to synthesize the fauna were checklists of the snakes of the island by Stuebing (1991, 1994), resulting in a field guide by Stuebing and Inger (1999) that was revised a decade and half later (Stuebing *et al.* 2014). The volumes make available a remarkable fauna within reach of the general public in terms of appreciation and they form valuable tools for investigations into their biology. Expectedly, the intervening eight years have seen active work on the snake fauna, including not only species descriptions but also changes in the generic (and even familial) allocation of species. Table 1.1 provides the most current checklist, updating species names (with particular emphasis of authority and dates of publication). As noted by Chan and Grismer (2021), systematics dominated the field of reptile research in terms of indexed publications for Malaysia as a whole over the past two decades, and one can foresee the future directions in the field, at least for the decade ahead. Such knowledge is surely the first step toward an integrative knowledge of the biodiversity of Borneo and will facilitate research in the field of conservation biology and promotion of biodiversity knowledge in its entirety.

Acknowledgments

Foremost, I am grateful to Harvey Lillywhite and Marcio Roberto Costa Martins for inviting me to contribute to this important volume and keeping their faith in its delivery, albeit beyond the initial deadline. I thank the Institute of Biodiversity and Environmental Conservation, Universiti Malaysia Sarawak, for support of my research, and the Niche Research Grant Scheme, Ministry of Higher Education, Government of Malaysia (IA010200-0708-0007) for field support. Kraig Adler, Aaron Bauer, Patrick David, David Gower, the late Robert F. Inger, Charles M. U. Leh, Alan Leviton, Colin McCarthy, and Robert B. Stuebing are thanked for publications, images, and information. Finally, I thank Genevieve V. A. Gee for reading a draft of the manuscript.

References

Adler, K. K. 1989. Herpetologists of the past. In K. Adler (ed.), *Contributions to the History of Herpetology*, No. 5. Society for the Study of Amphibians and Reptiles, Contributions to Herpetology, pp. 5–141.

Ali, J. R. 2018. Islands as biological substrates: Continental. *Journal of Biogeography* 45:1003–1018.

Bauer, A. M. 1998. South Asian herpetological specimens of historical note in the Zoological Museum, Berlin. *Hamadryad* 23:133–149.

Bauer, A. M., and E. O. Lavilla. 2022. *J. G. Schneider's Historiae Amphibiorum. Herpetology at the Dawn of the 19th Century*. Society for the Study of Amphibians and Reptiles.

Bauer, A. M., and R. Wahlgren. 2013. On the Linck collection and specimens of snakes figured by Johann Jakob Scheuchzer (1735): The oldest fluid-preserved herpetological collection in the world? *Bonn Zoological Bulletin* 62:220–252.

Bickford, D., D. J. Lohman, N. S. Sodhi, P. K. L. Ng, R. Meier, K. Winker, K. Ingram, and I. Das. 2007. Cryptic species as a window on diversity and conservation. *Trends in Ecology and Evolution* 22:148–155.

Brongersma, L. D. 1972. On the "Histoire Naturelle de Serpens" by de La Cépède, 1789 and 1790, with a request to reject this work as a whole, and with proposals to place seven names of snakes, being nomina oblita, on the Official Index of Rejected and Invalid Names in Zoology, and to place three names of snakes on the Official List of Specific Names in Zoology (Class Reptilia). *Bulletin of Zoological Nomenclature* 29:44–61.

Chan, K. O., and L. L. Grismer. 2021. A review of reptile research in Malaysia in the 21st century. *Raffles Bulletin of Zoology* 69:364–376.

Cogger, H. G., E. E. Cameron, and H. M. Cogger. 1983. *Zoological Catalogue of Australia. Volume 1. Amphibia and Reptilia.* Australian Government Publications Service.

Conor, E. F., and E. D. McCoy. 2013. Species–area relationships. In S. A. Levin (ed.), *Encyclopedia of Biodiversity* (2nd ed.). Academic Press, pp. 640–650.

Croizat, L. 1958. *Panbiogeography, or An Introductory Synthesis of Zoogeography, Phytogeography and Geology* (vols. 1, 2a, and 2b). Caracas (Privately published).

Das, I. 2004a. Herpetology of an antique land: The history of herpetological explorations and knowledge in India and south Asia. *Herpetological Expeditions and Voyages. Bonner Zoologische Beiträge* 52(2):215–229.

Das, I. 2004b. Collecting in the "Land below the Wind": Herpetological explorations of Borneo. *Herpetological Expeditions and Voyages. Bonner Zoologische Beiträge* 52:231–243.

Das, I. 2012. Looking east: Carolus Linnaeus and his herpetological species from Asia. *Bibliotheca Herpetologica* 9:104–114.

Das, I. 2015. Patrick Russell (1726–1805), surgeon and polymath-naturalist. *Hamadryad* 37:1–11.

Das, I., and C. Leh. 2005. A legacy of Wallace: Sarawak Museum and the history of herpetological research in Sarawak. In A. A. Tuen, and I. Das (eds.), *Wallace in Sarawak—150 Years Later: Proceedings of an International Conference on Biogeography and Biodiversity*. Institute of Biodiversity and Environmental Conservation, University Malaysia Sarawak, pp. 57–65.

David, P., and G. Vogel. 1996. *The Snakes of Sumatra. An Annotated Checklist and Key with Natural History Notes*. Edition Chimaira.

de Lang, R. 2017. *The Snakes of Java, Bali and Surrounding Islands*. Edition Chimaira.

Drew, L. 2011. Are we losing the science of taxonomy? As need grows, numbers and training are failing to keep up. *BioScience* 61:942–946.

Earl, G. W. 1837. *The Eastern Seas, or Voyages and Adventures in 1832–33–34, Comprising a Tour of the Island of Java– Visits to Borneo, the Malay Peninsula, Siam & c., Also an Account of the Present State of Singapore, with Observations on the Commercial Resources of the Archipelago.* Wm H. Allen and Co.

Goldmann, T. S., M. Scholz, and F. Dross. 2021. Skulls on the road: Historical traces of anatomical connections between Erlangen and Tartu/Dorpat. *Papers on Anthropology* 30:116–127.

Griffin, L. E. 1909. A list of snakes found in Palawan. *Philippine Journal of Science* 4:595–601.

Haile, N. S. 1958. The snakes of Borneo, with a key to the species. *Sarawak Museum Journal,* New Series 8:743–771.

Hanebuth, T. J. J., K. Stattegger, and A. Bojanowski. 2009. Termination of the Last Glacial Maximum sea-level lowstand: The Sunda sea- level record revisited. *Records of Quaternary Sea-level Changes: Global and Planetary Change* 66:76–84.

Heaney, L. R. 1991. A synopsis of climatic and vegetational change in southeast Asia. *Climate Change* 19:53–61.

Hirschberg, J. 1913. Marcus Elieser Bloch (1723–1799). *'Deutsche Medizinische' Wochenschrift* 39(19):900.

Inger, R. F., and H. Marx. 1965. The systematics and evolution of the Oriental colubrid snakes of the genus *Calamaria*. *Fieldiana Zoology* 49:1–304.

Iskandar, D. T., and E. Colijn. 2001. *A Checklist of Southeast Asian and New Guinean Reptiles. Part I: Serpentes.* Japan International Cooperation Agency; The Ministry of Forestry, The Gibbon Foundation and Institut of Technology.

Keith, A. N. 1940. *Land Below the Wind.* Little Brown and Company.

Lees, A. C., and S. L. Pimm. 2015. Species, extinct before we know them? *Current Biology* 25:177–180.

Linnaeus, C. 1758. *Systema Naturae per Regna Tria Naturae, Secundum Classes, Ordines, Genera, Species, cum Characteribus, Differentiis, Synonymis, Locis. Tomus I. Editio Decima, Reformata.* Laurentii Salvii.

Linnaeus, C. 1766. *Systema Naturae per Regna Tria Naturae, Secundum Classes, Ordines, Genera, Species, cum Characteribus, Differentiis, Synonymis, Locis. Tomus I. Editio Duodecima, Reformata.* Laurentii Salvii.

McCarthy, C. 1993. *Hydrophis* Latreille, 1801. In P. Golay, H. M. Smith, D. G. Broadley, J. R. Dixon, C. McCarthy, J.-C. Rage, B. Schätti, and M. Toriba (eds.), *Endoglyphs and Other Major Venomous Snakes of the World: A Checklist.* Azemiops S. A. Herpetological Data Centre, pp. 229–242.

McDiarmid, R. W., J. A. Campbell, and T.'S. A. Touré. 1999. *Snake Species of the World. A Taxonomic and Geographic Reference.* volume 1. Herpetologists' League.

McDowell, S. B. 1979. A catalogue of the snakes of New Guinea and the Solomons, with special reference to those of the Bernice P. Bishop Museums. Part III. Boinae and Acrochordoidea (Reptilia, Serpentes). *Journal of Herpetology* 13:1–92.

Morley, R. J., and J. R. Flenley. 1987. Late Cenozoic vegetational and environmental changes in the Malay Archipelago. In T. C. Whitmore (ed.), *Biogeographic Evolution of the Malay Archipelago.* Oxford Monographs on Biogeography. Clarendon Press, pp. 50–59.

Ng, H. H., and M. Kottelat. 2008. The identity of *Clarias batrachus* (Linnaeus, 1758), with the designation of a neotype (Teleostei: Clariidae). *Zoological Journal of the Linnaean Society* 153:725–732.

Parsons, C. M., and K. S. Murphy. 2012. Ecosystems under sail: Specimen transport in the eighteenth-century French and British Atlantics. *Early American Studies. An Interdisciplinary Journal* 10:503–529.

Poe, E. A. 1844. The purloined letter. In *The Gift: A Christmas, New Year, and Birthday Present, 1845.* Carey & Hart, pp. 41–61.

Quah, E. S. H., L. L. Grismer, K. K. P. Lim, A. M. S. S. Anuar, and P. Y. Imbun. 2019. A taxonomic reappraisal of the Smooth Slug Snake *Asthenodipsas laevis* (Boie, 1827) (Squamata: Pareidae) in Borneo with the description of two new species. *Zootaxa* 4646:501–526.

Quah, E. S. H., L. L. Grismer, K. K. P. Lim, M. S. S. Anuar, and K. O. Chan. 2020. A taxonomic revision of *Asthenodipsas malaccana* Peters, 1864 (Squamata: Pareidae) with a description of a new species from Borneo. *Zootaxa* 4729:1–24.

Raes, N., M. C. Roos, J. W. F. Slik, E. E. van Loon, and H. ter Steege. 2009. Botanical richness and endemicity patterns of Borneo derived from species distribution models. *Ecography* 32:180–192.

Russell, P. 1796. *An Account of Indian Serpents Collected on the Coast of Coromandel; Containing Descriptions and Drawings of Each Species; Together with Experiments and Remarks on their Several Poisons.* George Nicol.

Sandermann Olsen, S.-E. 1997. Bibliographia discipuli Linnæi. Bibliographies of the 331 pupils of Linnaeus. *Bibliotheca Linnæana Danica, Copenhagen* (9):1–458.

Schneider, J. G. T. 1799. *Historiae Amphibiorum naturalis et literariae. Fasciculcus Primus continens Ranas, Calamitas, Bufones, Salamandras et Hydros in Genera et Species Descriptos Notisque Suis Distinctos.* Friederici Frommanni.

Schneider, J. G. T. 1801. *Historiae Amphibiorum Naturalis et Literariae. Fasciculcus Secundus continens Crocodilos, Scincos, Chamaesauras, Boas, Pseudoboas, Elapes, Angues, Amphisbaenas et Caecilias.* Friederici Frommann.

Seba, A. 1734. *Locupletissimi Rerum Naturalium Thesauri Accurata et Descriptio, et Iconibus Artificiosissimus Expressio, per Universam Physices Historiam.* Vol. I. Janssonio-Waesbergios, & J. Wetstenium, & Gul. Smith.

Seba, A. 1735. *Locupletissimi Rerum Naturalium Thesauri Accurata et Descriptio, et Iconibus Artificiosissimus Expressio, per Universam Physices Historiam.* Vol. II. Janssonio-Waesbergios, & J. Wetstenium, & Gul. Smith.

Shaw, G. 1802. *General Zoology or Systematic Natural History.* Volume 3. Part. 2. Amphibia. G. Kearsley.

Somaweera, R. 2020. *A Naturalist's Guide to the Reptiles and Amphibians of Bali* (2nd ed). John Beaufoy.

Stearns, R. P. 1952. James Petiver: Promoter of natural sciences. c. 1663–1718. *Proceedings of the American Antiquarian Society* 62:243–365.

Stuebing, R. B. 1991. A checklist of the snakes of Borneo. *Raffles Bulletin of Zoology* 39:323–362.

Stuebing, R. B. 1994. A checklist of the snakes of Borneo: Addenda and corrigenda. *Raffles Bulletin of Zoology* 42:931–936.

Stuebing, R. B., and R. F. Inger. 1999. *A Field Guide to the Snakes of Borneo.* Natural History Publications (Borneo) Sdn Bhd.

Stuebing, R. B., R. F. Inger, and B. Lardner. 2014. *A Field Guide to the Snakes of Borneo.* Second Edition. Natural History Publications (Borneo) Sdn Bhd.

Uetz, P., P. Freed, R. Aguilar, F. Reyes, and J. Hošek (eds.). 2022. The reptile database. Accessed February 4, 2023. http://www.reptile-database.org.

Valdecasas, A. G., V. Correia, and A. M. Correas. 2006. Museums at the crossroad: Contributing to dialogue and wonder in natural history museums. *Museum Management and Curatorship* 21:32–43.

Voris, H. K. 2000. Maps of Pleistocene sea levels in Southeast Asia: Shorelines, river systems and time durations. *Journal of Biogeography* 27:1153–1167.

Wang, L.-J., and P.-X. Wang. 1990. Late Quaternary paleoceanography of the South China Sea: Glacial-interglacial contrasts on an enclosed basin. *Paleoceanography* 5:77–90.

Wheeler, A. 1995. Zoological collections in the British Museum: The Linnean Society's Museum. *Archives of Natural History* 22:235–254.

Wilting, A., R. Sollmann, E. Meijaard, K. M. Helgen, and J. Fickel. 2012. Mentawai's endemic, relictual fauna: Is it evidence for Pleistocene extinctions on Sumatra? *Journal of Biogeography* 39:1608–1620.

Zaher, H., R. W. Murphy, J. C. Arredondo, R. Graboski, P. R. Machado-Filho, K. Mahlow, G. G. Montingelli, *et al.* 2019. Large-scale molecular phylogeny, morphology, divergence-time estimation, and the fossil record of advanced caenophidian snakes (Squamata: Serpentes). *PLoS One* 14:e0216148.

2

Serpents of Paradise

Biogeography of the Snake Fauna on New Guinean Islands

Mark O'Shea and Simon Maddock

Introduction

New Guinea is the world's largest tropical island (785,753 km^2), yet its herpetofauna has received substantially less attention than that of Borneo, the second largest tropical island (748,168 km^2). Of the 43 families of snakes currently recognized globally, only nine are present in New Guinea, whereas Borneo has 18. However, the absence of certain taxonomic groups does not equate to the absence of their niches but rather offers the opportunity for diversification of those taxonomic groups that are present. The niche of a short, stocky, nocturnal, sedentary sit-and-wait, venomous ambush predator with vertically elliptical pupils that uses caudal luring to attract vertebrate prey is occupied elsewhere by viperids, but in Australasia death adders (Elapidae: *Acanthophis*) occupy this niche, in a classic case of evolutionary convergence. The niche of a highly alert, visually oriented, fast-moving, active diurnal lizard or mammal predator, represented in the rest of the world by nonvenomous racers or whipsnakes (Colubridae), rear-fanged venomous sandsnakes, or Montpellier snakes (Psammophiidae) is occupied by the Australo-Papuan elapids, such as whipsnakes (*Demansia*), brownsnakes (*Pseudonaja*), and taipans (*Oxyuranus*) in Australasia. And similarly, the terrestrial or semi-fossorial, vermivorous (worm-eater) niche of *Calamaria* (Colubridae) in the Oriental region, or *Atractus* and *Geophis* (Dipsadidae) in the Neotropics, is occupied in New Guinea by the highly speciated elapid genus *Toxicocalamus*. Curiously the "goo-eater" (gastropod mollusc-eater) niche inhabited by *Dipsas, Sibon,* and *Tropidodipsas* (Dipsadidae) in Latin American rainforests; *Duberria* (Pseudoxyrhophiidae) in Africa; and *Aplopeltura, Asthenodipsas,* and *Pareas* (Pareidae) in tropical Asian forests, remains unoccupied in New Guinea despite the apparent abundance of small gastropod molluscs in the region (Slapcinsky 2005, 2006; Slapcinsky and Lasley 2007).

In this chapter, we provide an overview of the non-marine snakes inhabiting the satellite islands and archipelagos in the New Guinea region and investigate whether their biogeographic relationships are of Oriental, Pacific, or Australian origin and whether the islands fit the island species–area relationship.

Mark O'Shea and Simon Maddock, *Serpents of Paradise* In: *Islands and Snakes.* Edited by: Harvey B. Lillywhite and Marcio Martins, Oxford University Press. © Oxford University Press 2023. DOI: 10.1093/oso/9780197641521.003.0002

We define the New Guinea region as the sovereign country of Papua New Guinea (PNG), including its archipelagos to the east, and Western New Guinea (WNG) comprising six Indonesian provinces and their associated archipelagos and islands. This is more of a political rather than biogeographical definition because it includes the islands of Bougainville and Buka, biogeographically part of the Solomon Islands, and excludes the Aru Islands (which are part of the Moluccan islands of Indonesia but lie on the Sahul Shelf) and the northern Torres Strait islands (which are part of Australia). As such, our definition of New Guinea ranges across a distance of more than 3,000 km, from the Boo Islands (1.1757°S 129.3023°E), the western-most island group in the Raja Ampat Islands, to Bougainville (6.3908°S 155.3944°E) in the Autonomous Region of Bougainville, and we have located 96 islands within this region that have been documented to contain at least one native non-marine snake species (Appendix I, 77 in PNG and 19 in WNG). These islands range in size from New Britain (36,520 km^2), Bougainville (9,318 km^2), and New Ireland (7,404 km^2), to 3-hectare Sable Island, to the northeast of Bougainville (see Appendix I).

A Summarized Geological History of New Guinea and Its Satellite Archipelagos

The New Guinea region does not have one single geological origin but is the result of a long and complicated plate tectonic history. Different authorities provide conflicting orders and dates for the tectonic events occurring in the New Guinea region, and it is beyond the scope of this chapter to delve too deeply into the geological past or test conflicting hypotheses. It is sufficient to recognize that New Guinea and its satellite archipelagos experienced an extremely active and diverse volcanic and tectonic past, with component parts of the New Guinea orogen originating from geographically distant regions, as summarized below.

Southern New Guinea

Southern New Guinea lay at the northern edge of the Australian Craton (Figure 2.1), an East Gondwanan landmass also known as Meganesia, which also included Tasmania and the Aru Islands and was located on the shallow, continental Sahul Shelf. A large area of southern New Guinea emerged above sea level during the Eocene and Oligocene (53–24 mya), but this had largely submerged again by the Early Miocene (20 mya), leaving a series of small islands along the leading edge of the Australian Craton. More islands formed by the Middle Miocene (15 mya), until New Guinea existed as a low-lying island arc on the leading edge of the northward-moving Australian Craton (Flannery 1995a).

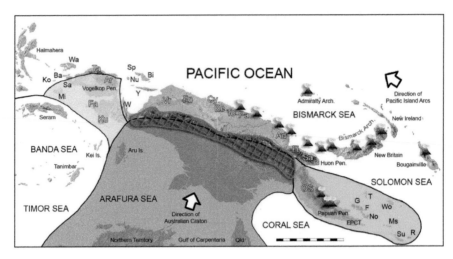

Figure 2.1 A summarized history of the plate tectonics of New Guinea. As the Australian Craton (orange) moved NNE, it collided with the Pacific islands arc terranes as they moved NNW. The collision produced the New Guinea Orogen, the Central Cordillera (orange hatched area). Later the East Papuan Composite Terrane (EPCT, blue) and Vogelkop microcontinent (yellow) collided with and accreted to the craton. Arrows indicate the direction of travel of the landmasses. Mountain ranges resulting from terrane aggregation (white text, west to east) are abbreviated as Ta (Tamrau), Ar (Arfak), Fa (Fakfak), Ku (Kumafa), Vr (Van Rees), Fo (Foja), Cy (Cyclops), Be (Bewani), To (Torricelli), Pa (Prince Alexander), Ad (Adelbert), Fi (Finisterre), Sa (Saruwaged), OS (Owen Stanley). Island abbreviations (black text) include Wa (Waigeo), Ba (Batanta), Sa (Salawati), and Mi (Misool) in the Raja Ampat Archipelago; Sp (Supiori), Bi (Biak), Nu (Numfor), and Y (Yapen) in the Schouten Islands; and G (Goodenough), F (Fergusson), No (Normanby), T (Trobriand), Wo (Woodlark), Ms (Misima), Su (Sudest), and R (Rossel) in Milne Bay. Additionally, the Wandamen Peninsula is abbreviated with the letter W. Volcano symbols indicate volcanic island arcs. Scale = 500 km.

One unusual feature existed on the otherwise east–west aligned leading edge of the craton, a ridge projecting northward located roughly where the border between PNG and WNG occurs today (Polhemus 2007). This ridge was subaerial during the Mesozoic, when much of the leading edge of the craton was submerged (Davies 1990), and it may have acted as a wedge, dividing the Pacific island arc terranes that began to collide with and accrete to the leading edge of the craton from the Late Cretaceous to the Oligocene (Polhemus 2007).

Northern New Guinea

As the Australian Craton moved northward it began to collide with a series of volcanic island arcs. One of the earliest orogenic events occurred when the Rouffer and Sepik Terranes accreted to the Australian Craton to initiate the orogenesis of the Central Cordillera of mainland New Guinea (Figure 2.1) (van Welzen *et al.* 2001). During the Miocene, the craton continued to collide with island arcs leading to further orogenesis along the Central Cordillera. The Coral Sea Basin began to open up in the Early Oligocene (62–56 mya) (Weissel and Watts 1979). The Papuan Peninsula (Bird's Tail Peninsula) was formed when the East Papuan Composite Terrane (EPCT; Figure 2.1) accreted to the Australian Craton in the Late Miocene (10–5 mya) (Pigram and Davies 1987). The Huon Peninsula was created when the Adelbert-Finisterre Terrane docked with New Guinea during the Pliocene (5 mya) (Davies *et al.* 1987; Polhemus 2007). At least 32 terranes collided with the Australian Craton, some of them already composite from earlier oceanic collisions (Pigram and Davies 1987), and the collision and accretion of later terranes with the ever-growing New Guinea Orogen led to the creation of the northern coastal mountain ranges between the Oligocene and the Pliocene (Figure 2.1, white text— east to west): Saruwaged, Finisterre, Adelbert, Prince Alexander, Torricelli, Bewani, Cyclops, Foja (Gauttier), Van Rees, and Arfak Ranges.

Vogelkop (Bird's Head) Peninsula and the Raja Ampat Islands

Since the Middle Jurassic (160 mya), portions of an emergent New Guinea existed as a series of isolated islands to the north of the Australian landmass (Flannery 1995a), with the Vogelkop a microcontinent which had detached from the Australian Craton during the Early Cretaceous (145–100 mya) (Pigram and Davies 1987; Polhemus 1996; Heads 2002), eventually becoming accreted to the New Guinea orogen in the region of the Wandamen Peninsula (Figure 2.1, black text W), during either the Middle Miocene (16–11 mya) or Pliocene (5–2.6 mya) (Polhemus and Polhemus 1998; Polhemus 2007).

Misool and most of Salawati in the Raja Ampat Archipelago (Figure 2.1, black text Mi and Sa) share their origin with the Vogelkop microcontinent, having rifted from the Australian Craton, but Waigao, Batanta, Kofiau (Figure 2.1, black text Wa, Ba and Ko), and the northern strip of Salawati lie on the Halmahera microplate (Figure 2.1), which arrived from the northeast between the Middle Oligocene (30 mya) and the Early Pliocene (5 mya) (Polhemus 2007).

Schouten Islands

The collisions of Pacific terranes with the New Guinea orogen did not always lead to accretion. In some areas, pieces rifted away from the orogen and were carried westward to either reaccrete in a different location (*e.g.*, the Tamrau Range on the Vogelkop, Figure 2.1, white text Ta), remain adrift, or be carried as far west as Sulawesi. The Schouten Islands—Yapen, Biak, Supiori, and Numfor (Figure 2.1., black text Y, Bi,

Sp and Nu)—lie on the 1,500-km Sorong-Yapen fault line. It has been hypothesized that the Schouten Islands originated as terranes accreted to the New Guinea orogen but were subsequently rifted from the landmass and carried westward into what is now Cenderwasih Bay (Figure 2.1) along the Sorong "strike-slip fault" (Polhemus 2007).

The Bismarck, Admiralty, and Solomons Archipelagos

The archipelagos east and northeast of New Guinea, namely the Bismarck, Admiralty, and Solomon archipelagos (Figure 2.1), are volcanic, oceanic island arcs that have yet to collide with mainland New Guinea. New Britain was 500 km away from New Guinea in the Early Pliocene (5 mya) and 300 km distant by the Middle Pliocene (3.5 mya) (Flannery 1995b), while today West New Britain is only 85 km from the Huon Peninsula and a series of islands (Umboi, Tolokiwa, Long, and Crown; Figure 2.2, east to west P18–15) are located in the Dampier and Vitiaz Straits between the two land-masses, providing routes for migration and faunal mixing.

The Milne Bay Archipelagos

The islands of Milne Bay east to southeast of New Guinea comprise the d'Entre-casteaux Archipelago, the Trobriand Islands, Woodlark Island, and the Louisiade Archipelago (Figure 2.2, P28–44). The d'Entrecasteaux Archipelago is linked to the Trobriand Islands by a shallow platform that may have been exposed as land during the Pleistocene. Similarly Woodlark Island lies on the Woodlark Rise while the rest of the Louisiade Archipelago (Misima, Sudest, Rossel) is located on the Pocklington Rise (Baldwin et al. 2012). These islands were previously much closer together be-fore the spreading of the Woodlark Basin between Woodlark and Misima Islands, which began in the Miocene (20–2 mya) (Pigram and Davies 1987; van Welzen et al. 2001). Faunal affinities are with the Papuan Peninsula, as part of the EPCT (Figure 2.1), rather than the Bismarck Archipelago or the Solomon Islands, both of which are separated from the Milne Bay islands by the deep Solomon Sea.

The Arafura Sea and Torres Strait during the Pleistocene

While the Coral Sea, southeast of New Guinea, is very deep, averaging 2,394 m with a maximum depth of 9,140 m below current sea level, the Arafura Sea to the southwest of New Guinea, is shallow (50–80 m) and did not exist for much of the Pleistocene (Figure 2.1). During the Pleistocene glacial maxima (several periods between 2.6 mya and 11,700 years ago), the Sahul Shelf between Australia and New Guinea, including the Aru Islands, would have been above sea level for prolonged periods, permitting the dispersal of faunal elements between the larger landmasses.

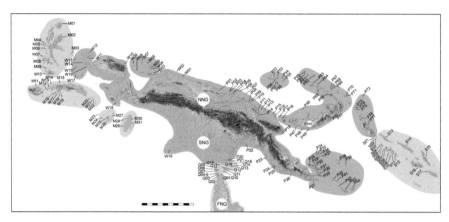

Figure 2.2   Islands with snakes included in this study. Archipelagos included in this study are shaded dark blue, whereas those in light blue are neighboring archipelagos included for comparison. The island of New Guinea is divided politically (as indicated by the yellow line in the center) into Papua New Guinea (PNG) and West New Guinea (WNG), and it is geologically separated by the Central Cordillera into southern New Guinea (SNG) and northern New Guinea (NNG). Islands of PNG are labeled using the letter P and a numeric code as follows: *Sandaun Province*: P01 Seleo, P02 Tarawai, P03 Walis, P04 Kairiru, P05 Muschu, P06 Vokeo, P07 Blup Blup; *Madang Province*: P08 Laing, P09 Boisa, P10 Manam, P11 Sarang, P12 Karkar, P13 Bagabag, P14 Kranket, P15 Crown, P16 Long; *Morobe Province*: P17 Tolokiwa, P18 Umboi, P19 Tami; *Western Province*: P20 Daru, P21 Bobo; *Gulf Province*: P22 Urama; *Central Province/NCD*: P23 Yule, P24 Loloata, P25 Motapure, P26 Abavi; *Milne Bay Province*: P27 Samarai, P28 Goodenough, P29 Wagabu, P30 Fergusson, P31 Dobu, P32 Normanby, P33 Boiaboiawaga, P34 Kula, P35 Kaileuna, P36 Kiriwina, P37 Kitava, P38 Woodlark, P39 Budi Budi Atoll, P40 Panaeati, P41 Misima, P42 Gulewa, P43 Sudest, P44 Rossel; P45 New Britain; *West New Britain Province*: P46 Garove, P47 Arawe, P48 Ganglo, P49 Gasmata; *East New Britain Province*: P50 Mioko, P51 Duke of York; *Manus Province*: P52 Manus, P53 Los Negros, P54 Pak, P55 Rambutyo, P56 Big Ndrova, P57 Lou, P58 Baluan; *New Ireland Province*: P59 Mussau, P60 Boliu, P61 Emirau, P62 Tench, P63 New Hanover, P64 Nusa, P65 Dyaul, P66 New Ireland, P67 Simberli, P68 Tatau, P59 Tabar, P70 Babase, P71 Ambitle, *Autonomous Region of Bougainville*: P72 Sable; P73 Pinipel, P74 Sirot, P75 Nissan Atoll, P76 Buka, P77 Bougainville. Islands of WNG are labeled using the letter W and a numeric code as follows: W01 Djamna, W02 Liki, W03 Yapen, W04 Mios Woendi, W05 Padaido, W06 Biak, W07 Supiori, W08 Mios Num, W09 Numfor; *West Papua Province*: W10 Mansinam, W11 Saonek, W12 Waigeo, W13 Gag, W14 Batanta, W15 Kofiau, W16 Salawati, W17 Misool, W18 Adi; *Papua Province*: W19 Yos Sudarso. Only the Far North of Queensland (FNQ) is of relevance to our discussion, and we label islands using the letter Q and a numeric code as follows: Q01 Eborac, Q02 Great Woody, Q03 Prince of Wales (Muralug), Q04 Thursday, Q05 Hammond, Q06 Horn, Q07 Moa, Q08 Badu, Q09 Jervis (Mabuiag), Q10 Warraber, Q11 Yam, Q12 Gabba, Q13 Murray, Q14 Yorke, Q15 Darnley, Q16 Saibai,

Figure 2.2 Continued

Q17 Dauan, Q18 Boigu. In the Solomon Islands, we use the letter S and a numeric code: *Western Province*: S01 Shortlands, S02 Pirumeri, S03 Fauro, S04 Vella Lavella, S05 Ranggona, S06 Simbo, S07 Gizo, S08 Kolombangara, S09 Rendova, S10 Tetepare, S11 New Georgia, S12 Vangunu; *Choiseul Province*: S13 Choiseul; *Isabel Province*: S14 Isabel; *Malaita Province*: S15 Malaita; *Guadalcanal Province*: S16 Guadacanal; *Makira-Ulawa Province*: S17 Makira. In the Moluccas, the code uses the letter M and a number as follows: *North Maluku Province*: M01 Morotai, M02 Halmahera, M03 Gebe, M04 Ternate, M05 Tidore, M06 Moti, M07 Kasiruta, M08 Bacan, M09 Bisa, M10 Obi; *Maluku Province*: M11 Buru, M12 Manipa, M13 Kelang, M14 Boano, M15 Seram, M16 Ambon, M17 Haruku, M18 Saparua, M19 Nusa Laut, M20 Panjang, M21 Manawoka, M22 Goram, M23 Kur, M24 Taam, M25 Kei Duluh, M26 Kei Kecil, M27 Kei Besar, M28 Wamar, M29 Trangan, M30 Wokam, M31 Kobroor.

The Gulf of Carpentaria was the 30,000 km^2 Lake Carpentaria (Torgersen *et al.* 1988), fed by rivers from southern New Guinea, Queensland, and the Northern Territory, as evidenced by the commonality of freshwater fish in the three regions (Allen 1991; Woinarski *et al.* 2007; Allen *et al.* 2008). The outflow traveled west to run into the deep-water Banda Sea, southwest of the Aru Islands.

The Torres Strait Islands lie between the Cape York Peninsula and southern PNG. The strait, which contains approximately 200 islands, measures only 152 km at its narrowest point, with a depth varying between 7 m in the north to 15 m in the south. Apart from Daru and Bobo, most of the islands in the Torres Strait are Australian territory today, with three of these (Boigu, Dauan, Saibai), lying only 7.6, 10.2, and 3.8 km, respectively, off the coast of southern Western Province, PNG. During the glacial maxima, much of the strait would have been dry land, while in the interglacial periods these islands would have provided a viable route for island-hopping dispersal.

Island Species Relationships

The *Island Species Area Relationship* (ISAR) has become a widespread phenomenon that has been supported by a significant number of studies in ecology and island biogeography (Matthews *et al.* 2021). ISAR, which states that if other abiotic factors are constant then the larger an island's area, the more species occur on it, is so widespread that it is possibly the closest to a "rule" one can find in ecology/biogeography. For this chapter, we obtained snake occurrence data found across New Guinea islands and surrounding archipelagos using museum records, scientific and non-scientific articles, books, and our own fieldwork, totaling 162 islands, plus the New Guinea mainland and Far North Queensland (Figure 2.2; Appendices I and II). Introduced species (*Indotyphlops braminus* and *Lycodon capucinus*), sea kraits (*Laticauda* spp.), and true sea snakes (*Aipysurus* spp., *Emydocephalus annulatus, Hydrelaps darwiniensis, Hydrophis* spp., and *Parahydrophis mertoni*), were omitted from our analyses as non-native or highly vagile, respectively, and essentially unrepresentative of any specific island. Our study therefore included the native snake faunas of 96 islands from New Guinea waters and 66 islands from neighboring Indonesia (Moluccas), Australia, and the Solomon Islands, for comparative purposes. In this section, we test ISAR on islands of the New Guinea region (Figure 2.3). However, we emphasize that

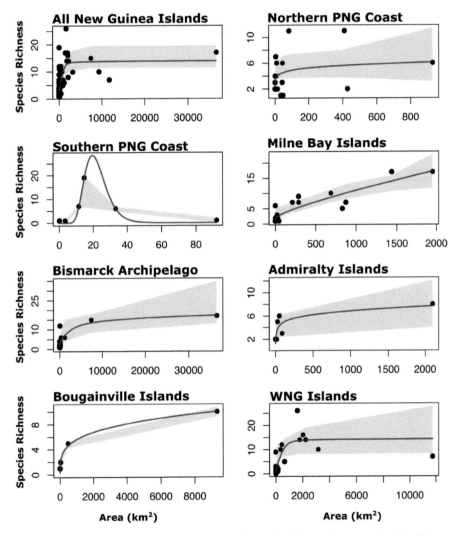

Figure 2.3 Multimodel-averaged ISAR curves for each of the regions examined in this section. Black dots represent islands, red lines are the average curve and grayed out area represents the confidence intervals. The number of islands containing snakes used for each plot: All New Guinea Islands = 96, Northern PNG Coast = 19, Southern PNG Coast = 7, Milne Bay Islands = 18, Bismarck Archipelago = 20, Admirality Islands = 7, Bougainville Islands = 6, WNG Islands = 19.

many islands may have been incompletely surveyed or sampled and figures obtained may be underestimates of their true snake faunas.

We divided the New Guinean islands and neighboring island groups into 15 archipelagos and landmasses—Northern Moluccas, Central Moluccas, Kei Islands, Aru Islands, Raja Ampat Islands, Schouten Islands, Admiralty Islands, Bismarck Archipelago, Solomon Islands, Milne Bay Islands, and Torres Strait Islands, plus New Guinea divided into Northern New Guinea, Southern New Guinea, the Vogelkop, and also Far North Queensland (Figure 2.4)—and compared their serpentine richness at family, genus, and species levels (Figures 2.4–2.6).

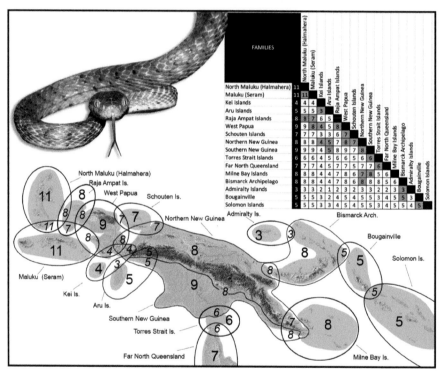

Figure 2.4 Representation of the snake families present in New Guinea, on satellite archipelagos, and neighboring regions of Indonesia, Australia, and the Solomon Islands. Figures in large roman script indicate the number of families present in each mainland block (West Papua, Northern New Guinea, Southern New Guinea) or archipelago ellipsoid, as annotated. The small numbers in italics indicate the number of families common to both areas enclosed by neighboring ellipsoids. Inset table: White figures on black are total families present in each area; figures in grid are the numbers of families common to both areas, figures black on gray indicate neighboring comparisons illustrated on the map. The inset is the Brown Treesnake (*Boiga irregularis*), a common species found in every block and archipelago.
Photograph by Mark O'Shea.

To test ISAR on a subset of the islands surrounding New Guinea, but omitting New Guinea itself, we gathered data for each of the islands that contained at least one species of native terrestrial, freshwater, or brackish snakes in our study region. Island area was obtained from the literature, online databases, or by drawing polygons around the coastline of each island on Google Earth, and then compared with the number of species (species richness hereafter). We tested ISAR using the R package *sars* (Matthews *et al.* 2019). First, we investigated ISAR patterns for all New Guinea islands. Second, we assessed New Guinea's inshore islands as three separate entities: Northern PNG Coast Islands (Seleo, Sandaun Province [P01], to Tami, Morobe Province [P19]); Southern PNG Coast Islands (Daru, Western Province [P20], to Abavi, Central Province [P26]); Western New Guinea (WNG) Coast Islands (Djamna, Papua Province [W01], including the Schouten and Raja Ampat Islands, to Yos Sudarso, Papua Province [W19]). Finally, we tested ISAR for the Milne Bay Islands, Bismarck Archipelago, Admiralty Islands, and Bougainville Islands (including Buka and islands to the north). On the raw data, we used a multimodal approach by attempting to fit all 20 models employed in the *sars* package using the "sar_average" function and averaged all models to create a multimodel-averaged ISAR curve for each of these eight island groups using information criteria weights (Guilhaumon *et al.* 2010). We also fitted the data to log-log power models using the "lin_pow" function.

Most island groups support a positive ISAR, indicating that, as island size increases, so does overall species richness (Figure 2.3, Table 2.1). There is a significant positive correlation in all island groups except for the "southern PNG coast" and "northern PNG coast" groups. On the southern PNG coast islands, Urama Island (92 km²), Gulf Province, PNG, located in a mangrove-dominated river delta, is seemingly the outlier with only the widely distributed Little Filesnake *Acrochordus granulatus* recorded there. In the northern PNG coast islands group, Long Island is considerably lacking in diversity for such a large island (428 km²) with only *Candoia carinata carinata* and

Table 2.1 Model fitting for the log-log ISAR models for each of the island groups assessed: adjusted r-squared (R^2_{adj}) = amount of variation in richness explained by area, *LogC* = coefficient of correlation, (z) = slope coefficient of the log-log species area relationship.

Island group	R^2_{adj}	*LogC*	*z*
All New Guinea islands	0.38	0.58	0.18
WNG Coast	0.48	0.93	0.18
Bismarck Archipelago	0.37	0.31	0.19
Southern PNG Coast	0.06	0.61	0.23
Northern PNG Coast	−0.01	1.00	0.06
Admiralty Islands	0.57	0.56	0.19
Bougainville Islands	0.79	0.19	0.20
Milne Bay Islands	0.58	0.80	0.18

C. paulsoni mcdowelli occurring on the island. However, most of Long Island is occupied by the large caldera lake, Lake Wisdom, which itself contains the small, seemingly snake-free Motmot Island (Cook *et al.* 2001). Similar to *A. granulatus*, *Candoia* species are widespread and, based on their island distribution (see below for specific details), can likely traverse marine barriers with relative ease.

Despite the positive trend in our ISAR results, as island size increases, the trend is less consistent in most island groups and could indicate incomplete recording. The exception to this pattern is seen in the islands of Milne Bay Province, an area with high endemicity, which show no evidence of leveling. The trend in ISAR is much

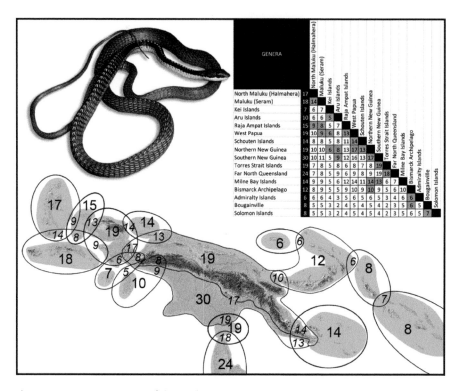

Figure 2.5 Representation of the snake genera present in New Guinea, on satellite archipelagos, and neighboring regions of Indonesia, Australia, and the Solomon Islands. Figures in large roman script indicate the number of genera present in each mainland block (West Papua, Northern NG, or Southern NG) or archipelago ellipsoid. The small numbers in italics indicate the number of genera common to both areas enclosed by neighboring ellipsoids. Inset table: White figures on black are total genera present in each area; figures in grid are the numbers of genera common to both areas, figures black on gray indicate neighboring comparisons illustrated on the map. The inset is the Coconut Treesnake (*Dendrelaphis calligastra*), a common species found in every block and archipelago except the Moluccas. For island group names see Figure 2.4. Photograph by Mark O'Shea.

more pronounced for smaller islands, with many larger islands having equal or fewer species compared to the smaller islands. Compared to other island groups in the New Guinea region, the islands of Milne Bay have received considerable attention from naturalists since the early to mid 20th century (Meek 1913; Brass 1956, 1959) and more recently from herpetologists (Kraus and Allison 2004; Kraus 2005, 2009, 2017a, 2017b, 2020; Roberts and Austin 2020).

A clear biogeographic story is observed in the snakes on New Guinea, especially when compared to surrounding islands (Figures 2.4–2.6). The Moluccas have more families (11) compared to other island groups discussed in this chapter, including

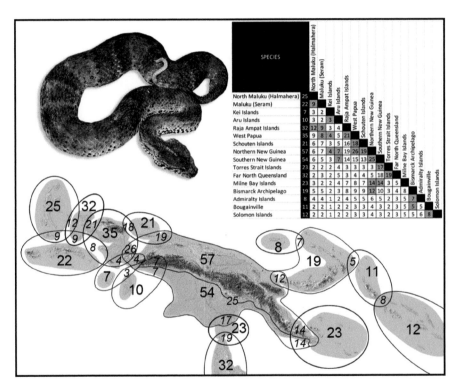

SPECIES	North Maluku (Halmahera)	Maluku (Seram)	Kei Islands	Aru Islands	Raja Ampat Islands	West Papua	Schouten Islands	Northern New Guinea	Southern New Guinea	Torres Strait Islands	Far North Queensland	Milne Bay Islands	Bismarck Archipelago	Admiralty Islands	Bougainville	Solomon Islands
North Maluku (Halmahera)	25															
Maluku (Seram)	22	9														
Kei Islands	7	3	2													
Aru Islands	10	3	2	3												
Raja Ampat Islands	32	12	9	3	4											
West Papua	35	9	8	4	5	21										
Schouten Islands	21	6	7	3	5	16	18									
Northern New Guinea	57	6	7	4	7	19	26	19								
Southern New Guinea	54	6	5	3	7	14	15	13	25							
Torres Strait Islands	23	2	2	2	4	3	3	3	3	17						
Far North Queensland	32	2	3	2	5	3	4	4	5	18	19					
Milne Bay Islands	23	3	2	2	4	7	8	7	14	14	3	5				
Bismarck Archipelago	19	5	5	2	3	8	9	9	12	10	3	4	8			
Admiralty Islands	8	4	4	1	2	4	5	5	6	5	2	3	5	7		
Bougainville	11	2	2	1	2	3	3	4	3	2	3	5	5	5		
Solomon Islands	12	2	2	1	2	3	3	3	4	3	2	3	5	5	6	8

Figure 2.6 Representation of the snake species present in New Guinea, on satellite archipelagos, and neighboring regions of Indonesia, Australia, and the Solomon Islands. Figures in large roman script indicate the number of species in each mainland block (West Papua, Northern NG, or Southern NG) or archipelago ellipsoid. The small numbers in italics indicate the number of species common to both areas enclosed by neighboring ellipsoids. Inset table: White figures on black are total species present in each area; figures in grid are the numbers of species common to both areas, figures black on gray indicate neighboring comparisons illustrated on the map. The inset is the Smooth-scaled Death Adder (*Acanthophis laevis*), a widely distributed species in New Guinea and the Moluccas. For island group names see Figure 2.4.
Photograph by Mark O'Shea.

mainland New Guinea (9) and Australia (7), with those only present on the Moluccas of Oriental origin. Genera and species richness is unsurprisingly highest on mainland New Guinea but with relatively high diversity in the Moluccas, the Western New Guinea islands, Queensland's Torres Strait islands, and Milne Bay islands. The Milne Bay islands are interesting because they are relatively small islands but demonstrate high endemicity.

This project could be expanded in order to compare the ophidiofaunas of individual islands of similar size using the *coefficient of biogeographical resemblance* (CBR) (Duellman 1990; O'Shea 1998). Factors that could be taken into consideration would include island factors, including island surface area, highest elevation, geological origin, vegetation cover, human impact and land use, closeness to next largest island, the direction and strength of oceanic currents, and the depth of the seabed between the islands. This could also be expanded further by looking at snake species characteristics (*i.e.*, maximum size, habitat preference, diel activity, dietary preference, and reproductive strategy).

Snake Distribution on the Islands off New Guinea

In this chapter, we cover a large area with a complex geological history and, consequently, complex evolutionary origins of the snake fauna that occurs in the region. Here we have grouped the New Guinea islands together into their various archipelagos—from west to east (a) the Raja Ampat Islands, (b) the Schouten Islands, (c) the Admiralty Islands, (d) Milne Bay islands, (e) the Bismarck Archipelago, and (f) Bougainville and its satellite islands. The snake faunas of these island groups are compared with those of archipelagos that could influence the composition of the New Guinea snake fauna (Figures 2.4–2.6), to the west (the North and Central Moluccas, Indonesia), the south (the southern Moluccas; *i.e.*, Kei and Aru Islands, Indonesia, as well as the Torres Strait Islands, Queensland, Australia), and southeast (the Solomon Islands). In this section, we provide an overview of the snake families occurring within our study system. We recognize that some authors consider Natricidae and Calamariidae as the subfamilies Natricinae and Calamarinae within the Colubridae, and continue to recognize Ahaetuliinae rather than Chrysopeleinae, *fide* Figueroa (2022), also within the Colubridae.

Acrochordidae (Filesnakes)

Acrochordidae is an Asian-Australasian family with a single genus (*Acrochordus*) comprising three extant and one extinct species. The freshwater species *A. javanicus* is confined to the rivers and swamps of mainland Southeast Asia and Sundaland, while a second freshwater species (*A. arafurae*) inhabits billabongs and slow rivers in northern Australia and southern New Guinea. The third species, *A. granulatus*, is the smallest and most widely distributed of the three, primarily inhabiting estuaries, mud flats, and mangrove swamps from southern India to Hainan Island, China, and southeast to northern Australia and the southwest Pacific. Despite this wide distribution, *A. granulatus* constitutes a single homogenous species (Sanders *et al.* 2010). An extinct species (*A. dehmi*), the sister taxon to *A. javanicus*, is known from several locations in South and Southeast Asia (Hoffstetter 1964; Head 2005) so it is presumed that *Acrochordus* is of Oriental origin.

Calamariidae (Reedsnakes)

The Calamariidae is an Oriental family that was recently resurrected from within the Colubridae (Zaher *et al.* 2019). It contains 96 species in seven genera, with 18 species and four genera present in Wallacea, the zoogeographical transition zone between the Oriental and Australasian zones comprising the Moluccan Islands, Sulawesi, and Lesser Sunda Islands. Their Wallacean distribution is centered over Sulawesi, and only two species occur in the Moluccas, *Calamaria ceramensis* on Seram, Ambon, and Saparu (Inger and Marx 1965; de Lang 2013) and *Calamorhabdium kuekenthali*, which is endemic to Bacan (de Lang 2013). Snakes of this group do not cross the Lydekker's Line, the imaginary divider separating the eastern edge of Wallacea from New Guinea, and they are absent from the Raja Ampat Islands and New Guinea (O'Shea 2021).

Candoiidae (Pacific Boas)

The Candoiidae comprises five species in the genus *Candoia*, three of which, *C. aspera*, *C. carinata*, and *C. paulsoni*, occur within the New Guinea region. However, the origins of *Candoia* (and the Fijian-Tongan iguanid genus *Brachylophus*) have long been the subject of discussion. Gibbons (1981) proposed that *Brachylophus* colonized the Fijian and Tongan islands, having arrived from the Caribbean aboard rafts of red mangrove (*Rhizophora mangle*) via the westward-flowing Southern Equatorial Current at some time prior to the closure of the Panamanian isthmus and the joining of North and South America (5.7 mya). McDowell (1979) considered West Indian *Epicrates* (= *Chilabothrus*) boas morphologically most similar to *Candoia*. These two hypotheses may have led others to consider a neotropical origin for *Candoia* as well. In favor of this argument are the placement of the easternmost species (*C. bibroni*) as the sister taxon to all other *Candoia* (Austin 2000) and the knowledge that *C. bibroni* occasionally inhabits mangroves, at least on Reef Island in the eastern Solomon Islands (McCoy 1980, 2007). Such a long journey is possible, as evidenced by Heyerdahl (1950), but it would require considerable physiological hardiness and specific adaptations in any terrestrial vertebrates making the journey, including low nutritional requirements and the ability to survive prolonged periods without freshwater (rain), both strategies for which boas and iguanas are better adapted than mammals. However, this route of transoceanic dispersal is now largely discounted for boas (Noonan and Sites 2010) and alternative theories have being proposed, or found renewed support, to state that *Candoia* is relictual and exists in the New Guinea region through Gondwanan vicariance (Underwood and Stimson 1990; Kluge 1991; Austin 2000), with associated extinctions in Australia. The absence of booid fossils in Australia may be an artifact explained by the continent's poor fossil record and the fact that most snake fossils consist of disassembled vertebrae, and it is difficult to distinguish between booid vertebrae and those that belong to pythons or extinct madtsoiids (Noonan and Sites 2010), both of which are present in the Australian fossil record. Other alternatives include a Southeast Asian origin (Noonan and Chippindale 2006; Noonan and Sites 2010) based on a proposed sister taxon relationship between Afro-Indian *Eryx* and *Candoia* (Vidal and Hedges 2004), though this may no longer be as strongly supported, with *Eryx* being genetically closer to traditional booid genera than to *Candoia* (Pyron *et al.* 2013). Therefore, the origins of *Candoia* cannot currently be determined with certainty.

Candoia paulsoni is the easternmost of the three New Guinea species (Smith *et al.* 2001), with the populations occurring in the Solomon Islands and on the islands to the north of Bougainville and east of New Ireland attributed to the nominate form (*C. p. paulsoni*); on Bougainville Island they are called *C. p. vindumi*, in the Louisiade Archipelago *C. p. mcdowelli*, and there are island endemics on Woodlark (*C. p. sadlieri*) and Misima (*C. p. rosadoi*). *Candoia p. mcdowelli* then extends throughout eastern and northern mainland Papua New Guinea. The species is not known from New Britain or New Ireland, and a record of *C. p. mcdowelli* from the Admiralty Islands is suspect. *Candoia* is also unreported from WNG, although a sixth subspecies (*C. p. tasmai*) occurs in the northern Moluccas 1,400 km west of the closest record for *C. p. mcdowelli*. This taxon is also reported from Gebeh Island, east of Halmahera and less than 40 km northwest of Gag Island in the Raja Ampat Islands, so it may yet be added to the herpetofauna of New Guinea.

The stocky-bodied, terrestrial *C. aspera* is distributed throughout mainland New Guinea, its northern coastal islands, the Admiralty and Bismarck Archipelagos (not Bougainville or the Louisiade Archipelago inhabited by *C. paulsoni*), and west to the Schouten Islands of Cenderwasih Bay and the Raja Ampat Islands. There are occasional records from the Moluccas. Two subspecies are recognized, the bulk of the range being occupied by *C. a. schmidti*, while the nominate *C. a. aspera* inhabits Duke of York Island and possibly also New Ireland. Occurring in sympatry with *C. aspera* throughout much of its range is the extremely slender and arboreal *C. carinata*, which is absent from the Papuan Peninsula and the Louisiade Archipelago where *C. paulsoni* is present. Two subspecies are recognized, *C. c. carinata* from New Guinea and *C. c. tepedeleni* from the Bismarck and Admiralty Archipelagos, and also, oddly, from Liki Island off northwestern New Guinea (Smith *et al.* 2001), 880 km west of the easternmost Admiralty Islands record. *Candoia carinata* is also found in the central

Moluccas. *Candoia aspera* and *C. carinata* occur in the southern Trans-Fly of New Guinea but neither is reported from the Torres Strait Islands or mainland Australia.

Colubridae (Typical Snakes)

Globally, the Colubridae is one of the most omnipresent higher snake taxa, but it is less well represented in the Australasian region. Both currently recognized subfamilies, Colubrinae and Chrysopeleinae, following Figueroa (2022), are present. The most frequently encountered colubrine is *Boiga irregularis*, the only Australasian member of an otherwise South and Southeast Asian genus comprising 37 species. *Boiga irregularis* is found throughout New Guinea and all its satellite archipelagos from where it occurs as far west as Sulawesi, as far south as New South Wales in Australia, as far east as Ugi Island in the southern Solomon Islands, and as far north as the Hermit Islands in the Admiralty Archipelago.

The niche of diurnal, arboreal predator of small frogs and lizards is occupied by the chrysopeleine genus *Dendrelaphis*. The Australasian region is occupied by a species group of nine *Dendrelaphis*, the *calligastra-punctulatus* group, which share a number of characteristics that separate them from their more Oriental congenerics (van Rooijen *et al.* 2015). Of these nine species, one is endemic to the Kei Islands, one to the Palau Islands, and one to Australia, including the Torres Strait Islands. The most widely distributed species is *D. calligastra*, which occurs from the Raja Ampat Islands to Queensland, the Admiralty, Louisiade, and Bismarck Archipelagos, Bougainville, and the entire Solomon Islands southeast to Utupua Island. The smallest distribution of the six New Guinea species is that of *D. papuensis*, which is endemic to the Trobriand Islands of Milne Bay Province. *Dendrelaphis modestus* is a representative of the Southeast Asian *D. caudolineatus* group, which inhabits the northern Moluccas (van Rooijen and Vogel 2012) but also the Raja Ampat Islands, which includes it in the New Guinea herpetofauna.

The Chrysopeleinae is also represented on Ternate Island in the northern Moluccas by *Ahaetulla prasina*, but with Ternate located west of Halmahera there seems little opportunity for this species to also occur in the Raja Ampat Islands, unlike *Chrysopelea rhodopleuron*, which inhabits Seram, Ambon, Buru, Manipa, and Nusa Laut in the central Moluccas, Bacan in the northern Moluccas, Aru in the southern Moluccas, and Misool in the Raja Ampat Islands.

The terrestrial nocturnal colubrid niche is occupied in New Guinea by *Stegonotus*, a large genus containing at least 25 species that occurs as far north as the Philippines and Borneo but demonstrates its greatest diversity (ca. 15 species) in New Guinea (Ruane *et al.* 2017; Kaiser *et al.* 2018, 2019a, 2019b, 2020, 2021; O'Shea and Richards 2021). The taxonomy of *Stegonotus* is complex, but the current distribution of the genus in the New Guinea region includes the Raja Ampat Islands (three species), the Vogelkop Peninsula (two), the Schouten Islands (three), Northern NG (three),

Southern NG (three), Admiralty Islands (two), Bismarck Archipelago (two), and Milne Bay Islands (two).

Cylindrophiidae (Asian Pipesnakes)

The family Cylindrophiidae, which occurs in Southeast Asia, southern China, and in the Greater and Lesser Sundas, reaches its eastern limit in the Moluccas. *Cylindrophis melanotus* occurs on Bacan, south of Halmahara in the northern Moluccas, and may also be present on Halmahera, Ternate, and Sanana in the Sula Islands, whereas *C. osheai* inhabits Boano, west of Seram in the central Moluccas. Four species are found in the southern Moluccas: *C. aruensis* on Damar, although the type locality for the two syntypes on the Aru Islands is considered questionable (*fide* Iskandar and Colijn 2002); *C. yamdena* on Yamdena; *Cylindrophis* cf. *boulengeri* on Babar; and *C. boulengeri* on Wetar. Like the Calamariidae, the Cylindrophiidae does not appear to cross the Lydekker Line.

Elapidae (Fixed Front-Fanged Venomous Snakes)

The only front-fanged venomous snake family in the Australasian region is the Elapidae, more specifically the subfamily Hydrophiinae which contains the terrestrial elapids of Australia, New Guinea, the Solomon Islands, and Fiji; the sea kraits (*Laticauda*); and the true sea snakes.

The Hydrophiinae is believed to have evolved from an Oriental *Calliophis*-like precursor, with *Laticauda* the sister genus to all other hydrophiines, followed by *Ogmodon* of Fiji, *Salomonelaps* + *Loveridgelaps* of the Solomons, *Micropechis* + *Aspidomorphus* + *Toxicocalamus* of New Guinea, and *Cacophis* in Australia (Sanders *et al.* 2008; Strickland *et al.* 2016).

Most of the terrestrial elapids of southern New Guinea (*Acanthophis*, *Cryptophis*, *Demansia*, *Furina*, *Glyphodon*, *Oxyuranus*, *Pseudechis*, and *Pseudonaja*) are likely of Australian origin and arrived via reinvasions of New Guinea from the larger southern continent (Wüster *et al.* 2005). Of these, only one (*Acanthophis laevis*) has ventured beyond southern New Guinea to occupy almost the entire island of New Guinea except the southeastern Papuan Peninsula and seemingly most of the Vogelkop and the Raja Ampat Islands. *Acanthophis laevis* has also radiated further east and southeast into the Moluccan islands, including Obi, Seram, Haruku, Saparua, Goram, and both the Kei and the Aru Islands. The southwestern Trans-Fly region of New Guinea and the Tanimbar Islands of the southern Moluccas are occupied by *A. rugosus*, which may also inhabit the Australian mainland.

New Guinea is also home to three endemic elapid genera that did not radiate north from Australia. *Micropechis*, *Aspidomorphus*, and *Toxicocalamus* are primarily distributed across those parts of New Guinea that were formed by the collisions of the

Pacific island arcs in the north, east, and west. *Micropechis ikaheka* occurs throughout New Guinea—except the southern Trans-Fly region of southern New Guinea—and into the Raja Ampat Islands. *Aspidomorphus* (three species) also occurs throughout New Guinea, the Raja Ampat Islands, the Milne Bay Islands, and the Bismarck Archipelago, but the most widely distributed species, *A. muelleri*, is the species that occurs on the Bismarck Archipelago, south of the Central Cordillera of New Guinea, and as far west as Seram. The genus *Toxicocalamus* is an endemic New Guinea genus which has recently received a great deal of taxonomic attention (Kraus 2009; Metzger *et al.* 2010; O'Shea *et al.* 2015; Kraus 2017b; O'Shea *et al.* 2018; Kraus 2020; Roberts and Austin 2020; Kraus *et al.* 2022). It occurs all across New Guinea island and also on offshore islands north of New Guinea and in the Milne Bay Islands, with six endemic island species.

The remaining Melanesian elapid genera are the three monotypic genera from the Solomon Islands. *Loveridgelaps elapoides* is endemic to the Solomon Islands, and although it is found on Shortlands Island only 10 km south of Bougainville it has not been reported from that large island. *Salomonelaps par* has also been found on Shortlands Island and Fauro Island and also on Buka Island to the north of Bougainville. The only terrestrial elapid known from Bougainville is the diminutive *Parapistocalamus hedigeri* which appears to be confined to the Empress Augusta Bay region in the west of the island.

Gerrhopilidae (Papillate Blindsnakes)

The Gerrhopilidae is a family of blindsnakes that diverged from the Typhlopidae on Indigascar (India + Madagascar) approximately 100 mya (Vidal *et al.* 2010) and is therefore likely of Asian origin. The genus *Gerrhopilus* contains 22 species, 5 of which have been described since the family's erection and 10 of which occur in Melanesia, on the Raja Ampats, New Guinea, the Bismarck Archipelago, and the islands of Milne Bay Province. *Gerrhopilis hades*, on Rossel Island in the Louisiade Archipelago, is currently the southernmost and easternmost record for the family.

Homalopsidae (Mudsnakes)

The Homalopsidae is an Asian and Australian family currently containing 29 genera and 57 species (Murphy and Voris 2014; Quah *et al.* 2017; Köhler *et al.* 2021), most of which are mildly venomous, rear-fanged, marine, brackish or freshwater snakes that prey on fish, frogs, and crustaceans, but the family also contains three genera of poorly known, fangless, terrestrial to fossorial vermivorous snakes. Of the mangrove-dwelling genera occurring in the New Guinea region, *Cerberus* and *Fordonia* are widely distributed in southern New Guinea and northern Australia, and thence northwestward through Indonesia and the Philippines to the Asian mainland. The freshwater *Pseudoferania* and the mangrove-dwelling *Myron* are confined to southern New Guinea and northern Australia, the latter with an endemic Aru Island species, while the brackish-water genus *Djokoiskandarus* and the freshwater *Huernia* are endemic to New Guinea island, the former southern coastal and the latter northwestern riverine.

Of the terrestrial fangless genera, *Brachyorrhos* is a Moluccan genus of five species (Murphy *et al.* 2012; Murphy and Voris 2020), but there is a single record from Kofiau Island in the Raja Ampats that makes *Brachyorrhos* a member of the New Guinea herpetofauna. *Calamophis* contains four species on the Vogelkop and Yapen Island in the Schouten Islands of Cenderawasih Bay (Murphy 2012; O'Shea and Kaiser 2016). There are no southern homalopsid records from further east than Port Moresby on the Papuan Peninsula and no northern records east of the Mamberamo River, Western New Guinea.

Natricidae (Keelbacks)

The large family Natricidae contains 37 genera and 266 species of freshwater aquatic watersnakes and keelbacks, which are primarily distributed through the Northern Hemisphere. The Australasian element comprises the genus *Tropidonophis*, containing 20 species (Malnate and Underwood 1988; Kraus and Allison 2004). *Tropidonophis* includes three Philippine endemics, two Moluccan endemics, two species in the Moluccas and WNG, four NG endemics, two species in NG and the Raja Ampat Islands, one species in NG and the Schouten Islands, one species in NG and the Aru Islands, one species in NG and Australia, one species in NG and the Milne Bay Islands, one Milne Bay Island endemic, and two Bismarck Archipelago endemics.

Pythonidae (Pythons)

The Pythonidae is well represented in Wallacea, New Guinea, and Australia, and the biogeographic links across the Torres Strait are readily apparent. The *Simalia amethistina/kinghorni* complex (Harvey *et al.* 2000) occurs throughout New Guinea, the Torres Strait, and northern Queensland, though where *S. kinghorni* of northern Queensland transitions into *S. amethistina* of New Guinea has yet to be determined. *Simalia amethistina* also inhabits the Aru Islands and is also reported from Gebe Island in the northern Moluccas. The islands of the northern, central, and southern Moluccas each contain a single endemic member of the *S. amethistina* complex (Harvey *et al.* 2000) and on some Moluccan islands the *S. amethistina*-complex species occurs in sympatry with *Malayopython reticulatus*, which is widely distributed across Indonesia to the Moluccas. *Antaresia* and *Liasis* are Australian genera represented by *A. papuensis* and *L. fuscus* in the Torres Strait and extreme southern New Guinea (O'Shea *et al.* 2004; Rawlings *et al.* 2004; Esquerré *et al.* 2021).

Both *Morelia spilota* and *M. viridis* occur in Queensland and New Guinea though only the former has been collected from the Torres Strait Islands while the latter is present on the Aru Islands, its type locality. In northern New Guinea, including the Schouten and Raja Ampat Islands, *M. viridis* is replaced by *M. azurea* (Rawlings and Donnellan 2003), with its type locality on Biak in the Schouten Islands. Within New Guinea, *Leiopython* exhibits a north–south partitioning similar to that of *M. viridis/azurea*, with *L. fredparkeri* in the south and *L. albertisi* in the north (Schleip 2008, 2014; Natusch 2021) including the Schouten and Raja Ampat Islands. The genus is replaced in the Bismarck Archipelago by *Bothrochilus boa*, but a population of *L. albertisi* is present in the Mussau Islands to the north of New Ireland.

Typhlopidae (Cosmopolitan Blindsnakes)

The blindsnakes present in the Australasian region belong to the subfamily Asiatyphlopinae of Hedges *et al.* (2014). Apart from *Indotyphlops* (see below), four genera are represented in the New Guinea region. *Acutotyphlops* is a genus of five relatively large species that prey on earthworms. Apart from a single Philippine species (Wallach *et al.* 2007) the distribution of the genus is centered over the islands east of New Guinea, with one species occurring through the Bismarck Archipelago and three species on Bougainville, one of which is also widely distributed in the Solomon Islands (Wallach 1995).

The largest typhlopid genus in Australasia is *Anilios*. Australia is home to 45 endemic *Anilios* and also two species that also occur in the Trans-Fly region of southern New Guinea. A third New Guinea species, and the only non-Australian member of the genus, occurs in northern New Guinea. The genus *Argyrophis* is an Asian genus of 13 species occurring from Nepal to Java and Taiwan, but with a potential species from the northeast Vogelkop of WNG. Brongersma (1934) described this taxon as *Typhlops fusconotus* on the basis of three specimens, but other authors synonymize it with *Argyrophis diardi* from South Asia or with *A. muelleri* from Southeast Asia.

The genus *Ramphotyphlops* occurs from Southeast Asia eastward into the Pacific, to Fiji, and the Caroline and Loyalty Islands. Twenty-two species are recognized but it is likely some of the more widely distributed taxa are species complexes. Five species are distributed across the Raja Ampat Islands, Vogelkop, and Moluccas, four of which occur in New Guinea. Two other species inhabit Bougainville and the Solomon Islands, one of which also occurs in the Bismarck, Admiralty, and Milne Bay archipelagos of PNG. The Solomon Islands is also home to two endemic *Ramphoyphlops* species. The genus *Malayotyphlops* is a Philippine genus of 12 species, with two Indonesian endemics, one of which, *M. kraali*, occurs on Seram and the Kei Islands, but this genus is not known from New Guinea. The commonly encountered parthenogenetic blindsnake *Indotyphlops braminus* (*fide* Wallach, 2020, 2021) is frequently encountered across the entire region, but because it has been distributed worldwide through the agencies of humans it has not been included in the species counts and calculations related to this project. See Figure 2.7.

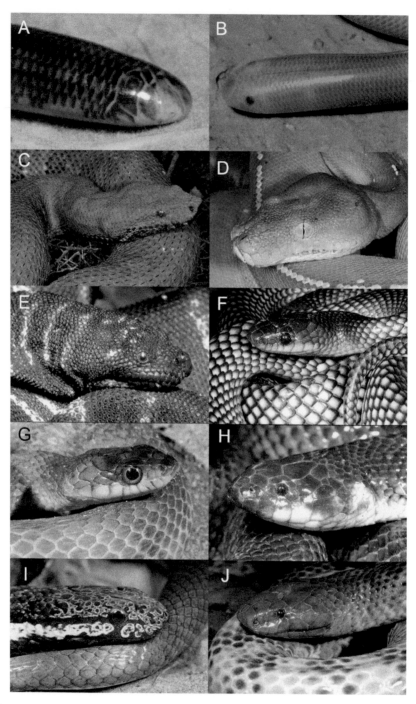

Figure 2.7 A montage of snake species found on the islands of New Guinea: (A) Typhlopidae: Eastern Yellow-bellied Blindsnake (*Ramphotyphlops depressus*); (B) Gerrhopilidae: Normanby Island Blindsnake (*Gerrhopilus persephone*);

Figure 2.7 Continued

(C) Candoiidae: New Guinea Ground Boa (*Candoia aspera*) New Guinea, Raja Ampat Islands, Admiralty Islands, and Bismarck Archipelago; (D) Pythonidae: Southern Green Tree Python (*Morelia viridis*) southern New Guinea, Milne Bay Islands, Aru Islands, and Queensland; (E) Little Filesnake (*Acrochordus granulatus*) throughout the region; (F) Colubridae: Reticulated Groundsnake (*Stegonotus reticulatus*) eastern New Guinea and Milne Bay islands; (G) Natricidae: Doria's Barred Keelback (*Tropidonophis doriae*) New Guinea and Aru Islands; (H) Crab-eating Mangrove Watersnake (*Fordonia leucobalia*) southern New Guinea, Kei Islands, and Lesser Sunda Islands; (I) Elapidae: Müller's Crowned Snake (*Aspidomorphus muelleri*) New Guinea, Bismarck Archipelago, and Raja Ampat Islands; (J) Elapidae: New Guinea Small-eyed Snake (*Micropechis ikaheka*) New Guinea and Raja Ampat Islands.
Photographs A–B courtesy of Fred Kraus, C–J by Mark O'Shea.

Conclusion

New Guinea and its surrounding archipelagos are complex in their geological history, and this has enabled at least 140 species of terrestrial or freshwater snakes to thrive on them. These island systems are important for global biodiversity, housing nine families and many endemic genera and species. New Guinean islands and surrounding archipelagos generally support the ISAR in that as island size increases so do the number of species occurring there. It is important to note that as island size increases the correlation becomes less pronounced, with the exception of the Milne Bay Islands. The islands of the d'Entrecasteaux Archipelago, the Trobriand Islands, Woodlark Island, and the Louisiade Archipelago (Milne Bay) exhibit high reptile and amphibian endemism.

It is currently believed that snakes, both scolecophdians (blindsnakes) and alethinophidian ("true snakes," those above the Scolecophidia) arose from a Gondwanan origin (Vidal *et al.* 2009). The Typhlopidae and Gerrhopilidae belong in the Typhlopoidea, an East Gondwana superfamily that evolved on Indigascar and diverged from the West Gondawa Leptotyphlopidae and Anomalepididae over 100 mya (Vidal *et al.* 2010). The alethinophidian families are also East Gondwanan, belonging to the Afrophidia, the Out of Africa clade containing all but two families of more advanced snakes (Vidal *et al.* 2009).

At the family level, most of the snakes of New Guinea appear to have arrived from Asia, although the Hydrophiinae has arrived via a circuitous route, from the Pacific, into New Guinea, into Australia, and back into New Guinea. There are two hypotheses for the origin of the Pythonidae comprising Gondwanan and Laurasian dispersal events, with one recent paper proposing the latter (Zaher and Smith 2020), that the pythons radiated southward from Lauriasia. The origins of the Candoiidae have still to be determined.

At the generic level, *Argyrophis, Ramphotyphlops, Gerrhopilus, Acrochordus, Boiga, Dendrelaphis, Cerberus,* and *Fordonia* are distributed widely across the Oriental region, although *Ramphotyphlops, Gerrhopilus,* and *Dendrelaphis* have also speciated considerably in and around New Guinea. The genera *Anilios, Antaresia, Liasis, Morelia, Simalia, Myron, Acanthophis, Cryptophis, Demansia, Furina, Glyphodon, Pseudechis, Pseudonaja,* and *Oxyuranus* have their origins on the Australian continent but the degree of their radiation in New Guinea varies enormously from southern coastal to New Guinea–wide, with *Simalia* and *Acanthophis* ranging westward into the Moluccas and *Liasis* occurring in the Lesser Sunda islands, most likely from Western Australia rather than New Guinea. *Apodora, Bothrochilus, Leiopython, Djokoiskandarus, Huernia, Aspidomorphus, Micropechis,* and *Toxicocalamus* are endemic to New Guinea and its satellite archipelagos. *Brachyorrhos* would appear to be a Wallacean genus with its close relative *Calamophis* a WNG endemic. The genera *Stegonotus* and *Tropidonophis* have their centers of greatest speciation on New Guinea, but both are also found south to Australia and west and northwest to Indonesia and the Philippines. *Salomonelaps* and *Pseudopistocalamus* are really Solomons endemics, like *Loveridgelaps,* but they are recorded from the New Guinea region because of the political status of the Autonomous Region of Bougainville within PNG. It is therefore clear that New Guinea has a snake fauna with elements from far-distant origins to east, west, and south.

We wish to note that the islands of New Guinea are almost certainly home to many taxonomically undocumented species, due in part to a lack of rigorous scientific exploration on many of these islands. Many of the earliest natural history collectors, such as Luigi Maria d'Albertis (1841–1901) or Albert Stewart Meek (1871–1943) were interested primarily in beetles, butterflies, and birds (and, in d'Albertis's case, human body parts), and although they collected reptiles and amphibians when opportunities presented themselves, they were much more a "by-catch." Even large-scale, organized zoological expeditions to New Guinea, such as the German Kaiserin-Augusta-Fluß Expedition (1912–1913) up the Sepik River, or the seven American Museum of Natural History Archbold Expeditions to New Guinea (1933–1939, 1953–1964) did not include a herpetologist on their staff, with the collection of "lower vertebrates" often left to the expedition doctor or a general collector. Since insects, birds, and mammals were the primary interests, reptiles and amphibians at the time may not have been as diligently reported in the field notes as the more popular taxa. Often the original collector was long dead by the time an obscure specimen they had collected was identified as something important by a herpetologist from a later age. Thankfully this scenario is changing and now expeditions go to New Guinea with herpetology in the forefront of their collective minds.

Acknowledgments

The authors would like to thank Hinrich Kaiser for casting his expert critical eye over this manuscript; we feel it has been greatly enhanced by his comments and suggestions. We would also like to thank Colin Dubreuil for providing feedback on the statistical analyses.

References

Allen, G. R. 1991. *Field Guide to the Freshwater Fishes of New Guinea*. Christensen Research Institute.

Allen, G. R., A. W. Storey, and M. Yarrao. 2008. *Freshwater Fishes of the Fly River, Papua New Guinea*. Ok Tedi Mining.

Austin, C. C. 2000. Molecular phylogeny and historical biogeography of Pacific island boas (*Candoia*). *Copeia* 2000(2):341–352.

Baldwin, S. L., P. G. Fitzpatrick, and L. E. Webb. 2012. Tectonics of New Guinea. *Annual Review of Earth and Planetary Sciences* 40:495–520.

Brass, L. J. 1956. Results of the Archbold Expeditions. No. 75. Summary of the fourth Archbold Expedition to New Guinea (1953). *Bulletin of the American Museum of Natural History* 111(2):77–152.

Brass, L. J. 1959. Results of the Archbold Expeditions. No. 79. Summary of the fifth Archbold Expedition to New Guinea (1956–1957). *Bulletin of the American Museum of Natural History* 118(1):1–70.

Brongersma, L. D. 1934. Contribution to Indo-Australian herpetology. *Zoologische Mededelingen* 17(3–4):161–251.

Cook, S., R. Singidan, and I. W. B. Thornton. 2001. Colonization of an island volcano, Long Island, Papua New Guinea, with an erergent island, Motmot, in its caldera lake. IV. Colonization by non-avian vertebrates. *Journal of Biogeography* 28:1353–1363.

Davies, H. L. 1990. Structure and evolution of the border region of Papua New Guinea. In G. J. Garmen and Z. Carmen (eds.), *Petroleum Exploration in Papua New Guinea*. PNG Chamber of Mines and Petroleum, pp. 245–270.

Davies, H. L., J. Lock, D. L. Tiffin, E. Honza, Y. Okuda, F. Murakami, and K. Kisimoto. 1987. Convergent tectonics in the Huon Peninsula region, Papua New Guinea. *Geo-Marine Letters* 7:143–152.

de Lang, R. 2013. *The Snakes of the Moluccas (Maluku), Indonesia: A Field Guide to the Land and Non-Marine Aquatic Snakes of the Moluccas with Identification Key*. Chimaira.

Duellman, W. E. 1990. Herpetofaunas in neotropical rainforests: Comparative composition, history, and resource use. In A. H. Gentry (ed.), *Four Neotropical Rainforests*. Yale University Press, pp. 455–505.

Esquerré, D., S. C. Donnellan, C. J. Pavón-Vásquez, J. Fenker, and J. S. Keogh. 2021. Phylogeography, historical demography and systematics of the world's smallest pythons (Pythonidae, *Antaresia*). *Molecular Phylogenetics and Evolution* 161(107181):1–19.

Figueroa, A. 2022. Chrysopeleinae Cope, 1893 (Squamata: Colubridae) as the valid name for the family-group currently known as Ahaetuliinae Figueroa, McKelvy, Grismer, Bell and Lailvaux, 2016 (Squamata: Colubridae). *Zootaxa* 5120(4):595–597.

Flannery, T. F. 1995a. *Mammals of New Guinea*. Reed Books.

Flannery, T. F. 1995b. *Mammals of the South-West Pacific and Moluccan Islands*. Reed Books.

Gibbons, J. R. H. 1981. The biogeography of *Brachylophus* (Iguanidae) including the description of a new species, *Brachylophus vitiensis*, from Fiji. *Journal of Herpetology* 15(3):255–273.

Guilhaumon, F., D. Mouillot, and O. Gimenez. 2010. mmSAR: An r-package for multimodel species–area relationship inference. *Ecography* 33(2):420–424.

Harvey, M. B., D. G. Barker, L. K. Ammerman, and P. T. Chippindale. 2000. Systematics of pythons of the *Morelia amethistina* complex (Serpentes: Boidae) with the description of three new species. *Herpetological Monographs* 14:139–185.

Head, J. J. 2005. Snakes of the Siwalik Group (Miocene of Pakistan): Systematics and relationship to environmental change. *Palaeontologia Electronica*. 8(1):1–33.

Heads, M. 2002. Birds of paradise, vicariance biogeography and terrane tectonics in New Guinea. *Journal of Biogeography* 29:261–283.

Hedges, S. B., A. B. Marion, K. M. Lipp, J. Marin, and N. Vidal. 2014. A taxonomic framework for typhlopid snakes from the Caribbean and other regions (Reptilia, Squamata). *Caribbean Herpetology* 49:1–61.

Heyerdahl, T. 1950. *The Kon-Tiki Expedition*. Allen and Unwin.

Hoffstetter, R. 1964. Les serpents du Neogene du Pakistan (couches des Siwaliks). *Bulletin de la Societe Geologique de France* 6:467–474.

Inger, R. F., and H. Marx. 1965. The systematics and evolution of the oriental colubrid snakes of the genus *Calamaria*. *Fieldiana Zoology* 49:1–304.

Iskandar, D. T., and E. Colijn. 2002. *A Checklist of Southeast Asian and New Guinean Reptiles Part 1. Serpentes*. Department of Biology, Institute of Technology.

Kaiser, C., H. Kaiser, and M. O'Shea. 2018. The taxonomic history of Indo-Papuan groundsnakes, genus *Stegonotus* Duméril et al. 1854 (Serpentes, Colubridae), with some taxonomic revisions and the designation of a neotype for *S. parvus* (Meyer, 1874). *Zootaxa* 4512(1):1–73.

Kaiser, C., J. Lapin, M. O'Shea, and H. Kaiser. 2020. Carefully examining Bornean *Stegonotus* (Serpentes, Colubridae): The montane groundsnake population in Sabah is a new and distinct species. *Zootaxa* 4894(1):58–68.

Kaiser, C., M. O'Shea, and H. Kaiser. 2019a. Corrigenda and addenda to "The taxonomic history of Indo-Papuan groundsnakes, genus *Stegonotus* D. B. and D., 1854 with some taxonomic revisions and the designation of a neotype for *S. parvus*" by Kaiser et al. (2018). *Zootaxa* 4615:392–394.

Kaiser, C. M. O'Shea, and H. Kaiser. 2019b. A new species of Indo-Papuan groundsnake, genus *Stegonotus* Duméril et al. 1854 (Serpentes, Colubridae), from the Bird's Head Peninsula of West Papua, Indonesia, with comments on differentiating morphological characters. *Zootaxa* 4590(2):201–230.

Kaiser, H., C. Kaiser, S. Mecke, and M. O'Shea. 2021. A new species of *Stegonotus* (Serpentes: Colubridae) from the remnant coastal forests of southern Timor-Leste. *Zootaxa* 5027(4):489–514.

Kluge, A. G. 1991. Boine snake phylogeny and research cycles. *Miscellaneous Publications of the Museum of Zoology, University of Michigan* 178:1–58.

Köhler, G., K. P. P. Khaing, N. L. Than, D. Baranski, T. Schell, C. Greve, A. Janke, and S. U. Pauls. 2021. A new genus and species of mud snake from Myanmar (Reptilia, Squamata, Homalopsidae). *Zootaxa* 4915(3):301–325.

Kraus, F. 2005. New species of blindsnake from Rossel Island, Papua New Guinea. *Journal of Herpetology* 39(4):591–595.

Kraus, F. 2009. New species of *Toxicocalamus* (Squamata: Elapidae) from Papua New Guinea. *Journal of Herpetology* 65(4):460–467.

Kraus, F. 2017a. New species of blindsnakes (Squamata: Gerrhopilidae) from the offshore island of Papua New Guinea. *Zootaxa* 4299(1):75–94.

Kraus, F. 2017b. Two new species of *Toxicocalamus* (Squamata: Elapidae) from Papua New Guinea. *Journal of Herpetology* 51(4):574–581.

Kraus, F. 2020. A new species of *Toxicocalamus* (Squamata: Elapidae) from Papua New Guinea. *Zootaxa* 4859(1):127–137.

Kraus, F. 2023. A revision of *Gerrhopilus inornatus* (Squamata: Gerrhopilidae) reveals a multi-species complex. *Zootaxa* 5231(1):1–23.

Kraus, F., and A. Allison. 2004. A new species of *Tropidonophis* (Serpentes: Colubridae: Natricinae) from the D'Entrecasteaux Islands, Papua New Guinea. *Proceedings of the Biological Society of Washington* 117(3):303–310.

Kraus, F., H. Kaiser, and M. O'Shea. 2022. Hidden diversity in semi-fossorial Melanesian forest snakes: A revision of the *Toxicocalamus loriae* complex (Squamata, Elapidae) from New Guinea. *Vertebrate Zoology* 72:997–1034.

Malnate, E. V., and G. Underwood. 1988. Australasian natricine snakes of the genus *Tropidonophis*. *Proceedings of the Academy of Natural Sciences of Philadelphia* 140(1):59–201.

Matthews, T. J., K. Triantis, and R. J. Whitaker. 2021. *The Species-Area Relationship: Theory and Application*. Cambridge University Press.

Matthews, T. J., K. Triantis, R. J. Whittaker, and F. Guilhaumon. 2019. Sars: An R package for fitting, evaluating and comparing species–area relationship models. *Ecography* 42(8):1446–1455.

McCoy, M. 1980. *Reptiles of the Solomon Islands*. Wau Ecology Institute Handbook.

McCoy, M. 2007. *Reptiles of the Solomon Islands*. Pensoft.

McDowell, S. B. 1979. A catalogue of the snakes of New Guinea and the Solomons, with special reference to those in the Bernice P. Bishop Museum. Part III. Boinae and Acrochordidae. *Journal of Herpetology* 13(1):1–92.

Meek, A. S. 1913. *A Naturalist in Cannibal Land*. T. Fisher Unwin.

Metzger, G. A., F. Kraus, A. Allison, and C. L. Parkinson. 2010. Uncovering cryptic diversity in *Aspidomorphus* (Serpentes: Elapidae): Evidence from mitochondrial and nuclear markers. *Molecular Phylogenetics and Evolution* 54:405–416.

Murphy, J. C. 2012. Synonymised and forgotten, the Bird's Head stout-tailed snakes, *Calamophis* Meyer (Squamata: Serpentes: Homalopsidae). *Raffles Bulletin of Zoology* 60(2):515–523.

Murphy, J. C., Mumpuni, R. De Lang, D. J. Gower, and K. L. Sanders. 2012. The Moluccan short-tailed snakes of the genus *Brachyorrhus* Kuhl (Squamata: Serpentes: Homalopsidae, and the status of *Calamophis* Meyer. *Raffles Bulletin of Zoology* 60(2):501–514.

Murphy, J. C., and H. K. Voris. 2014. A checklist and key to the homalopsid snakes (Reptilia, Squamata, Serpentes), with the description of new genera. *Fieldiana Life and Earth Sciences* 8:1–43.

Murphy, J. C., and H. K. Voris. 2020. A new species of *Brachyorrhos* from Seram, Indonesia and notes on fangless homalopsids (Squamata, Serpentes). *Philippine Journal of Systematic Biology* 14(2):1–8.

Natusch, D. 2021. Phylogenomics, biogeography and taxonomic revision of New Guinean pythons (Pythonidae, *Leiopython*) harvested for international trade. *Molecular Phylogenetics and Evolution* 158(106960):1–10.

Noonan, B. P., and P. T. Chippindale. 2006. Dispersal and vicariance: The complex evolutionary history of boid snakes. *Molecular Phylogenetics and Evolution* 40:347–358.

Noonan, B. P., and J. W. Jr. Sites. 2010. Tracing the origins of iguanid lizards and boine snakes of the Pacific. *American Naturalist* 175(1):61–72.

O'Shea, M. 1998. The Reptilian Herpetofauna of the Ilha de Maracá. In W. Milliken and J. A. Ratter (eds.), *Maracá: The Biodiversity and Environment of an Amazonian Rainforest*. Wiley, pp. 231–262.

O'Shea, M. 2021. Wallacea: A hotspot of snake diversity (foreword). *Biodiversity Biogeography and Nature Conservation in Wallacea and New Guinea* 4:7–17.

O'Shea, M., A. Allison, and H. Kaiser. 2018. The taxonomic history of the enigmatic Papuan snake genus *Toxicocalamus* (Elapidae: Hydrophiinae) with the description of a new species from the Managalas Plateau of Oro Province, Papua New Guinea, and a revised dichotomous key. *Amphibia-Reptilia* 39(4):403–433.

O'Shea, M., and H. Kaiser. 2016. The first female specimen of the poorly known Arfak Stout-tailed Snake, *Calamophis sharonbrooksae* Murphy, 2012, from the Vogelkop Peninsula of Indonesian West New Guinea, with comments on the taxonomic history of primitive homalopsids. *Amphibian and Reptile Conservation* 10(2):1–10.

O'Shea, M., F. Parker, and H. Kaiser. 2015. A new species of New Guinea worm-eating snake, genus *Toxicalamus* (Serpentes: Elapidae), from the Star Mountains of Western Province, Papua New Guinea, with a revised dichotomous key to the genus. *Bulletin of the Museum of Comparative Zoology* 161(6):241–264.

O'Shea, M., and S. R. Richards. 2021. A striking new species of Papuan groundsnake (*Stegonotus*: Colubridae) from southern Papua New Guinea, with a dichotomous key to the genus in New Guinea. *Zootaxa* 4926(1):26–42.

O'Shea, M., R. G. Sprackland, and I. H. Bigilale. 2004. First record of the genus *Antaresia* (Squamata: Pythonidae) from Papua New Guinea. *Herpetological Review* 35(3):225–227.

Pigram, C. J., and H. L. Davies. 1987. Terranes and the accretion history of the New Guinea orogen. *BMR Journal of Australian Geology and Geophysics* 10:193–211.

Polhemus, D. A. 1996. Island arcs, and their influence on Indo-Pacific biogeography. In A. Keast and S. E. Miller (eds.), *The Origin and Evolution of Pacific Island Biotas, New Guinea to Eastern Polynesia: Patterns and Processes*. SPB Academic Publishing, pp. 51–66.

Polhemus, D. A. 2007. Tectonic geology of Papua. In A. J. Marshall and B. M. Beehler (eds.), *The Ecology of Papua (Part One)*. Periplus Editions, pp. 137–164.

Polhemus, D. A., and J. T. Polhemus. 1998. Assembling New Guinea: 40 millions years of island arc accretion as indicated by the distributions of aquatic Heteroptera (Insecta). In R. Hall and J. D. Holloway (eds.), *Biogeography and Geological Evolution of SE Asia*. Backhuys Publishers, pp. 327–340.

Pyron, R. A., F. T. Burbrink, and J. J. Wiens. 2013. A phylogeny and revised classification of Squamata, including 4161 species of lizards and snakes. *BMC Evolutionary Biology* 13:93–146.

Quah, E. S. H., L. L. Grismer, P. L. Wood Jr., M. K. Thura, T. Zin, H. Kyaw, N. Lwin, *et al.* 2017. A new species of mud snake (Serpentes, Homalopsidae, *Gylophis* Murphy and Voris, 2014) from Myanmar with a first molecular phylogenetic asessment of the genus. *Zootaxa* 4238(4):571–582.

Rawlings, L. H., D. G. Barker, and S. C. Donnellan. 2004. Phylogenetic relationships of the Australo-Papuan *Liasis* pythons (Reptilia: Macrostomata) based on mitochondrial DNA. *Australian Journal of Zoology* 52:215–227.

Rawlings, L. H., and S. C. Donnellan. 2003. Phylogeographic analysis of the green python, *Morelia viridis*, reveals cryptic diversity. *Molecular Phylogenetics and Evolution* 27:36–44.

Roberts, J. R., and C. C. Austin. 2020. A new species of New Guinea worm-eating snake (Elapidae: *Toxicalamus* Boulenger, 1896), with comments on postfrontal bone variation based on micro-computed tomography. *Journal of Herpetology* 54(4):446–459.

Ruane, S., S. J. Richards, J. D. McVay, B. Tjaturadi, K. Krey, and C. C. Austin. 2017. Cryptic and non-cryptic diversity in New Guinea ground snakes of the genus *Stegonotus* Duméril, Bibron and Duméril, 1854: A description of four new species (Squamata: Colubridae). *Journal of Natural History* 52(1–2):1–28.

Sanders, K. L., M. S. Y. Lee, R. Leys, R. Foster, and J. S. Keogh. 2008. Molecular phylogeny and divergence dates for Australasian elapids and sea snakes (Hydrophiinae): Evidence from seven genes for rapid evolutionary radiations. *Journal of Evolutionary Biology* 21:682–695.

Sanders, K. L., Mumpuni, A. Hamidy, J. J. Head, and D. J. Gower. 2010. Phylogeny and divergence times of filesnakes (*Acrochordus*): Inferences from morphology, fossils and three molecular loci. *Molecular Phylogenetics and Evolution* 56:857–867.

Schleip, W. D. 2008. Revision of the genus *Leiopython* Hubrecht 1879 (Serpentes, Pythonidae) with the redescription of taxa recently described by Hoser (2000) and the description of new species. *Journal of Herpetology* 42(4):645–667.

Schleip, W. D. 2014. Two new species of *Leiopython* Hubecht, 1879 (Pythonidae: Serpentes): Non-compliance with the International Code of Zoological Nomenclature leads to unavailable names in zoological nomenclature. *Journal of Herpetology* 48(2):272–275.

Slapcinsky, J. 2005. Six new species of *Paryphantophis* (Gastropoda, Pulmonata, Charopidae) from the Papuan Peninsula of New Guinea. *Nautilus* 119(1):27–42.

Slapcinsky, J. 2006. *Paryphantophis* (Gastropoda, Pulmonata, Charopidae) from the Louisiade Archipelago of New Guinea. *Nautilus* 120(4):119–130.

Slapcinsky, J., and R. Lasley. 2007. Three new species of *Paryphantopsis* (Gastropoda, Pulmonata, Charopidae) from the Nakanai Mountains, New Britain, Papua New Guinea. *Nautilus* 121(4):182–190.

Smith, H. M., D. Chiszar, K. Tepedelen, and F. van Breukelen. 2001. A revision of the bevel-nosed boas (*Candoia carinata* complex) (Reptilia: Serpentes). *Hamadryad* 26(2):283–315.

Strickland, J. L., S. Carter, F. Kraus, and C. L. Parkinson. 2016. Snake evolution in Melanesia: Origin of the Hydrophiinae (Serpentes, Elapidae), and the evolutionary history of the enigmatic New Guinean elapid *Toxicocalamus*. *Zoological Journal of the Linnean Society* 177:1–16.

Torgersen, T., J. Luly, P. de Deckker, M. R. Jones, D. E. Searle, A. R. Chivas, and W. J. Hullman. 1988. Late Quaternary environments of the Carpentaria Basin, Australia. *Palaeogeography Palaeoclimatology Palaeoecology* 67:245–261.

Underwood, G., and A. F. Stimson. 1990. A classification of pythons (Serpentes, Pythonidae). *Journal of Zoology* 221(4):565–603.

van Rooijen, J., and G. Vogel. 2012. A revision of the taxonomy of *Dendrelaphis caudolineatus* (Gray, 1834) (Serpentes: Colubridae). *Zootaxa* 3272:1–25.

van Rooijen, J., G. Vogel, and R. Somaweera. 2015. A revised taxonomy of the Australo-Papuan species of the colubrid genus *Dendrelaphis* (Serpentes: Colubridae). *Salamandra* 51(1):33–56.

van Welzen, P. C., H. Turner, and M. C. Roos. 2001. New Guinea: A correlation between accreting areas and dispersing Sapindaceae. *Cladistics* 17:242–247.

Vidal, N., and S. B. Hedges. 2004. Molecular evidence for a terrestrial origin of snakes. *Proceedings of the Royal Society Biological Sciences Series B* 271: S226–S229.

Vidal, N., J. Marin, M. Morini, S. Donnellan, W. R. Branch, R. Thomas, M. Vences, *et al.* 2010. Blindsnake evolutionary tree reveals long history of Gondwana. *Biology Letters* 6:558–561.

Vidal, N., J.-C. Rage, A. Coloux, and S. B. Hedges. 2009. Snakes (Serpentes). In S. B. Hedges and S. Kumar (eds.). *The Timetree of Life*. Oxford University Press, pp. 390–397.

Wallach, V. 1995. A new genus for the *Ramphotyphlops subocularis* species group (Serpentes: Typhlopidae), with description of a new species. *Asiatic Herpetological Research* 6:132–150.

Wallach, V., R. M. Brown, A. C. Diesmos, and G. V. A. Gee. 2007. An enigmatic new species of blind snake from Luzon Island, northern Philippines, with a synopsis of the genus *Acutotyphlops* (Serpentes, Typhlopidae). *Journal of Herpetology* 41(4):690–702.

Weissel, J. K., and A. B. Watts. 1979. Tectonic evolution of the Coral Sea Basin. *Journal of Geophysical Research* 84(B9):4572–4582.

Woinarski, J., B. Mackey, H. Nix, and B. Traill. 2007. *The Nature of Northern Australia: Natural Values, Ecological Processes and Future Prospects*. Australian National University Press.

Wüster, W., A. J. Dumbrell, C. Hay, C. E. Pook, D. J. Williams, and B. G. Fry. 2005. Snakes across the Strait: Trans-Torresian phylogeographic relationships in three genera of Australasian snakes (Serpentes: Elapidae: *Acanthophis*, *Oxyuranus*, and *Pseudechis*). *Molecular Phylogenetics and Evolution* 34:1–14.

Zaher, H., R. W. Murphy, J. C. Arredondo, R. Graboski, P. R. Machado-Filho, K. Mahlow, G. G. Montingelli, *et al.* 2019. Large-scale molecular phylogeny, morphology, divergence-time estimation, and the fossil record of advanced caenophidian snakes (Squamata: Serpentes). *Plos ONE* 14(5):e0216148.

Zaher, H., and K. T. Smith. 2020. Pythons in the Eocene of Europe reveal a much older divergence of the group in sympatry with boas. *Biology Letters* 16(20200735):1–6.

Appendix I New Guinea islands included in study with coordinates, areas, and snake species count

Island name	Province	Figure 2.2 annotation	Latitude south	Longitude east	Area km²	Snake species
New Guinea					785,753	99
		PAPUA NEW GUINEA				
Seleo	Sandaun	P01	3.14643	142.48637	0.78	3
Tarawai	East Sepik	P02	3.20955	143.26026	4.3	7
Walis	East Sepik	P03	3.22803	143.29720	10.2	6
Kairiru	East Sepik	P04	3.35142	143.56212	44.4	6
Muschu	East Sepik	P05	3.41272	143.59439	42.3	3
Vokeo	East Sepik	P06	3.22411	144.09570	34.5	1
Blupblup	East Sepik	P07	3.50743	144.60423	6.7	4
Laing	Madang	P08	4.17276	144.87338	0.05	3
Boisa	Madang	P09	3.99633	144.96541	1.4	4
Manam	Madang	P10	4.05013	145.03769	83	11
Sarang	Madang	P11	4.76975	145.70220	0.08	2
Karkar	Madang	P12	4.58028	145.97546	410	11
Bagabag	Madang	P13	4.80182	146.22150	36.5	1
Kranket	Madang	P14	5.19917	145.81920	1.24	2
Crown	Madang	P15	5.11513	146.96626	14	2
Long	Madang	P16	5.29924	147.06591	428	2
Tolokiwa	Morobe	P17	5.31547	147.59001	47.7	1
Umboi	Morobe	P18	5.65517	147.93491	930	6
Tami	Morobe	P19	6.76247	147.90948	0.21	2
Bobo	Western	P20	9.11029	143.24255	33.2	6
Daru	Western	P21	9.08042	143.19945	14.7	19
Urama	Gulf	P22	7.58564	144.61774	92	1
Yule	Central	P23	8.82833	146.53699	11.5	7
Loloata	Nat. Cap. Dist.	P24	9.53815	147.29059	0.12	1
Motupore	Nat. Cap. Dist.	P25	9.52379	147.28478	0.19	1
Abavi	Central	P26	10.18268	148.71689	3.5	1
Samarai	Milne Bay	P27	10.61187	150.66411	0.29	6
Goodenough	Milne Bay	P28	9.35274	150.23531	687	10
Wagabu	Milne Bay	P29	9.33641	150.68921	0.26	1
Fergusson	Milne Bay	P30	9.52503	150.58757	1,437	17
Dobu	Milne Bay	P31	9.75377	150.86729	0.08	1
Normanby	Milne Bay	P32	10.04848	151.01100	1,949	17
Boiaboiawaga	Milne Bay	P33	10.20917	150.90536	0.08	1
Kuia	Milne Bay	P34	8.59928	150.86857	1.7	2

Appendix I Continued

Island name	Province	Figure 2.2 annotation	Latitude south	Longitude east	Area km²	Snake species
Kaileuna	Milne Bay	P35	8.52760	150.95805	45.5	1
Kiriwina	Milne Bay	P36	8.48290	151.06591	290.5	9
Kitava	Milne Bay	P37	8.62381	151.33631	24	3
Woodlark	Milne Bay	P38	9.10461	152.75498	874	7
Budi Budi	Milne Bay	P39	9.28924	153.69469	2	1
Panaeti	Milne Bay	P40	10.68352	152.37423	30.3	2
Misima	Milne Bay	P41	10.66125	152.78661	215	7
Gulewa	Milne Bay	P42	11.06706	152.53372	0.59	1
Sudest	Milne Bay	P43	11.51456	153.46302	830	5
Rossel	Milne Bay	P44	11.36433	154.23697	292.5	7
New Britain		P45	5.74929	150.76723	36,520	17
Garove	W New Britain	P46	4.65625	149.51241	57.8	1
Arawe	W New Britain	P47	6.16757	149.02221	0.99	4
Ganglo	W New Britain	P48	6.22844	149.53894	1.31	1
Gasmata	W New Britain	P49	6.29186	150.28701	0.93	2
Mioko	E New Britain	P50	4.23151	152.46089	1.5	4
Duke of York	E New Britain	P51	4.18291	152.45592	50	12
Manus	Manus	P52	2.06635	147.13594	2,100	7
Los Negros	Manus	P53	2.02552	147.41012	54	6
Pak	Manus	P54	2.07806	147.64644	9.2	2
Rambutyo	Manus	P55	2.30062	147.80089	88	3
Big Ndrova	Manus	P56	2.22011	147.23462	0.81	2
Lou	Manus	P57	2.37758	147.34123	29	5
Baluan	Manus	P58	2.55724	147.28378	14	2
Mussau	New Ireland	P59	1.44960	149.62075	350	6
Boliu	New Ireland	P60	1.53459	149.65203	1	2
Emirau	New Ireland	P61	1.65918	149.96719	34	1
Tench	New Ireland	P62	1.64874	150.67304	0.39	1
New Hanover	New Ireland	P63	2.54153	150.23990	1,186	6
Nusa	New Ireland	P64	2.56809	150.78243	0.93	1
Dyaul	New Ireland	P65	2.94370	150.89496	100	2
New Ireland	New Ireland	P66	4.32432	152.86228	7,404	15
Simberi	New Ireland	P67	2.62719	151.98956	61.7	1
Tabar	New Ireland	P68	2.93960	152.03103	116	2
Tatau	New Ireland	P69	2.75285	151.95877	138.2	4
Babase	New Ireland	P70	4.01559	153.70767	30	1
Ambitle	New Ireland	P71	4.08215	153.63249	107	3

(continued)

Appendix I Continued

Island name	Province	Figure 2.2 annotation	Latitude south	Longitude east	Area km^2	Snake species
Sable	Bougainville	P72	3.67199	154.65204	0.03	1
Pinipel	Bougainville	P73	4.41226	154.12031	7.7	1
Sirot	Bougainville	P74	4.46610	154.16172	0.93	1
Nissan Atoll	Bougainville	P75	4.49338	154.22716	37	2
Buka	Bougainville	P76	5.30781	154.63822	492	5
Bougainville	Bougainville	P77	6.37608	155.38213	9,318	10
WESTERN NEW GUINEA (INDONESIA)						
Djamna	Papua	W01	2.00673	139.24847	1	2
Liki	Papua	W02	1.60368	138.72962	13.7	2
Yapen	Papua	W03	1.78000	136.19085	2,227	14
Mios Woendi	Papua	W04	1.25514	136.37434	0.72	3
Padaido	Papua	W05	1.23786	136.51923	5.16	1
Biak	Papua	W06	1.08246	136.02071	1,786	14
Supiori	Papua	W07	0.77848	135.54849	659	5
Mios Num	Papua	W08	1.49924	135.18424	106	1
Numfoor	Papua	W09	1.02319	134.88359	355	10
Mansinam	West Papua	W10	0.90563	134.09998	4.6	9
Saonek	West Papua	W11	1.06064	130.89935	0.31	2
Weigao	West Papua	W12	1.75774	131.14499	3,155	10
Gag	West Papua	W13	1.18757	129.83618	61.8	3
Batanta	West Papua	W14	0.84423	130.66739	453	12
Kofiau	West Papua	W15	1.90058	130.15470	144	1
Salawati	West Papua	W16	0.18070	130.91069	1,623	26
Misool	West Papua	W17	0.44269	129.88699	2,034	16
Adi	West Papua	W18	4.21778	133.46911	158	2
Yos Sudarso	Papua	W19	7.90132	138.44948	11,742	7

Appendix II Islands of the Moluccas, Queensland, and Solomon Islands compared with New Guinea islands included in study, with coordinates, areas, and snake species count

Island name	Province or State	Figure 2.2 annotation	Latitude south	Longitude east	Area km²	Snake species
			INDONESIA			
Morotai	North Maluku	M01	2.28429	128.38691	2,266	7
Halmahera	North Maluku	M02	0.59758	127.76721	18,040	14
Gebe	North Maluku	M03	0.12260	129.49672	157	1
Ternate	North Maluku	M04	0.79553	127.36138	76	13
Tidore	North Maluku	M05	0.67401	127.40409	127	8
Moti	North Maluku	M06	0.45831	127.41410	24.8	1
Kasiruta	North Maluku	M07	0.37208	127.19057	472	2
Bacan	North Maluku	M08	0.56240	127.51023	1,900	13
Bisa	North Maluku	M09	1.21521	127.53926	190	3
Obi	North Maluku	M10	1.48904	127.75735	2,542	4
Buru	Maluku	M11	3.36176	126.73884	8,473	6
Manipa	Maluku	M12	3.30552	127.57675	159.7	3
Kelang	Maluku	M13	3.20529	127.73290	147	1
Boano	Maluku	M14	2.98065	127.92004	134	5
Seram	Maluku	M15	3.06055	129.48702	17,454	17
Ambon	Maluku	M16	3.62255	128.11273	806	15
Haruku	Maluku	M17	3.55906	128.46546	150	5
Saparua	Maluku	M18	3.57045	128.66598	168.1	7
Nusa Laut	Maluku	M19	3.66684	128.78217	28	5
Panjang	Maluku	M20	4.00721	131.22628	20.5	2
Manawoka	Maluku	M21	4.10951	131.31701	42	2
Goram	Maluku	M22	4.02399	131.41731	66	4
Kur	Maluku	M23	5.34228	131.99326	38	3
Taam	Maluku	M24	5.73786	132.18483	14	2
Kei Duluh	Maluku	M25	5.62245	132.78245	109	4
Kei Kecil	Maluku	M26	5.72038	132.71820	453	5
Kei Besar	Maluku	M27	5.63270	133.03761	555	7
Wokam	Maluku	M28	5.69265	134.51506	1604	6
Kobroor	Maluku	M29	6.19757	134.53896	1723	8
Trangan	Maluku	M30	6.53314	134.34812	2149	6
Wamar	Maluku	M31	5.79175	134.21659	56	4
			AUSTRALIA			
Eborac	Queensland	Q01	10.68259	142.53350	0.06	1
Great Woody	Queensland	Q02	10.70426	142.35268	0.46	1
Prince of Wales	Queensland	Q03	10.67906	142.17002	204.6	9

(continued)

Appendix II Continued

Island name	Province or State	Figure 2.2 annotation	Latitude south	Longitude east	Area km²	Snake species
Thursday	Queensland	Q04	10.57992	142.21940	3.5	7
Hammond	Queensland	Q05	10.55865	142.20714	14.5	2
Horn	Queensland	Q06	10.61158	142.28694	53	4
Moa	Queensland	Q07	10.17184	142.25567	170.2	3
Badu	Queensland	Q08	10.11879	142.14568	101	9
Jervis	Queensland	Q09	9.95533	142.18330	7.5	2
Warraber	Queensland	Q10	10.20828	142.82310	0.73	1
Yam	Queensland	Q11	9.89596	142.77340	2	3
Gabba	Queensland	Q12	9.76668	142.63370	4.3	1
Murray	Queensland	Q13	9.91859	144.04938	5.7	7
Yorke	Queensland	Q14	9.75224	143.41090	1.65	1
Darnley	Queensland	Q15	9.57970	143.77896	5.5	4
Saibai	Queensland	Q16	9.40513	142.70877	107.9	11
Dauan	Queensland	Q17	9.42175	142.53623	3.4	4
Boigu	Queensland	Q18	9.27405	142.22309	89.6	3
SOLOMON ISLANDS						
Shortlands	Western	S01	7.04642	155.73985	248	7
Pirumeri	Western	S02	7.11039	155.88280	3.7	1
Fauro	Western	S04	6.91514	156.08123	75	5
Vella Lavella	Western	S04	7.75219	156.65644	629	6
Simbo	Western	S06	8.28482	156.53056	14.7	3
Gizo	Western	S07	8.08213	156.80170	1.5	4
Kolombangara	Western	S08	7.99334	157.06315	688	5
Rendova	Western	S09	8.54908	157.30575	411	4
Tetepare	Western	S10	8.74089	157.56094	118	3
New Georgia	Western	S11	8.14305	157.50884	2,037	6
Vangunu	Western	S12	8.66348	157.99631	509	4
Choiseul	Choiseul	S13	7.05027	156.95127	2,971	6
Santa Isabel	Isabel	S14	7.66654	158.62679	2,999	7
Malaita	Malaita	S15	8.94513	160.14570	4,307	9
Guadalcanal	Guadalcanal	S16	9.57766	160.14570	5,336	10
Makira	Makira-Ulawa	S17	10.57361	161.80970	3,190	7

3

Reproductive Strategies of the Golden Lancehead, *Bothrops insularis*, from Queimada Grande Island

Constraints and Challenges

Karina N. Kasperoviczus, Henrique B. Braz, Ligia G. S. Amorim, and Selma M. Almeida-Santos

Introduction

Queimada Grande is a small island (0.43 km^2) located approximately 33 km off the coast of São Paulo state, southeastern Brazil (24°30′ S, 43°42′ W). It is a rocky, craggy island with no beaches and, therefore, difficult to access when the sea is rough (Amaral 1921; Martins *et al.* 2019). Most of the island is covered by vegetation typical of the Atlantic Forest. The climate is humid subtropical. Warmer temperatures occur from spring (October–December) to summer (January–March) and are associated with higher rainfall, whereas lower temperatures occur from autumn (April–June) to winter (July–September) and are associated with lower rainfall (Martins *et al.* 2019). The island has no source of fresh water besides rainfall (Amaral 1921).

Several seabirds and 41 perching birds are seen on the island, but most of them are seasonal migrants for short periods (Marques *et al.* 2012). Only two passerine birds seem to reside on the island: the House Wren, *Troglodytes aedon*, and the Bananaquit, *Coereba flaveola* (Marques *et al.* 2012). The island also harbors two species of bats (the molossids *Nyctinomops laticaudatus* and *N. macrotis*) but no terrestrial mammals. Other vertebrates inhabiting the island include two anuran amphibians (*Haddadus binotatus* and the endemic *Scinax peixotoi*), three small lizards (the gymnophthalmid *Colobodactylus taunay*, the scincid *Mabuya macrorhyncha*, and the introduced gekkonid *Hemidactylus mabouia*), two worm lizards (the amphisbaenids *Amphisbaena hogei* and *Leposternon microcephalum*), and two snakes (the dipsadid *Dipsas albifrons* and the viperid *Bothrops insularis*).

Bothrops insularis, commonly known as Golden Lancehead, is endemic to Queimada Grande. This species has been studied *in situ* and *ex situ* for more than 100 years, and some peculiarities of its biology were observed already in the first studies (Kasperoviczus and Almeida-Santos 2012). One intriguing peculiarity of this species

Karina N. Kasperoviczus, Henrique B. Braz, Ligia G. S. Amorim, and Selma M. Almeida-Santos, *Reproductive Strategies of the Golden Lancehead,* Bothrops insularis, *from Queimada Grande Island* In: *Islands and Snakes*. Edited by: Harvey B. Lillywhite and Marcio Martins, Oxford University Press. © Oxford University Press 2023. DOI: 10.1093/oso/9780197641521.003.0003

is that most, if not all, females exhibit hemipenes and associated retractor muscles (Hoge *et al.* 1959; Kasperoviczus 2009). The level of development of these structures varies among individual females (Kasperoviczus 2009; Garcia *et al.* 2022). Whereas some females have vestigial or malformed hemipenes and retractor muscles, others have hemipenes and retractor muscles similar to males (Figure 3.1A–E), suggesting that at least some females can evert and retract the hemipenes. Indeed, we recently observed a captive female everting the hemipenis and moving it from side to side (Figure 3.1F). It remains unclear whether the female hemipenes have been selected for some function or fixed in the population by genetic drift. Nevertheless, female

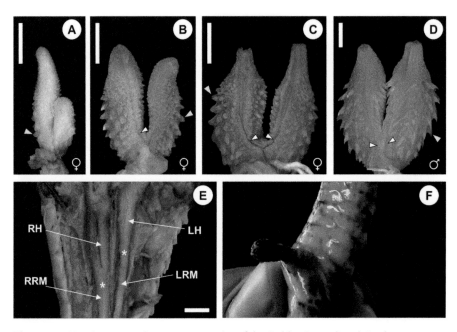

Figure 3.1 Hemipenes and retractor muscles of the Golden Lancehead, *Bothrops insularis*. (A) Female hemipenis with a malformation in one of the lobes. Note the absence of spermatic grooves and fully formed apices and the presence of spines with blunt and poorly formed tips. (B) Female hemipenis with slightly more developed spines and incomplete spermatic grooves but still showing malformation in one of the lobes. (C) A more developed female hemipenis, with complete spermatic grooves but also showing blunt spines. (D) Male hemipenis with complete spermatic sulcus, spines in greater number, and sharp spines. (E) Dissection of the tail of a female showing the hemipenes and retractor muscles. Asterisks indicate the point of origin of the retractor muscles. (F) Hemipenial eversion in a captive adult female. White arrowheads: Spermatic sulcus. Yellow arrowheads: Spines. RH: Right hemipenis. LH: Left hemipenis. RRM: Right retractor muscle. LRM: Left retractor muscle. Scale bar = 3 mm.

Photographs by Karina Kasperoviczus (A–E) and Kelly Kishi (F).

Golden Lanceheads have ZW sex chromosomes and are considered genetically female (Beçak 1965; Beçak et al. 1990).

Bothrops insularis exhibits many biological, ecological, and morphological divergences compared to its closest mainland relative, the Jararaca Lancehead, B. jararaca. Many of these divergences are consequences of insularity and are predicted by the island syndrome (Novosolov and Meiri 2013; Novosolov et al. 2013). For example, B. insularis occurs at a much higher population density than B. jararaca, showing one of the highest population densities recorded for snakes in the world (Marques et al. 2002; Martins et al. 2008; Guimarães et al. 2014; Abrahão et al. 2021). This high population density presumably reflects their coexistence with no interspecific competitors and few potential predators. Another difference from its mainland relative is the adult body size, which is, on average, much smaller (620 mm in males and 721 mm in females; Marques et al. 2013) than that of B. jararaca (760 mm in males and 1,115 mm in females; Almeida-Santos 2005).

Island environments also impose changes in the dietary niche and may even lead to the use of novel resources. Juvenile B. jararaca and B. insularis feed on small ectothermic prey (Martins et al. 2002). Adult B. jararaca feed primarily on small terrestrial mammals (Martins et al. 2002). Inasmuch as such prey type is lacking on the island, adult B. insularis have adapted to feed almost exclusively on passerine birds (Marques et al. 2002, 2012; Martins et al. 2002). Accordingly, B. insularis has evolved not only arboreal and diurnal habits but also morphological traits commonly found in arboreal snakes, such as long tails and more cranially positioned hearts than B. jararaca (Wüster et al. 2005). However, the island's resident passerine birds seemingly have evolved predation avoidance, and adult B. insularis feeds almost exclusively on two species of migratory passerine birds (Elaenia chilensis and Turdus flavipes) that visit and spend a few days on the island (Marques et al. 2012). In other words, adult B. insularis rely heavily on a highly seasonal resource to obtain energy to fuel reproduction. How male and female B. insularis manage such a seasonal and scarce energy source to reproduce is a great challenge that we describe here.

Temporal Dynamics of Food Availability and Feeding Activity

The White-crested Elaenia (Elaenia chilensis) and the Yellow-legged Thrush (Turdus flavipes) compose approximately 95% of the items consumed by the Golden Lancehead (based on a sample composed almost exclusively by adults; Marques et al. 2012). These migratory birds reach Queimada Grande Island in different seasons. The White-crested Elaenia visits the island in late summer, and the Yellow-legged Thrush visits the island during winter (Marques et al. 2012). However, adult B. insularis feed mainly in late summer on the White-crested Elaenia and to a lesser extent in winter on the Yellow-legged Thrush (Marques et al. 2012). Outside the migratory seasons of these birds, adult Golden Lanceheads may face a fasting period, as no migratory birds appear on the island in substantial numbers and other prey types compose a minor portion of the adult diet (Marques *et al.* 2012).

To gain insights into sex differences in feeding activity, we revisited the dietary data from Marques *et al.* (2012), separating individuals by sex and considering only adults. Both male and female Golden Lanceheads prey on the White-crested Elaenia during late summer (Figure 3.2). Males seemingly stop feeding in autumn-winter, whereas some females prey on the Yellow-legged Thrush and other unidentified birds in late autumn and early winter (Figure 3.2). It is noteworthy, however, that even during the summer, when the feeding peak occurs, the frequency of males and females that consume some food seems relatively low, ranging from 20–29% of the individuals sampled (Figure 3.2). Moreover, only two of the 47 (*i.e.*, 4.3%) individual *B. insularis* that were found to contain food items by Marques *et al.* (2012) had eaten more than one bird. Both snakes were females; one had eaten two White-crested Elaenia in late summer (March), and the other had eaten two Yellow-legged Thrushes in early winter (July). These observations suggest that annual energy acquisition by adult individuals of *B. insularis* is quite low, such that a substantial number of individuals may acquire only 1–2 food items each year.

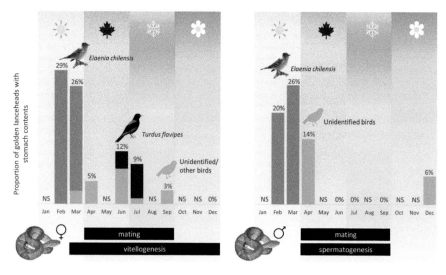

Figure 3.2 Seasonal timing of feeding and reproductive phenology of males (left) and females (right) of the Golden Lancehead, *Bothrops insularis*, from Queimada Grande Island. Green bars indicate the White-crested Elaenia (*Elaenia chilensis*), dark blue bars the Yellow-legged Thrush (*Turdus flavipes*), and gray bars other or unidentified birds. Values above the bars indicate the percentage of individuals found with stomach contents per month. NS: No sampling done in the month.

Photographs of the birds by Arthur Macarrão (*Elaenia chilensis*) and Claudia Brasileiro (*Turdus flavipes*). Photograph of *Bothrops insularis* by Ricardo Sawaya.

The Reproductive Phenology of the Golden Lancehead is Largely Driven by the Seasonal Timing of Prey Availability

The reproductive phenology is relatively conserved phylogenetically across *Bothrops*. Male *Bothrops* studied thus far show one of two seasonal patterns of spermatogenesis: spring–summer or summer–autumn (reviewed in Silva *et al.* 2020). Mainland *B. jararaca*, in particular, shows a spring–summer spermatogenesis pattern (Almeida-Santos and Salomão 2002; Almeida-Santos 2005; Kasperoviczus 2013). Regardless of the timing of spermatogenesis, mating season occurs in autumn in most *Bothrops* species, synchronously with the onset of vitellogenesis. Therefore, female *Bothrops* must store sperm in their oviduct until ovarian follicles reach ovulatory sizes, which occurs primarily in spring. Parturition occurs between summer and early autumn in most *Bothrops* species (Almeida-Santos and Salomão 2002).

Bothrops insularis also reproduce seasonally, like the mainland *B. jararaca* and most congeners. However, the timing of many reproductive events of *B. insularis* differs widely from its congeners. In spring–summer, when male *Bothrops* have testes producing spermatozoa, male *B. insularis* have inactive testes (Kasperoviczus 2009). Spermatogenesis in *B. insularis* starts in autumn and proceeds until early spring and is therefore delayed compared to its congeners (Figure 3.2; Kasperoviczus 2009). In early autumn, viable sperm can already be found in the ductus deferentia (Kasperoviczus 2009; Silva *et al.* 2015). As with other congeners, mating occurs in autumn, when some females are at early vitellogenesis. Therefore, female *B. insularis* must also store sperm in their oviduct until ovarian follicles reach ovulatory sizes later in the year. However, the mating season of *B. insularis* extends to late winter (Figure 3.2) and is, therefore, much longer than the autumnal mating season observed in most congeners. Some female *B. insularis* have enlarged, preovulatory-sized follicles in early winter and ovulate in late winter (Marques *et al.* 2013), much earlier than female *B. jararaca*, which have enlarged, preovulatory-sized follicles in early spring and ovulate in mid-spring (Almeida-Santos 2005). However, between late winter and late spring, many female *B. insularis* still have small to mid-sized (10–15 mm diameter) vitellogenic follicles (Marques *et al.* 2013), while female *B. jararaca* have no small to mid-sized follicles since mid-autumn (Almeida-Santos 2005). Parturitions, however, have been recorded between mid-summer and early autumn (Marques *et al.* 2013), as in most *Bothrops* species (Almeida-Santos and Salomão 2002).

Why does *B. insularis* show this distinctive reproductive pattern compared with other congeners? We suggest that the reproductive strategies of male and female Golden Lanceheads are largely driven by the seasonal timing of bird availability. Energy intake and storage are required to initiate sperm production and vitellogenesis (Olsson *et al.* 1997; Shine 2003). Because male *B. insularis* feed essentially in late summer (when spermatogenesis begins or peaks in other congeners), spermatogenesis is pushed to begin in autumn (Figure 3.2), only after males have acquired energy reserves. Histological analyses have detected sperm in the posterior oviduct of early vitellogenic females in April (Kasperoviczus 2009), thus confirming that mating occurs as soon as early autumn. Between late autumn and early winter, males are still actively searching for females to copulate. In an expedition in late autumn, for example, 40% of the males we observed were crawling, a proportion much higher than the maximum of 9% observed in other periods (Marques *et al.* 2013). In fact, our group has already observed seven interindividual interactions on the island between late autumn and early winter (Figure 3.3), most of them lasting 3–4 hours (Marques *et al.* 2013). In one of the records (sex unrecorded), the courting individuals had similar body sizes (Figure 3.3B), contrasting with other courtship observations in which the male was much smaller than the female (Figure 3.3C). Interestingly, none of the interactions recorded between late autumn and early winter resulted in mating (Marques *et al.* 2013).

All females that courted during late autumn and early winter observations had no vitellogenic follicles detectable by palpation (Marques *et al.* 2013), suggesting they

Figure 3.3 Courtship behavior in the Golden Lancehead, *Bothrops insularis*, on Queimada Grande Island. (A) A pair observed on a tree branch in mid-autumn (May 30, 2018). (B) A pair (sex unrecorded) found in early winter (July 1995). Note that, unlike in C, both individuals are similar-sized. (C) A male inspecting a female using tongue flicking while aligning his body to hers (late autumn, July 2007). No copulation was observed.
Photographs by Ligia Grazziely dos Santos Amorim (A), Paulo Lara (B), and Otavio Augusto Vuolo Marques (C).

were non-reproductive. Given the high cost required to sustain vitellogenesis and the relatively low proportion of female *B. insularis* that feed in summer (26–29% of the sampled females; Figure 3.2), it is plausible that only a few females are able to obtain enough energy reserves during the feeding activity in summer. We suspect that most females acquire insufficient (if any) energy stores in summer feeding activity to fuel vitellogenesis, and, consequently, they are not receptive to mate between autumn and early winter. Moreover, females had stomach contents in three of the six male–female interactions observed between late autumn and early winter, indicating recent feeding (Marques *et al.* 2013). Indeed, in this period, some females feed on Yellow-legged Thrushes visiting the island (Figure 3.2). Non-receptiveness after feeding has been reported in an arboreal congener, the Two-striped Forest Pitviper *B. bilineatus* (Turci *et al.* 2009). Thus, some female Golden Lanceheads may devote their time during late autumn and early winter to foraging rather than mating. Between mid- and late winter (August–September), mating has been recorded in free-ranging or recently captured *B. insularis* (Amaral 1921; Amorim *et al.* 2019). Therefore, by feeding on thrushes in late autumn–early winter, some females may acquire the required energy, or even gain "extra" energy, to fuel vitellogenesis and increase their chances of reproducing. This feeding activity in late autumn–early winter may also explain why many female *B. insularis* are still at early to mid-vitellogenesis between late winter and spring. Follicular growth beginning and proceeding between late winter and early spring is also suggestive that mating stimulates ovarian folliculogenesis (Whittier and Crews 1986).

To increase reproductive success, males may have evolved an extended mating season (compared with mainland relatives) to track this irregular and asynchronous vitellogenesis in females. As there may be few females accessible (*i.e.*, in estrus) during the mating season, it may be advantageous for males to employ some mechanism to ensure their paternity once they manage to copulate. On some occasions, males have been found on top of females during the mating season (Figure 3.4). Unfortunately, we cannot assure whether these individuals did mate, but on one occasion, the female had an enlarged cloaca (Figure 3.4A–B), suggesting recent mating. Nevertheless, the finding of males coiled on top of females is highly suggestive of mating guarding or vigilance behavior to prevent rival males from approaching and attempting to mate with the female (Luiselli 1993; Almeida-Santos *et al.* 1999). A similar interaction has been described in the Two-striped Forest Pitviper, *B. bilineatus*, in which a male coiled over a female after courtship on a tree trunk (Turci *et al.* 2009).

Reproductive Output

The reproductive output of *B. insularis* differs enormously from that of *B. jararaca*. Potential litter size (estimated by the number of enlarged vitellogenic follicles) is significantly smaller in *B. insularis* (mean = 8.2 ± 4.2, range = 3–20) than in *B. jararaca* (mean 19.6 ± 6.3, range = 11–36) (Marques *et al.* 2013). Moreover, female *B. insularis* exhibit a

Figure 3.4 Potential mate guarding or post-mating vigilance behavior by male Golden Lanceheads, *Bothrops insularis*, during the mating season on Queimada Grande Island. (A) A male coiled on top of a female on a tree in late autumn (June). (B) A closer photo of the couple depicted in A showing the female's enlarged cloaca (arrowhead), suggesting recent mating. (C) A male on top of a female found on the ground in early winter (July).
Photographs by Flora Roncolatto Ortiz (A–B) and Karina Rodrigues da Silva Banci (C).

high (~45%) incidence of undeveloped oviductal eggs (Marques *et al.* 2013), presumably decreasing their reproductive output. However, the small litter size of *B. insularis* seems to reflect the smaller body size it reaches relative to *B. jararaca* because the interspecific difference in potential litter size disappears after the effect of maternal body size on litter size is taken into account (Marques *et al.* 2013). Data from free-ranging individuals brought to our laboratory show that female *B. insularis* produce smaller offspring than female *B. jararaca* (unpaired *t*-test: $t = 8.22$, df = 221, p < 0.0001); offspring size averages 229.0 ± 19.5 mm (range = 190–275 mm) in *B. insularis* and 254.3 ± 19.9 mm (range = 181–290 mm) in *B. jararaca*. However, unlike what was observed for litter size, the offspring size in *B. insularis* is still smaller than in *B. jararaca* after accounting for the interspecific variation in maternal body size (ANCOVA with maternal SVL as the covariate: $F_{1,20}$ = 6.67, p = 0.018). The frequency of reproductive female *B. insularis* observed per year averages 15%, varying from 0% to 50% (Marques *et al.* 2013). These values suggest that females show triennial or longer cycles. In contrast, approximately half of the females of mainland populations of *B. jararaca* reproduce annually, indicating a biennial reproductive frequency (Sazima 1992; Almeida-Santos 2005). Moreover, the proportion of newborns sampled in the population of *B. insularis* (0.6%) is much lower than that of mainland *B. jararaca* (17.8%), suggesting a lower rate of newborn recruitment (Marques *et al.* 2013). Therefore, female *B. insularis* produce smaller litters of smaller offspring at much longer intervals than their mainland closest relative, *B. jararaca*. Even with this low reproductive output, *B. insularis* shows a much higher population density than its sister species (*B. jararaca*), presumably reflecting the occurrence of few potential predators on the island.

The divergence in reproductive output between *B. insularis* and *B. jararaca* has been suggested to result from physiological constraints on gravid females or the effects of prolonged inbreeding in *B. insularis* (Wüster *et al.* 2005). Alternatively, the low reproductive output of *B. insularis* may directly or indirectly reflect its low annual rate of food intake (see also Marques *et al.* 2012). The limited energetic resources available on the island may have favored the evolution of smaller body sizes in *B. insularis* compared with *B. jararaca* (see Chapter 4, this volume), which in turn influences their smaller litter size. As female snakes usually do not feed during pregnancy, female Golden Lanceheads likely miss the feeding peak in summer and have little opportunity to feed during the years they are pregnant. Additionally, the low feeding rate of Golden Lanceheads may lengthen the time required to accumulate sufficient energy reserves for reproduction. Unfortunately, we do not know whether and to what extent the number of migrating birds varies annually, but a high resource fluctuation among years may be the source of the great year-to-year variation in the reproductive frequency of *B. insularis*. Female Golden Lanceheads kept in our laboratory as part of a breeding and conservation program have been fed mice monthly and observed to reproduce every other year (S. M. Almeida-Santos, unpublished data). Thus, the low reproductive frequency of *B. insularis* may result from the low annual food intake on the island.

Conclusion

Animals living on islands often show divergent traits compared to their mainland relatives. Despite its relatively recent origin, the island endemic Golden Lancehead, *B. insularis*, also differs in many traits from its mainland relatives, including its reproductive phenology and reproductive output. We argue here that the divergence in reproductive traits is largely due to the low and highly seasonal feeding activity of Golden Lanceheads on the island.

The discussion herein raises several questions for future studies. Answering these questions requires experimental tests in nature, captivity, or a combination of both. Future studies should manipulate food intake by Golden Lanceheads in captivity to test if the reproductive divergences described here reflect phenotypic plasticity or local adaptation. Preliminary observations indicate that, under a constant food intake regime, female Golden Lanceheads are able to reproduce biennially, like their mainland relative, *B. jararaca*. It is unclear, however, how litter size and offspring size will respond to a constant food intake regime. The same is true for male reproductive phenology. Would the timing of spermatogenesis change (resembling that of mainland relatives) in individual males kept on a constant food intake regime but mimicking the climate conditions on the island?

The potential tradeoff between feeding and mating in winter also warrants further investigation. Specifically, do males trade feeding opportunities for mating, while females trade mating opportunities for feeding? Or does this pattern merely reflect a limitation of male Golden Lanceheads to consume the Yellow-legged Thrush, which is larger than the White-crested Elaenia? An intriguing feature of *B. insularis* that has yet to be explained is why females produce so many nonviable eggs compared to their congeners and other snakes. Does the species' low annual feeding rate also influence the high frequency of nonviable eggs? Does the high frequency of nonviable eggs reflect the deleterious effects of prolonged inbreeding?

In addition, many questions on the reproductive behavior of the Golden Lancehead remain to be answered. For example, our suggestion of mating guarding or vigilance behavior by males requires confirmation. Do males and females mate once or more than once during the extended mating season? Moreover, does the female hemipenis show some role during reproductive interactions? These questions may also be assessed *in situ* and *ex situ*; however, given the high population density and extended mating season, *in situ* observations should be relatively straightforward to make and have the advantage of eliminating any captivity-induced behavioral anomaly. Ultimately, a comprehensive understanding of the reproductive biology of the Golden Lancehead is crucial to implementing informed and successful conservation strategies.

References

Abrahão, C. R., L. G. Amorim, A. M. Magalhães, C. R. Azevedo, J. H. H. Grisi-Filho, and R. A. Dias. 2021. Extinction risk evaluation and population size estimation of *Bothrops insularis* (Serpentes: Viperidae), a critically endangered insular pitviper species of Brazil. *South American Journal of Herpetology* 19:32–39.

Almeida-Santos, S. M. 2005. Modelos reprodutivos em serpentes: Estocagem de esperma e placentação em *Crotalus durissus* e *Bothrops jararaca* (Serpentes: Viperidae). PhD thesis, Universidade de São Paulo.

Almeida-Santos, S. M., E. A. Peneti, M. G. Salomão, E. S. Guimarães, and P. S. Sena. 1999. Predatory combat and tail wrestling in hierarchical contests of the Neotropical rattlesnake *Crotalus durissus terrificus* (Serpentes: Viperidae). *Amphibia-Reptilia* 20:88–96.

Almeida-Santos, S. M., and M. G. Salomão. 2002. Reproduction in neotropical pitvipers, with emphasis on species of the genus *Bothrops*. In G. W. Schuett, M. Höggren, M. E. Douglas and H. W. Greene (eds.), *Biology of the Vipers*. Eagle Mountain Publishing, pp. 445–462.

Amaral, A. 1921. Contribution towards the knowledge of snakes in Brazil. Part I. Four new species of Brazilian snakes. *Anexos das Memórias do Instituto Butantan* 1:46–81.

Amorim, L. G. S., C. R. Azevedo, W. S. Azevedo, and S. M. Almeida-Santos. 2019. First record of mating of *Bothrops insularis* (Serpentes: Viperidae) in nature, with comments on sexual behaviour. *Herpetology Notes* 12:225–227.

Beçak, M. L., M. N. Rabello-Gay, W. Beçak, M. Soma, R. F. Batistic, and I. Trajtengertz. 1990. The W chromosome during the evolution and in sex abnormalities of snakes. In E. Olmo (ed.), *Cytogenetics of Amphibians and Reptiles*. Birkhauser Verlag, pp. 221–240.

Beçak, W. 1965. Constituição cromossômica e mecanismo de determinação do sexo em ofídios sulamericanos. I. Aspectos cariotípicos. *Memórias do Instituto Butantan* 32:37–78.

Garcia, V. C., L. G. S. Amorim, and S. M. Almeida-Santos. 2022. Morphological and structural differences between the hemipenes and hemiclitores of golden lancehead snakes, *Bothrops insularis* (Amaral, 1922), revealed by radiography. *Anatomia Histologia Embryologia* 51:557–560.

Guimarães, M., R. Munguía-Steyer, P. F. Doherty, M. Martins, and R. J. Sawaya. 2014. Population dynamics of the critically endangered golden lancehead pitviper, *Bothrops insularis*: Stability or decline? *PLoS One* 9:e95203.

Hoge, A. R., H. E. Belluomini, G. Schreiber, and A. M. Penha. 1959. Sexual abnormalities in *Bothrops insularis* (Amaral) 1921 (Serpentes). *Memórias do Instituto Butantan* 29:17–88.

Kasperoviczus, K. N. 2009. Biologia reprodutiva da jararaca ilhoa, *Bothrops insularis* (Serpentes: Viperidae), da Ilha da Queimada Grande, São Paulo. Master's dissertation, Universidade de São Paulo.

Kasperoviczus, K. N. 2013. Evolução das estratégias reprodutivas de *Bothrops jararaca* (Serpentes: Viperidae). PhD thesis, Universidade de São Paulo.

Kasperoviczus, K. N., and S. M. Almeida-Santos. 2012. Instituto Butantan e a jararaca-ilhoa: Cem anos de história, mitos e ciência. *Cadernos de História da Ciência - Instituto Butantan* 8:255–269

Luiselli, L. 1993. Are sperm storage and within-season multiple mating important components of the adder reproductive biology? *Acta Oecologica (Montrouge)* 14:705–710.

Marques, O. A. V., K. N. Kasperoviczus, and S. M. Almeida-Santos. 2013. Reproductive ecology of the threatened pitviper *Bothrops insularis* from Queimada Grande Island, southeast Brazil. *Journal of Herpetology* 47:393–399.

Marques, O. A. V., M. Martins, P. F. Develey, A. Macarrão, and I. Sazima. 2012. The golden lancehead *Bothrops insularis* (Serpentes: Viperidae) relies on two seasonally plentiful bird species visiting its island habitat. *Journal of Natural History* 46:885–895.

Marques, O. A. V., M. Martins, and I. Sazima. 2002. A jararaca da ilha da Queimada Grande. *Ciência Hoje* 31:56–59

Martins, M., O. A. V. Marques, and I. Sazima. 2002. Ecological and phylogenetic correlates of feeding habits in neotropical pitvipers of the genus *Bothrops*. In G. W. Schuett, M. Höggren, M. E. Douglas, and H. W. Greene (eds.), *Biology of the Vipers*. Eagle Mountain Publishing, pp. 307–328.

Martins, M., R. J. Sawaya, S. M. Almeida-Santos, and O. A. V. Marques. 2019. The Queimada Grande Island and its biological treasure. In H. B. Lillywhite and , M. Martins (eds.), *Islands and Snakes: Isolation and Adaptive Evolution*. Oxford University Press, pp. 117–137.

Martins, M., R. J. Sawaya, and O. A. V. Marques. 2008. A first estimate of the population size of the critically endangered lancehead, *Bothrops insularis*. *South American Journal of Herpetology* 3:168–174.

Novosolov, M., and S. Meiri. 2013. The effect of island type on lizard reproductive traits. *Journal of Biogeography* 40:2385–2395.

Novosolov, M., P. Raia, and S. Meiri. 2013. The island syndrome in lizards. *Global Ecology and Biogeography* 22:184–191.

Olsson, M., T. Madsen, and R. Shine. 1997. Is sperm really so cheap? Costs of reproduction in male adders, Vipera berus. *Proceedings of the Royal Society of London B* 264:455–459.

Sazima, I. 1992. Natural history of the Jararaca pitviper, *Bothrops jararaca*, in southeastern Brazil. In J. A. Campbell and E. D. Brodie, Jr (eds.), *Biology of the Pitvipers*. Selva Press, pp. 199–216.

Shine, R. 2003. Reproductive strategies in snakes. *Proceedings of the Royal Society of London B* 270:995–1004.

Silva, K. B., M. A. Zogno, A. B. Camillo, R. J. G. Pereira, and S. M. Almeida-Santos. 2015. Annual changes in seminal variables of golden lancehead pitvipers (*Bothrops insularis*) maintained in captivity. *Animal Reproduction Science* 163:144–150.

Silva, K. M. P., H. B. Braz, K. N. Kasperoviczus, O. A. V. Marques, and S. M. Almeida-Santos. 2020. Reproduction in the pitviper *Bothrops jararacussu*: Large females increase their reproductive output while small males increase their potential to mate. *Zoology* 142:125816.

Turci, L. C. B., S. Albuquerque, P. S. Bernarde, and D. B. Miranda. 2009. Uso do hábitat, atividade e comportamento de *Bothriopsis bilineatus* e de *Bothrops atrox* (Serpentes: Viperidae) na floresta do Rio Moa, Acre, Brasil. *Biota Neotropica* 9:197–206.

Whittier, J. M., and D. Crews. 1986. Ovarian development in red-sided garter snakes, *Thamnophis sirtalis parietalis*: Relationship to mating. *General and Comparative Endocrinology* 61:5–12.

Wüster, W., M. R. Duarte, and M. G. Salomão. 2005. Morphological correlates of incipient arboreality and ornithophagy in island pitvipers, and the phylogenetic position of *Bothrops insularis*. *Journal of Zoology* 266:1–10.

4

Lanceheads in Land-Bridge Islands of Brazil

Repeated and Parallel Evolution of Dwarf Pitvipers

*Ricardo J. Sawaya, Fausto E. Barbo, Felipe G. Grazziotin,
Otavio A. V. Marques, and Marcio Martins*

Introduction

The structure, function, and behavior of insular populations are the outcome of evolutionary changes in response to insular environments (Martins and Lillywhite 2019). A well-known example of such responses is the change in body size observed in many insular vertebrates, with populations becoming smaller or larger in comparison with their mainland relatives. Indeed, there are several examples of both "gigantism" in small animals and "dwarfism" in large animals on islands, what has been called the "island rule" (Foster 1964; Van Valen 1973; Case 1978; Lomolino 1985, 2005; Whittaker and Fernández-Palacios 2007). Note that here "gigantism" and "dwarfism" refer only to the larger or smaller body size of insular populations in relation to the size of its mainland, ancestral populations. The island rule results from a combination of factors affecting animals of different sizes in different ways (*e.g.*, Lomolino 2005; Benítez-López *et al.* 2021). Although pervasive among vertebrates, especially mammals, birds, and reptiles, dwarfism and gigantism could be mediated by climate, resource availability and island features like size and isolation (Case 1978; Lomolino 2005; Benítez-López *et al.* 2021). The most pronounced effects of these factors have been demonstrated for mammals and reptiles in smaller and more remote islands (Benítez-López *et al.* 2021). Common examples are the extinct dwarf elephant from the Mediterranean (Stock 1935) and the giant tortoises of the Galapagos Islands (see Carlquist 1974).

Snakes are a relatively well-studied group of vertebrates regarding body size changes on islands. In a review on body size evolution in snakes, Boback (2003) found that island features including area, age, and isolation were apparently not involved in body size changes in insular snakes. On the other hand, this author found consistent evidence of body size changes in insular snakes related to prey size availability on islands, the so-called *diet alteration hypothesis*: snakes feeding on smaller prey than those found on the mainland tend to become smaller, while those feeding

Ricardo J. Sawaya, Fausto E. Barbo, Felipe G. Grazziotin, Otavio A. V. Marques, and Marcio Martins, *Lanceheads in Land-Bridge Islands of Brazil* In: *Islands and Snakes*. Edited by: Harvey B. Lillywhite and Marcio Martins, Oxford University Press.

upon larger prey tend to become larger (see figure 1.8 in Martins and Lillywhite 2019). Furthermore, Boback (2003) associated the trends found in body size of insular snakes to their foraging strategies: the typically ambush (sit-and-wait) vipers tend to become dwarfs in islands while the active-foraging colubrids (and also an elapid) tend to become giants. In addition, colubrids and elapids frequently feed on nestling seabirds while dwarf vipers frequently feed on squamate reptiles.

Elapids (genus *Notechis*) from Australia are a good example of body size changes in insular populations owing to differences in prey size availability. Different populations of *Notechis* colonized land-bridge islands in coastal Australia. Most of them feed on chicks of nesting seabirds, a relatively large prey compared to the original prey size of the mainland populations from which the insular populations originated. Consequently, populations of *Notechis* inhabiting islands containing seabird colonies became "giants" compared to their mainland ancestors (see review in Aubret 2019). The shifts in body size observed in insular populations of *Notechis* occurred in a relatively short period of time: these populations are estimated to be less than 10,000 years old (Keogh *et al.* 2005; they became isolated from mainland populations since the Holocene climatic optimum).

In addition to the size of prey available on islands, resource availability may interact with growth, also resulting in body size changes in snakes (Martins and Lillywhite 2019; Hasegawa and Mori, Chapter 6 of this volume). The larger body size of the insular population of Florida cottonmouth snakes (*Agkistrodon conanti*) at Seahorse Key may have been the outcome of a higher abundance of food at the island, compared to mainland habitats (Lillywhite and Sheehy 2019). On the other hand, dwarfism observed in a population of *Vipera ammodytes* in Golem Grad Island (Macedonia) has been interpreted as the outcome of lower energy intake of their available ectothermic prey (lizards and centipedes) when compared to mainland populations preying mainly upon rodents (Tomović *et al.* 2022; see also Chapter 12, this volume). These authors suggest that both plasticity and adaptation may influence the phenotype of this island viper.

Among the allometric processes (*i.e.*, processes related to changes in body proportions during development) affecting populations, heterochrony is known to induce dramatic changes in the phenotype of different organisms, including body size (Gould 1977; Gould and MacFadden 2004; Keyte and Smith 2014). Although there are several definitions for heterochrony (reviewed in Smith 2001), it can be generally understood as changes in rate or timing of some particular ontogenetic processes (*i.e.*, processes occurring during the development and growth of an individual; Rice 1997). Heterochrony can generate dwarf or giant organisms through several distinct processes as pedomorphosis, peramorphosis, proportional dwarfism, and proportional gigantism (Gould 1977; Alberch *et al.* 1979; Smith 2001). Progenesis—the attainment of sexual maturity by an organism in an early developmental stage—is one of the processes generating pedomorphic populations, resulting in adults exhibiting juvenile characters including smaller body size (Gould 1977; Alberch *et al.* 1979). In dwarfism by progenesis, although maturation happens earlier, the relationship between shape and size remains constant (Smith 2001). Other heterochronic processes that can generate dwarf organisms are proportional dwarfism (Gould 1977; Alberch *et al.* 1979). This process is not defined by changes in maturation time or shape development as in pedomorphism. Instead, they are defined by the decrease of the body size in relation to shape and maturation, which remain constant. However, the process involved in each of the cases of body size change in insular snakes is still poorly known and deserve further studies. In any case, these changes in body size most probably facilitated the persistence of snakes in insular habitats (McNab 2002).

In this chapter, we explore the biology of lanceheads of the *Bothrops jararaca* species complex inhabiting land-bridge islands in Brazil, highlighting the repeated and parallel evolution of body size in some species, and we discuss the diet alteration hypothesis as the most probable cause of such changes.

The Geographical Setting and the *Bothrops jararaca* Species Complex

The Brazilian coast is 7,367 km long (Hudson 1998), extending along more than 37 degrees of latitude, from 33°44′ South to 04°26′ North. The Atlantic Forest covers most of the Brazilian coast, from 3° to 30° South, and from 35° to 60° West, from the coast to inland areas, reaching Argentina and Paraguay (Galindo-Leal and Câmara 2003). There are hundreds of land-bridge islands along the Brazilian coast, which must have been isolated from the mainland during sea-level oscillations from the late Pleistocene to the early Holocene (Furtado 2013; see also figure 9 in Barbo *et al.* 2022b). It has been proposed that portions of the Atlantic Forest coastal populations have been trapped on land-bridge islands off the Brazilian coast during these sea-level oscillations of the last hundreds of thousands of years, resulting in the isolation and diversification of many insular populations and species (Barbo *et al.* 2022b).

The genus *Bothrops* is one of the most speciose genera among crotaline snakes (Campbell and Lamar 2004; Carrasco *et al.* 2012) consisting of 48 currently recognized species inhabiting a wide diversity of ecosystems in Central and South America (Uetz *et al.* 2022). Species of *Bothrops* are generally separated into seven monophyletic groups: *B. alternatus, B. atrox, B. jabrensis, B. jararaca, B. jararacussu, B. neuwiedi,* and *B. taeniatus* group (Salomão *et al.* 1997; Campbell and Lamar 2004; Wüster *et al.* 2002; Carrasco *et al.* 2012; Alencar *et al.* 2016; Barbo *et al.* 2022a).

The *B. jararaca* species group has been well studied in recent years (*e.g.*, Sazima 1992; Martins *et al.* 2001, 2002), including the description of four new island species in the past 20 years (Marques *et al.* 2002; Barbo *et al.* 2012, 2016, 2022b). The current taxonomic composition of this group consists of six species (Figure 4.1): *B. jararaca*, a widespread mainland species, considered a species complex including two

Figure 4.1 Lanceheads of the *Bothrops jararaca* species group. The mainland species *B. jararaca* has been divided by molecular studies in the South Clade (A) and North Clade (B) populations (see Barbo *et al.* 2022a for details). Five insular species have been described so far: (C) *B. alcatraz*; (D) *B. germanoi*; (E) *B. insularis*; (F) *B. otavioi*; and (G) *B. sazimai*.

Photographs: (A) M. B. Martins (Rio Grande do Sul state); (B) R. J. Sawaya (Cunha Municipality, São Paulo state); (C) O. A. V. Marques (Alcatrazes Island, São Paulo state); (D) M. R. Duarte (Moela Island, São Paulo state); (E) M. Martins (Queimada Grande Island, São Paulo state); (F) F. C. Centeno (Vitória Island, São Paulo state); (G) R. J. Sawaya (Franceses Island, São Paulo state).

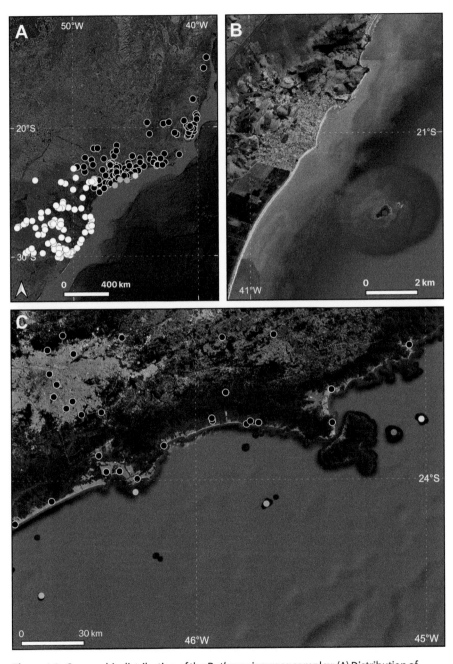

Figure 4.2 Geographic distribution of the *Bothrops jararaca* complex: (A) Distribution of the South Clade (SC: white dots) and the North Clade (NC: black dots) of the mainland populations of *B. jararaca*; (B) distribution of the insular species *B. sazimai* in Ilha dos Franceses, Espírito Santo state; (C) São Paulo state coast, highlighting the distribution of the island dwellers, from West to East: *B. insularis* (green dot), *B. germanoi* (orange dot), *B. alcatraz* (blue dot), *B.* aff. *jararaca* from Búzios Island (pink dot), and *B. otavioi* (yellow dot).

main clades in southern and northern regions of the Atlantic Forest (Figures 4.1A,B, Figure 4.2; see Grazziotin *et al.* 2006; Barbo *et al.* 2022b), and five island endemics on the southeastern Brazilian coast: *B. alcatraz* from Alcatrazes Island (Marques *et al.*, 2002) (Figure 4.1C); *B. germanoi* from Moela Island (Barbo *et al.* 2022b) (Figure 4.1D); *B. insularis* from Queimada Grande Island (Amaral 1921) (Figure 4.1E); *B. otavioi* from Vitória Island (Barbo *et al.* 2012) (Figure 4.1F); and *B. sazimai* from Franceses Island (Barbo *et al.* 2016) (Figure 4.1G). Another insular population in the State of São Paulo, from Búzios Island, has also been considered as another possibly endemic species, but it is still under formal description (F. E. Barbo and R. J. Sawaya, unpublished data).

The Mainland Species and the Forms from Land-Bridge Islands

Adults of the mainland lancehead *B. jararaca* feed mainly on rodents. This endothermic prey represents more than 70% of their diet (Martins *et al.* 2002; Table 4.1). On the other hand, ectothermic prey corresponds to most of the diet of juvenile *B. jararaca*. Thus, similarly to several snakes, there is an evident ontogenetic shift in the diet of this pitviper, with juveniles preying on ectothermic animals (centipedes, anurans, and lizards) and adults preying on endothermic prey (rodents and birds; Martins *et al.* 2002). Juveniles of *B. jararaca* usually have a pale tail tip associated with caudal luring (Sazima 1991). Both juveniles and adults are ambush (sit-and-wait) predators and hunt mostly at night (Sazima 1992).

We have recorded a considerable variation in body size (snout-vent length [SVL]) among the North and South clades of *B. jararaca* in the mainland species as well as in the six island forms (Figure 4.3; Table 4.1). Mainland populations of *B. jararaca* from the South and North clades reach more than 1,100 and 1,400 mm SVL, respectively. The ranges of body size of the four island populations feeding upon ectothermic prey (*B. alcatraz*, *B. germanoi*, *B. otavioi*, and *B. sazimai*) are less than 700 mm of SVL, or 53.7–85.3% of SVL of their mainland counterparts (Table 4.1). The only two island forms that feed on endothermic prey, *B. aff. jararaca* from Búzios Island and the Golden Lancehead, *B. insularis*, have more than 1,000 mm of SVL, or 105.4–106.3% and 92.2–97.0% of mainland forms, respectively (Figure 4.3; Table 4.1).

Body sizes of mature individuals, given by the smallest mature male recorded (males are smaller than females; Figure 4.3; Table 4.1), indicate that insular lanceheads feeding upon ectothermic prey are pedomorphic. The smallest mature males of *B. alcatraz* and *B. otavioi* have between 300 and 400 mm of SVL, and *B. sazimai* between 400 and 500 mm of SVL (Figure 4.3). Comparatively, the smallest males of the mainland *B. jararaca* recorded have between 500 and 600 mm SVL in the South Clade and 700 to 800 mm in the North Clade. On the other hand, the insular forms preying upon endothermic prey had the smallest mature males, between 500 and

Table 4.1 Body size range and diet of mainland and insular forms of lanceheads of the *Bothrops jararaca* complex

Species / population	SVL range (mm) (Median; N)	% size (Median SVL) from Clade South and Clade North	Smallest adult male (SVL, mm)	Diet
B. jararaca (Clade South)	197.0 – 1158 (674.0; N = 241)	- 99.2%	506[a]	endothermic: mammals (N = 32); birds (N = 3);
B. jararaca (Clade North)	202 – 1461 (679.5; N = 468)	100.8% -	613[a]	ectothermic: centipedes (N = 1); frogs (N = 8)*
B. alcatraz	363 – 523 (439.0; N = 24)	65.1% 64.6%	365[b]	ectothermic: centipedes (N = 9); frogs (N = 1); lizards (N = 3)[b]
B. germanoi	215 – 632 (575.0; N = 17)	85.3% 84.6%	-	ectothermic: centipedes (N = 4), lizards (N = 2)[c]
B. insularis	385 – 1016 (654.0; N = 374)	97.0 % 96.2%	505[d]	endothermic: birds (N = 30); ectothermic: centipedes (N = 2); frogs (N = 3); snake (N = 1)*
B. otavioi	203 – 692 (365.0; N = 31)	54.2% 53.7%	388[e]	ectothermic: frogs (N = 3)[e]
B. sazimai	306 – 700 (565.0; N = 57)	83.8% 83.1%	460[f]	ectothermic: centipedes (N = 6); lizards (N = 14); snake (N = 1)[f]
B. aff. jararaca (Búzios Island)	235 – 1011 (716.5; N = 82)	106.3% 105.4%	661[a]	endothermic: birds (N = 6); ectothermic: lizards (N = 3)*

Showing proportion of snout to vent length (SVL) from Clade South and Clade North populations of *Bothrops jararaca*, respectively, and smallest SVL of mature males. The categories of prey recorded are for both populations together.

[a] Kasperoviczus 2013.

[b] Marques *et al.* 2002.

[c] Barbo *et al.* 2022b.

[d] Marques *et al.* 2013.

[e] Barbo *et al.* 2012.

[f] Barbo *et al.* 2016.

* Unpublished data.

600 mm of SVL for *B. insularis*, and 600 and 700 mm of SVL for *B. aff. jararaca* from Búzios Island (Figure 4.3). Thus, our data on body size and diet support the diet alteration hypothesis (Boback 2003): species that feed on relatively small ectothermic prey (centipedes, frogs, and/or lizards) are dwarfs, compared to those feeding on

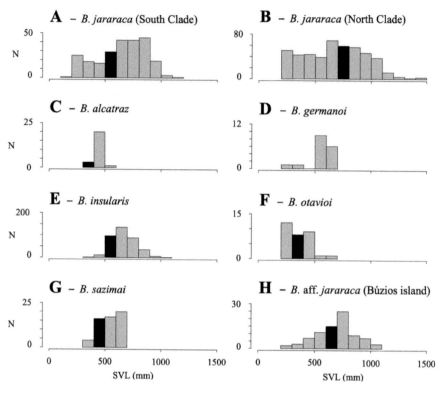

Figure 4.3 Body size variation of mainland and insular forms of lanceheads of the *Bothrops jararaca* complex. Frequency distributions and ranges of body size (snout to vent length [SVL] in millimeters) of (A) South and (B) North Clades of the mainland populations of *B. jararaca*, (C) *B. alcatraz*; (D) *B. germanoi*; (E) *B. insularis*; (F) *B. otavioi*; (G) *B. sazimai*; and (H) *B. aff. jararaca* from Búzios Island. Black bars correspond to the SVL class of the smallest mature male recorded. Histogram classes of 100 mm; *N* = number of individuals.

endothermic prey (birds and/or mammals; see Table 4.1). Furthermore, lanceheads are typically ambush (sit-and-wait) predators, which also supports the patterns found by Boback (2003).

Rodents have not been recorded in any of the six islands where the above-mentioned populations and species of the *B. jararaca* complex occur (Marques 2021). Dietary data obtained from the dissection of preserved specimens and from the regurgitation of live individuals sampled in the field allowed the characterization of the diet of these island pitvipers (Table 4.1). The diets of all island species and populations are different from the mainland populations of *B. jararaca*. As stated above, the six insular lanceheads may have originated through the isolation of populations of the mainland *B. jararaca*, and a character optimization analysis indicated that the diet of the mainland lancehead *B. jararaca* is the ancestral condition for this group of snakes (Martins *et al.* 2002).

Only two forms of insular lanceheads analyzed here (*B.* aff. *jararaca* from Búzios island and the Golden Lancehead, *B. insularis*) feed on endothermic prey, both of them feeding on birds (Table 4.1; Martins *et al.* 2002; Marques *et al.* 2012; F. E. Barbo, pers. obs.). Data obtained from regurgitations of live specimens of *B. insularis* in the field revealed that the Golden Lancehead essentially depends on two migratory species of birds (Marques *et al.* 2012; Marques 2021). The frequency of specimens of the Golden Lancehead (*B. insularis*) found on the vegetation is correlated with the abundance of birds in Queimada Grande Island. Most of the year there are more snakes on the ground, but at the end of rainy season (March), when the White-crested Elaenia (*Elaenia chilensis*) is abundant on the island, up to 70% of the island pitvipers have been recorded on the vegetation (Marques *et al.* 2012). The choice of higher substrates may favor the encounter of these birds by the snakes. The greater arboreality of the Golden Lancehead among other species of *Bothrops* (see Martins *et al.* 2001) may explain some of its derived morphological characters, including slender body, longer tail, thinner and more flexible scales, more anterior heart, and shorter fangs (Wüster *et al.* 2005; Marques 2021). Another typical feature of the Golden Lancehead is its black tail in both adults and juveniles (Marques *et al.* 2021). Caudal luring has been recorded in juveniles (Andrade *et al.* 2010), and postures of foraging in free-living animals, with the tail tip exposed lying over body coils and close to the head, indicate that caudal luring also probably occurs in *B. insularis* (Martins *et al.* 2002). In addition to the occurrence of bird remains in some individuals of *B.* aff. *jararaca* from Búzios island (F. E. Barbo, pers. obs.), the feeding biology of this form is still unknown.

A Hypothesis for the Dwarfism in Ectothermic Specialist Insular Species

The predominantly diurnal hunting habit of the Golden Lancehead has been related to ornithophagy (Marques 2021). All ectothermic-specialist insular species—*B. alcatraz*, *B. germanoi*, *B. otavioi*, and *B. sazimai*—feed mainly on nocturnal prey (centipedes, frogs, and lizards; Table 4.1; Barbo *et al.* 2012, 2016, 2022b). Two of these species have contrasting tail coloration (pale in *B. germanoi* and dark in *B. otavioi*), but all of them have smaller body sizes, being ecologically similar to the juveniles of the mainland *B. jararaca* (Marques 2021). Similarly to Queimada Grande Island, all other islands have several species of birds. Alcatrazes Island, for example, houses at least 82 bird species (Muscat *et al.* 2014). What, then, must have caused divergence of diet in the four ectothermic-specialist island pitvipers, and not in the Golden Lancehead and the Búzios island population? A plausible explanation would be the difference in snake predator pressure on each island as an outcome of their area and distance from the mainland. Islands that are larger and located closer to the mainland harbor greater species richness. Alcatrazes and Vitória islands are relatively far from the mainland (35 and 26 km from the coast) but have larger areas (1.3 and 2.2 km², respectively) than Queimada Grande Island (33 km from the coast; 0.4 km²). Alcatrazes Island has a greater richness of birds including hawks, in addition to the tegu lizard (*Tupinambis*), which can prey on snakes. Considering the size of Vitória Island, it is expected that it will also have a greater richness of birds. Although the Franceses and Moela Islands are smaller (0.15 and 0.2 km², respectively), they are much closer to the mainland (4.0 and 2.5 km, respectively), making it easy for continental birds to access the islands. The absence of a terrestrial predator, the tegu lizard, and the low richness of birds that are potential predators of snakes in Queimada Grande Island may have allowed the Golden Lancehead to explore a different niche, hunting birds on the vegetation during the day (Marques 2021). On the other hand, the greater number of diurnal predators (birds and/or tegu lizards) may have been an important predation pressure for the other four island pitvipers to keep secretive habits during the day, as in juveniles of *B. jararaca*. By hunting at night, they could only feed on the same prey as the juveniles of *B. jararaca*. As ectothermic prey (centipedes, frogs, and lizards) are generally smaller compared to endothermic prey, as well as provide less energy, these insular forms must have evolved to pedomorphic dwarf forms.

In a study of geographic variation on morphology, diet and reproduction of *B. jararaca* populations of the mainland, Moraes (2008) found that juveniles from mid-elevation areas feed on endothermic prey (mammals) at a higher frequency in relation to those from coastal lowland areas. Mid-elevation populations also consume relatively larger prey and have larger bodies, proportionally larger heads, and are stouter. The larger body size of those mid-elevation populations was also associated with a higher frequency of endothermic prey in their diets as an outcome of phenotypic or adaptive plasticity (Moraes 2008). Such plasticity may have played important roles in the evolution of body size in this snake complex, perhaps facilitating the persistence of insular forms in depauperate habitats of the land-bridge islands.

Conclusion

Future captive studies with dwarf lanceheads manipulating ectothermic and endothermic prey, and small and large prey (see Aubret *et al.* 2004), could clarify whether the body size patterns observed in each island species are due to plain phenotypic plasticity (the same gene pool producing different phenotypes in response to different environmental conditions; Travis 1994), adaptive plasticity (individuals showing a plastic response presenting higher fitness; Price *et al.* 2003; Ghalambor *et al.* 2007), or directional selection (extreme phenotype favored over other phenotypes and fixed by natural selection). Anecdotal observations on captive individuals of *B. alcatraz* indicate that maximum body size does not change in captivity even when fed with endothermic prey (A. Fellone and S. Travaglia-Cardoso, pers. comm.), suggesting that dwarfism is an adaptation to the new conditions in the Alcatrazes Island and not phenotypic plasticity. Monitoring the growth of these snakes in captivity and determining their size when sexually mature would also help to understand the process responsible for pedomorphosis in these species: neoteny or progenesis.

Finally, besides their usefulness to help in our understanding of the evolution of snake diet and body size in islands, by occurring in very small islands and being susceptible to stochastic events and anthropic threats like poaching and fires (see Marques *et al.* 2021; Silveira *et al.* 2021a, 2021b; Barbo *et al.* 2016, 2022b), the island lanceheads of coastal Brazil deserve special attention for conservation. Indeed, all three of the species that have had their conservation status assessed by the International Union for Conservation of Nature (IUCN) are threatened: *B. alcatraz* (Vulnerable), *B. insularis* (Critically Endangered), and *B. otavioi* (Critically Endangered). Furthermore, Barbo *et al.* (2016, 2022b) provided evidence that *B. sazimai* and *B. germanoi* are also threatened.

Acknowledgments

We thank Fundação de Amparo à Pesquisa do Estado de São Paulo (FAPESP) for grants to RJS, OAVM, and MM (#2020/12658-4); FEB (#2012/09156-0); and FGG (#2012/08661-3 and 2016/50127-5). We also thank Conselho Nacional de Desenvolvimento Científico e Tecnológico (CNPq) for research fellowships to RJS (#312795/2018-1), FGG (#312016/2021-2), OAVM (#312359/2020-9), and MM (#309772/2021-4).

References

Alberch, P., S. J. Gould, G. F. Oster, and D. B. Wake. 1979. Size and shape in ontogeny and phylogeny. *Paleobiology* 5:296–317.

Andrade, D. V., O. A. V. Marques, R. S. Gavira, F. E. Barbo, R. L. Zacariotti, and I. Sazima. 2010. Tail luring by the golden lancehead (*Bothrops insularis*), an island endemic snake from south-eastern Brazil. *South American Journal of Herpetology* 5:175–180.

Alencar, L. R. V., T. B. Quental, F. G. Grazziotin, M. L. Alfaro, M. Martins, M. Venzon, and H. Zaher. 2016. Diversification in vipers: Phylogenetic relationships, time of divergence and shifts in speciation rates. *Molecular Phylogenetics and Evolution* 105:50–62.

Amaral, A. 1921. Contribuição para o conhecimento dos ophidios do Brasil. Parte I. Quatro novas espécies de serpentes Brasileiras. *Anexos das Memórias do Instituto Butantan* 1:1–37.

Aubret, F. 2019. Pleasure and pain: Insular tiger snakes and seabirds in Australia. In H. L. Lillywhite and M. Martins (eds.), *Islands and Snakes: Isolation and Adaptive Evolution.* Oxford University Press, pp. 138–155.

Aubret, F., Shine, R., and X. Bonnet. 2004. Adaptive developmental plasticity in snakes. *Nature* 431:261–262.

Barbo, F. E., W. W. Booker, M. R. Duarte, B. Chaluppe, J. A. Portes-Junior, F. L. Franco, and F. G. Grazziotin. 2022b. Speciation process on Brazilian continental islands, with the description of a new insular lancehead of the genus *Bothrops* (Serpentes, Viperidae). *Systematics and Biodiversity* 20:1–25.

Barbo, F. E., J. L. Gasparini, A. P. Almeida, H. Zaher, F. G. Grazziotin, R. B. Gusmão, J. M. G. Ferrarini, and R. J. Sawaya. 2016. Another new and threatened species of lancehead genus *Bothrops* (Serpentes, Viperidae) from Ilha dos Franceses, Southeastern Brazil. *Zootaxa* 4097:511–529.

Barbo, F. E., F. G. Grazziotin, G. A. Pereira-Filho, M. A. Freitas, S. H. Abrantes, and M. N. C. Kokubum. 2022a. Isolated by dry lands: Integrative analyses unveil the existence of a new species and a previously unknown evolutionary lineage of Brazilian Lanceheads (Serpentes: Viperidae: *Bothrops*) from a Caatinga moist-forest enclave. *Canadian Journal of Zoology* 100:147–159.

Barbo, F. E., F. G. Grazziotin, I. Sazima, M. Martins, and R. J. Sawaya. 2012. A new and threatened species of lancehead from Southeastern Brazil. *Herpetologica* 68:418–429.

Benítez-López, A., L. Santini, J. Gallego-Zamorano, B. Milá, P. Walkden, M. A. Huijbregts, and J. A. Tobias. 2021. The island rule explains consistent patterns of body size evolution in terrestrial vertebrates. *Nature Ecology & Evolution* 5:768–786.

Boback, S. M. 2003. Body size evolution in snakes: Evidence from island populations. *Copeia* 1:81–94.

Campbell, J. A., and W. W. Lamar. 2004. *The Venomous Reptiles of the Western Hemisphere* (vol. 1). Cornell University Press.

Carlquist, S. 1974. *Island Biology*. Columbia University Press.

Carrasco, P. A., C. I. Mattoni, G. C. Leynaud, and G. J. Scrocchi. 2012. Morphology, phylogeny and taxonomy of South American bothropoid pitvipers (Serpentes,Viperidae). *Zoologica Scripta* 41:1–15.

Case, T. J. 1978. A general explanation for insular body size trends in terrestrial vertebrates. *Ecology* 59:1–18.

Foster, J. B. 1964. Evolution in mammals on islands. *Nature* 202:234–235.

Furtado, V. V. 2013. Upper quaternary sea level fluctuations and still stands on the continental shelf of São Paulo state, Brazil: A summary. *Revista Brasileira de Geofísica* 31:43–48.

Galindo-Leal, C., and I. G. Câmara. 2003. Atlantic forest hotspot status: An overview. In C. Galindo-Leal and I. G. Câmara (eds.), *The Atlantic Forest of South America: Biodiversity Status, Threats, and Outlook*. Island Press, pp. 3–11.

Ghalambor, C. K., J. K. McKay, S. P. Carroll, and D. N. Reznick. 2007. Adaptive versus nonadaptive phenotypic plasticity and the potential for contemporary adaptation in new environments. *Functional Ecology* 21:394–407.

Gould, S. J., 1977. *Ontogeny and Phylogeny*. Harvard University Press.

Gould, G. C., and B. J. MacFadden. 2004. Gigantism, dwarfism, and Cope's rule: "Nothing in evolution makes sense without a phylogeny." *Bulletin of the American Museum of Natural History* 285:219–237.

Grazziotin, F. G., M. Monzel, S. Echeverrigaray, and S. L. Bonato. 2006. Phylogeography of the *Bothrops jararaca* complex (Serpentes: Viperidae): Past fragmentation and island colonization in the Brazilian Atlantic Forest. *Molecular Ecology* 15:3969–3982.

Hudson, R. A. 1998. *Brazil: A Country Study* (5th ed.). Library of Congress.

Kasperoviczus, K. N. 2013. Evolução das estratégias reprodutivas de (Serpentes: Viperidae) *Bothrops jararaca*. Ph.D. Thesis. Faculdade de Medicina Veterinária e Zootecnia, Universidade de São Paulo. São Paulo. 134 pp. DOI 10.11606/T.10.2013.tde-10122013-103521

Keogh, J. S., I. A. Scott, and C. Hayes. 2005. Rapid and repeated origin of insular gigantism and dwarfism in Australian tiger snakes. *Evolution* 59:226–233.

Keyte, A. L., and K. K. Smith. 2014. Heterochrony and developmental timing mechanisms: Changing ontogenies in evolution. *Seminars in Cell & Developmental Biology* 34:99–107.

Lillywhite, H. B., and C. M. Sheehy III. 2019. The unique insular population of cottonmouth snakes at Seahorse Key. In H. L. Lillywhite and M. Martins (eds.), *Islands and Snakes: Isolation and Adaptive Evolution*. Oxford University Press, pp. 201–240.

Lomolino, M. V. 1985. Body size of mammals on islands: The island rule reexamined. *American Naturalist* 125:310–316.

Lomolino, M. V. 2005. Body size evolution in insular vertebrates: Generality of the island rule. *Journal of Biogeography* 32:1683–1699.

Marques, O. A. V. 2021. *A Ilha das cobras: Biologia, evolução e conservação da jararaca-ilhoa na Queimada Grande*. Editoria Ponto A, 72 pp.

Marques, O. A. V., M. Martins, P. F. Develey, A. Macarrão, and I. Sazima. 2012. The golden lancehead *Bothrops insularis* (Serpentes: Viperidae) relies on two seasonally plentiful bird species visiting its island habitat. *Journal of Natural History* 46:885–895.

Marques, O. A. V., Martins, M. R. C., and Sawaya, R. J. 2021. Bothrops alcatraz. The IUCN Red List of Threatened Species 2021: e.T46344A123179114. Accessed October 18, 2022. https://dx.doi.org/10.2305/IUCN.UK.2021-3.RLTS.T46344A123179114.en

Marques, O. A. V., M. Martins, and I. Sazima. 2002. A new species of pitviper from Brazil, with comments on evolutionary biology and conservation of the *Bothrops jararaca* group. *Herpetologica* 58:303–312.

Martins, M., and H. L. Lillywhite. 2019. Ecology of snakes on islands. In H. L. Lillywhite and M. Martins (eds.). *Islands and Snakes: Isolation and Adaptive Evolution*. Oxford University Press, pp. 1–44.

Martins M., M. S. Araujo, R. J. Sawaya, and R. Nunes. 2001. Diversity and evolution of macrohabitat use, body size and morphology in a monophyletic group of Neotropical pitvipers (*Bothrops*). *Journal of Zoology* 254:529–538.

Martins, M., O. A. V. Marques, and I. Sazima. 2002. Ecological and phylogenetic correlates of feeding habits in neotropical pitvipers (genus *Bothrops*). In G. W. Schuett, M. Höggren, M. E. Douglas, and H. W. Greene (eds.), *Biology of the Vipers*. Eagle Mountain Publishing, pp. 307–328.

McNab, B. K. 2022. Minimizing energy expenditure facilitates vertebrate persistence on oceanic islands. *Ecology Letters* 5:693–704.

Moraes, R. A. 2008. Variações em caracteres morfológicos e ecológicos em populações de Bothrops jararaca (Serpentes: Viperidae) no Estado de São Paulo. Unpublished master's dissertation, Universidade de São Paulo, São Paulo. https://teses.usp.br/teses/disponiveis/41/41134/tde-13062008-103811/en.php

Muscat, E., J. Y. Saviolli, A. Costa, C. A. Chagas, M. Eugênio, E. L. Rotenberg, and F. Olmos. 2014. Birds of the Alcatrazes archipelago and surrounding waters, São Paulo, southeastern Brazil. *Check List* 10:729–739.

Price, T. D., A. Qvarnström, and D. E. Irwin. 2003. The role of phenotypic plasticity in driving genetic evolution. *Proceedings of the Royal Society of London Series B* 270:1433–1440.

Rice, S. H. 1997. The analysis of ontogenetic trajectories: When a change in size or shape is not heterochrony. *Proceedings of the National Academy of Sciences* 94:907–912.

Salomão, M. G., W. Wüster, R. S. Thorpe, and BBBSP. 1997. DNA evolution of South American pitvipers of the genus *Bothrops*. In R. S. Thorpe, W. Wüster, and A. Malhotra (eds.), *Venomous Snakes: Ecology, Evolution, and Snakebite*. Clarendon Press, pp. 89–98.

Sazima, I. 1991. Caudal luring in two Neotropical pitvipers, *Bothrops jararaca and B. jararacussu*. *Copeia* 1:245–248.

Sazima, I. 1992. Natural history of the jararaca pitviper, *Bothrops jararaca*, in southeastern Brazil. In J. A. Campbell, and E. D. Brodie, Jr. (eds.), *Biology of the Pitvipers*. Selva, pp. 199–216.

Silveira, A. L., A. L. da C. Prudente, A. J. S. Argôlo, C. R. Abrahão, C. de C. Nogueira, F. E. Barbo, G. C. Costa, *et al.* 2021a. *Bothrops insularis*. The IUCN Red List of Threatened Species 2021: e.T2917A123180264. Accessed October 18, 2022. https://dx.doi.org/10.2305/IUCN.UK.2021-3.RLTS.T2917A123180264.en.

Silveira, A. L., A. L. da C. Prudente, A. J. S. Argôlo, C. R. Abrahão, C. de C. Nogueira, C. Strüssmann, D. Loebmann, *et al.* 2021b. *Bothrops otavioi*. The IUCN Red List of Threatened

Species 2021: e.T50957350A123739664. Accessed October 18, 2022. https: //dx.doi.org/ 10.2305/IUCN.UK.2021-3.RLTS.T50957350A123739664.en.

Smith, K. K. 2001. Heterochrony revisited: The evolution of developmental sequences. *Biological Journal of the Linnean Society* 73:169–186.

Stock, C. 1935. Exiled elephants of the Channel Islands. *Scientific Monthly* 41:205–214.

Travis, J. 1994. Evaluating the adaptive role of morphological plasticity. In P. C. Wainwright, and S. M. Reilly (eds.), *Ecological Morphology: Integrative Organismal Biology*. University of Chicago Press, pp. 99–122.

Tomović, L., M. Anđelković, A. Golubović, D. Arsovski, R. Ajtić, B. Sterijovski, S. Nikolić, *et al.* 2022. Dwarf vipers on a small island: Body size, diet and fecundity correlates. *Biological Journal of the Linnean Society* 20:1–13.

Uetz, P., P. Freed, and J. Hosek. 2022. The reptile database. http: //www.reptile-database.org

Van Valen, L. 1973. A new evolutionary law. *Evolutionary Theory* 1:1–33.

Whittaker, R. J., and J. M. Fernández-Palacios. 2007. *Island Biogeography: Ecology, Evolution, and Conservation*. Oxford University Press.

Wüster, W., M. R. Duarte, and M. Graça-Salomão. 2005. Morphological correlates of incipient arboreality and ornithophagy in island pitvipers, and the phylogenetic position of *Bothrops insularis*. *Journal of Zoology* 266:1–10.

Wüster, W., M. Graça-Salomão, J. A. Quijada-Mascareñas, R. S. Thorpe, and BBBSP. 2002. Origins and evolution of the South American pitvipers fauna: Evidence from mitochondrial DNA sequence analysis. In G. W. Schuett, M. Höggren, M. E. Douglas, and H. W. Greene (eds.), *Biology of the Vipers*. Eagle Mountain Publishing, pp. 111–129.

5

Two Islands, Two Origins

The Snakes of Trinidad and Tobago

John C. Murphy, John C. Weber, Michael J. Jowers,
and Robert C. Jadin

Background

The Trinidad and Tobago archipelago is in the southern Caribbean between the Caribbean Sea and the central Atlantic Ocean, northeast of Venezuela. The sister islands are often cited as the southernmost islands of the Lesser Antilles. However, Trinidad is a continental island with strong geologic and geomorphic connections to mainland South America, having separated approximately 4 Ma. Tobago, on the other hand, is a far-traveled, Cretaceous oceanic arc that formed at the leading edge of the Caribbean plate and was accreted onto the northern coast of South America. Consequently, neither of the two islands are part of the Neogene Lesser Antilles Island arc.

The first checklist of the Trinidad herpetofauna was compiled by Dr. J. Court, who recognized 10 snake species with the acknowledgment that more were present. The list was published in de Verteuil's 1858 volume *Trinidad: Its Geography, Natural Resources, Administration, Present Condition, and Prospects.* Reinhardt and Lütken (1862) first noted the Trinidad herpetofauna was more like the South America fauna than the Lesser Antilles fauna. Fredric Ober made a collection on Tobago that was reported by Cope (1879), including a specimen of a pitviper in the genus *Bothrops.* However, a re-examination by Tuck and Hardy (1973) found none of Ober's specimens originated on Tobago. Multiple papers by Mole and Urich (Mole and Urich 1891; Mole 1910, 1924) at the turn of the 20th century included collaboration with George Boulenger at the British Museum of Natural History. In the 1910 paper, Mole suggested Tobago had six or seven snake species. Mertens (1972) listed 20 species from Tobago, while Hardy (1982) later discussed the Tobago herpetofauna but did not attempt a complete list of Tobago snakes. The number of Trinidad snake species increased to 38 in Mole's 1924 paper, which were thereafter confirmed by Emsley (1977). Murphy (1997) and Murphy *et al.* (2018) list 45 species from Trinidad and 21 species from Tobago, though some of the species' names changed with updated taxonomy. Some species are considered endemics or near endemics, while others represent widespread neotropical taxa.

John C. Murphy, John C. Weber, Michael J. Jowers, and Robert C. Jadin, *Two Islands, Two Origins* In: *Islands and Snakes.* Edited by: Harvey B. Lillywhite and Marcio Martins, Oxford University Press. © Oxford University Press 2023.
DOI: 10.1093/oso/9780197641521.003.0005

Here, we discuss the origins of the islands and their respective snake faunas and review recent phylogenetic papers to assess any patterns possibly related to geologic influences.

Abbreviations used: AB, anthropogenic biomes; MF, mangrove forests; TTMF, Trinidad and Tobago moist forests; TTDF, Trinidad and Tobago dry forests; Ma, millions of years ago; kya, thousands of years ago; MIS, marine isotope stage or marine oxygen-isotope stages; OIS, oxygen isotope stage (alternating warm and cool periods in the Earth's paleoclimate).

Geography and Geology

The accreted Cretaceous oceanic Tobago forearc-arc terrane formed at the leading edge of the Caribbean plate several thousand kilometers west of its current location and then entered the gap between the Americas (*e.g.*, Weber *et al.* 2015). The entire Neogene history of the Caribbean–South American plate boundary may be due to the progressive eastward migration of a transform fault system that includes a bow wave of contractile deformation and a deep propagating lithospheric tear (*i.e.*, a STEP—subduction transform edge propagator—fault) (Arkle *et al.* 2021). Topographically, the Cordillera de la Costa-Paria-Northern Range forms the highest mountains in the region and is geologically a hinterland metamorphic belt formed from South American deposits that were pushed down, deformed, and exhumed back to the surface in the Cenozoic (Weber *et al.* 2001). Central and south Trinidad's geology include similar Mesozoic-Paleogene South American deposits and a cover of paleo-Orinoco deposits. Although the archipelago forms one country, the islands had very different geological origins. Today both support a continental flora and fauna.

Trinidad lies 11 km off the northeast coast of Venezuela and 130 km south of the Grenadines. The island is 4,768 km² in area with an average length of 80 km and an average width of 59 km. Tobago is 36 km northeast of Trinidad and measures about 298 km² in area, 41 km in length and 12 km at its greatest width. Additionally, the flora and fauna of both major islands are decidedly South American due to their proximity to mainland South America and the multiple dry land connections that existed during sea level low stands, with only minor contributions from the Antilles (Figure 5.1).

The position of Trinidad and Tobago in relationship to the mainland is shown in Figure 5.2 as well as the disjunct nature of the Caribbean Coastal ranges (5.2a), the location of Orinoco Basin (Fig. 5.2b), the Guiana Shield (Fig. 5.2c) and the Amazon Drainage Basin (5.2d).

Despite their shared life forms, the two large islands of the archipelago had entirely different geological origins, as outlined above. In addition, modern geomorphic processes variably affect the two main islands. Tobago's low southwestern plains and topographic asymmetry—markedly highest in the north—are due to both active tilting

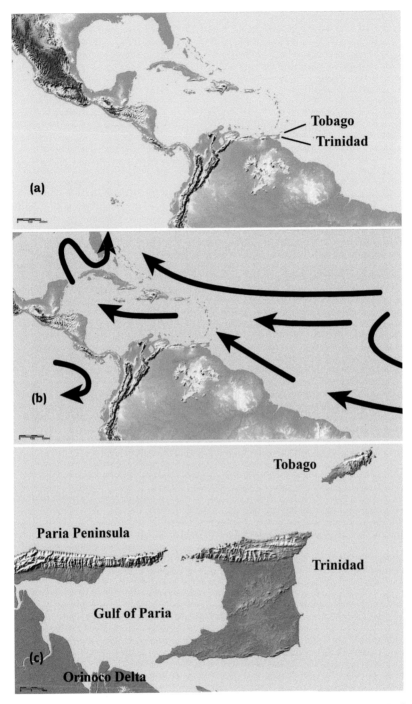

Figure 5.1 (a) Trinidad lies between 10°3' and 10°51' north latitudes and between 61°55' and 60°55' west longitudes. Tobago lies between 11°8' and 11°30' north latitudes and 61°5' and 60°28' west longitudes. (b) The direction of the major currents in the region; the significance of a west and northwest movement is a likely explanation for the near absence of Lesser Antilles species in Trinidad and Tobago. The Lesser Antilles amphibians and reptiles that are present in Trinidad and Tobago are the result of human-mediated dispersal. (c) The relationship between the South American mainland, Trinidad, and Tobago.

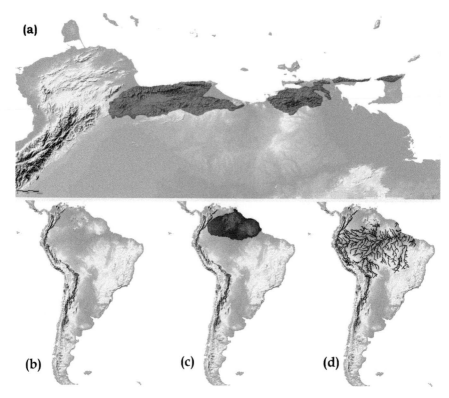

Figure 5.2 (a) Illustrates the relationship between mainland Venezuela and Trinidad and Tobago. The areas in red are Caribbean Coastal Ranges and Trinidad's Northern Range. (b) The blue is the Orinoco Basin and Llanos. (c) The Guiana Shield, (d) The Amazon Basin and its drainage pattern.

along the north coast and a combination of recent sinking and high sea level stands in the south. On the other hand, Trinidad's modern landscape is highly affected by westward tilting and sinking into the Gulf of Paria pull-apart basin (Ritter and Weber 2007; Arkle *et al.* 2017a, 2017b). In addition, during the Miocene and Pliocene, as well as during Pleistocene interglacials, much of Trinidad and parts of Tobago were covered with water.

The two islands are separated from each other by a channel that reaches a depth more than 91 m, but much of the channel is less than 72 m deep. The most prominent feature of Tobago (and the area of most herpetological interest) is the Main Ridge. The Main Ridge runs east–west for about 24 km, with the highest point reaching 549 m. The surrounding terrain is steep and divided, with ridges and deep gullies cut by fast-moving streams, and it is markedly asymmetric, with the steepest slopes and streams on the north. The southwest end of the island is a low lying, remnant coral platform that contains patches of mangrove.

Trinidad can be divided into five physiographic provinces. The Northern Range is a tall, east–west trending, metamorphic mountain range that is an extension of the Coastal Cordillera of Venezuela. Its highest peaks are Cerro del Aripo (940 m) and El Tucuche (936 m). The Northern Range is 11–16 km wide and about 88 km long. Most of the region lies between 150–456 m in elevation, but most peaks and ridge tops are between 456 and 760 m asl.

The Northern Basin at surface appears to be largely a dissected alluvial plain, but at depth exhibits a complex history of Neogene basin fill and tectonism. The basin averages 16 km wide but widens on its sunken western edge to about 25 km where it contains Caroni Swamp (an area of about 103 km^2 of mangroves).

The Central Range is an east-northeast–west-southwest trending belt of low-lying hills cored and geologically underlain by generally non-resistant Paleogene and Cretaceous rocks. The average width is 5–8 km, and the elevation varies between 60 and 300 m. The range runs diagonally across Trinidad from Manzanilla to Pointe-á-Pierre, about 60 km. Tamana Hill is the highest point, reaching 307 m in elevation. The active Central Range transform fault divides the island into a northern Caribbean plate and a southern South American plate. Drainage is mostly to the northwest into the Caroni Swamp and, to the southeast, into the Nariva Swamp and tributaries of the Ortoire River.

The Southern Basin consists of rolling hills below 60 m in elevation whereas the Southern Range, also known as the Trinity Hills, stands as a small, isolated range at 303 m asl. The northern edge of the Southern Basin is defined by a sharp escarpment and geological boundary that runs east-to-west between Mosquito Creek and Radix Point. Within the eastern Basin, to the east, streams drain into the Atlantic via the Nariva and Ortoire Rivers, and to the west, the streams flow via the Oropouche River into the Gulf of Paria. The Nariva swamp is a 258 km^2 complex of palm marsh, mangroves, and herbaceous swamp.

The Serpent's Mouth (*aka* the Columbus Channel)—a saltwater barrier between Trinidad and Venezuela—is a strait lying between Icacos Point in southwest Trinidad and the north coast and northern Orinoco delta edge of Venezuela. At its narrowest point it is 14 km wide, and it connects the Atlantic Ocean to the Gulf of Paria.

Zoogeographic Categories

The snakes of Trinidad and Tobago fit into the eight zoogeographic categories listed below and used to classify the Trinidad and Tobago fauna (Figure 5.3). Each species has been assigned to one of these categories, which are based on Murphy (1997) and Murphy *et al.* (2018) in addition to updated distributions available in the literature.

WIDESPREAD (W) taxa have distributions that extend into Middle or North America or occur on both sides of the Andes Mountains. Most of these

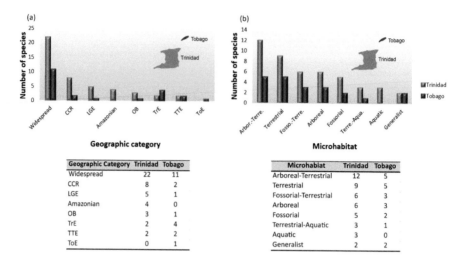

Geographic Category	Trinidad	Tobago
Widespread	22	11
CCR	8	2
LGE	5	1
Amazonian	4	0
OB	3	1
TrE	2	4
TTE	2	2
ToE	0	1

Microhabiat	Trinidad	Tobago
Arboreal-Terrestrial	12	5
Terrestrial	9	5
Fossorial-Terrestrial	6	3
Arboreal	6	3
Fossorial	5	2
Terrestrial-Aquatic	3	1
Aquatic	3	0
Generalist	2	2

Figure 5.3 (a) The graph shows the eight geographical categories used to organize the Trinidad snake fauna. The codes are W = widespread, A = Amazonian, CCR = Central Coastal Ranges, LGE = Lowland Guyana Endemics, OB = Orinoco Basin, TrE = Trinidad Endemics, ToE = Tobago Endemics, TTE = Trinidad and Tobago endemics (the endemic species occur on both main islands). (b) The graph shows the number of species on each island using one of eight microhabitats.

originated in Amazonia and dispersed northward after the closure of the Panamanian portal.

AMAZONIAN (A) taxa have their distribution centered in the greater Amazon Basin and have extended into Trinidad and Tobago and surrounding areas during periods of favorable habitat expansion. None of the taxa in this category is known to occur west of the Andes.

CARIBBEAN COASTAL RANGE (CCR) taxa are associated with a montane complex that extends from the Santa Marta region of Colombia, across northern Venezuela and Trinidad to Tobago. Most, but not all, of the taxa in this category reach the eastern limit of their distribution in Trinidad and Tobago.

LOWLAND GUIANA ENDEMICS (LGE) are taxa that are found at elevations below 1,000 m on the Guiana Shield and have extended their distribution to Trinidad and Tobago.

ORINOCO BASIN (OB) taxa are lowland species that have a distribution restricted to the Orinoco Basin.

TRINIDAD AND TOBAGO ENDEMICS (TTE) are present on Trinidad and Tobago, absent elsewhere.

TRINIDAD ENDEMICS (TrE) are present on Trinidad, unknown from elsewhere.

TOBAGO ENDEMICS (ToE) are present on Tobago, unknown from elsewhere.

Diversity of Trinidad and Tobago Snakes

Snakes are represented by eight, possibly nine families in the archipelago as follows, with the number of species represented in parentheses: Anomalepidiidae (1), Leptotyphlopidae (2), Typhlopidae (2), Boidae (4), Colubridae (13), Dipsadidae (24), Elapidae (2), and Viperidae (2). A member of the family Aniliidae has been reported from Trinidad but was based on one, now lost, specimen of *Anilius scaytale* (Boos 2001). Of the eight families reported from Trinidad, five of those families (Anomalepidiidae, Leptotyphlopidae, Aniliidae, Elapidae, Viperidae) are unknown from Tobago. A list of species known from Trinidad and Tobago, the habitats they use, and the zoogeographic categories is provided in the captions for Figures 5.3–5.8.

The portion of snakes from each zoogeographic category is broken down by island in Figure 5.3a. The island snake assemblages are composed of about 50% widespread taxa, and 9% of Trinidad taxa have Amazonian distributions, while there are no Tobago species with an Amazonian distribution. Both islands have species with a CCR distribution, Trinidad with 17% and Tobago with 10%. All endemics are associated with the CCR, and almost all are restricted to the Northern Range of Trinidad and the Main Ridge of Tobago. If we include those endemic species, CCR species make up 26% (12/46) of the Trinidad snake fauna and 38% (8/21) of the Tobago snake fauna.

The percentages of snakes that use the various microhabitats are remarkably similar for both Trinidad and Tobago (Figure 5.3b). Species that are arboreal (A) or arboreal and terrestrial (A-T) compose about 39% of the taxa on both islands. Aquatic (q) and aquatic-terrestrial (t-q) species comprise 14% of Trinidad snake species but only 5% of the Tobago snake fauna. This is likely the result of no large, natural bodies of fresh water on Tobago. The only Tobago snake that is likely to be associated with water is the endemic *Erythrolamprus pseudoreginae*, and it is associated with streams on the Main Ridge, though it is not highly aquatic.

A comment on the discovery and description of *E. pseudoreginae* is appropriate here because it illustrates the problem of finding secretive species even on small islands. The first specimen of *E. pseudoreginae* was collected in 1978, by Dave Stephens while he was hiking Pigeon Peak Trace, and it was deposited at the Smithsonian. In 1994, Kurt Auffenberg collected the second specimen on Gilpin Trace Trail. Thus, two specimens were collected over 37 years. During that time birders and hikers would occasionally obtain photographic evidence of its presence. Despite 10 visits to the island between 1984 and 2015, I (JCM) did not find the snake. In 2016, I visited Tobago's Main Ridge with Alvin Braswell (North Carolina State Museum) and Renoir Auguste (University of the West Indies). Alvin was a few steps behind me on the trail and a specimen crawled out of the leaf litter right in front of him. With so few accounts of this and other secretive taxa, the natural history of these species remains unexplored.

Perhaps the most interesting unresolved biogeographic question is, What are the causes of the species distributions shared between the Venezuelan mainland and Tobago but absent from Trinidad? Hypotheses include the extinction of intermediate populations, the attachment of the Tobago terrane to what is now the central coast of Venezuela, and overwater dispersal (Hardy 1982; Murphy 1997; Jowers *et al.* 2015, 2021). Many island and mainland snake taxa have not been taxonomically or phylogenetically assessed using molecular data. As more of them are sampled and analyzed, the patterns and processes involved in species distributions should become clearer.

Another question that remains unresolved is, Why are there so many species of snakes on Trinidad and Tobago per square kilometer? Trinidad is 4,768 km^2 and has 0.009437 snake species/km^2. Tobago is 300 km^2 and has 0.07 snake species/km^2. Guedes *et al.* (2017) looked at Neotropical snake species in ecoregions and found Trinidad mangroves to have the highest number of snake species per square kilometer in the Neotropics (0.053 snake species/km^2); Trinidad and Tobago moist forests ranked 10th (0.00847 snake species/km^2), and Trinidad and Tobago dry forests ranked 38th (0.002176 snake species/km^2). Why Trinidad and Tobago have such large numbers of snakes per square kilometer remains unknown. However, we suspect it might be a function of how well various areas have been studied. Trinidad and Tobago are relatively well known compared to other regions in the Neotropics, and many of these other areas may thus contain numerous unknown species.

Figures 5.4 to 5.8 serve as a checklist for the species of the islands, although some species are not illustrated for lack of a photo while being mentioned in the captions. The codes in brackets are geographic regions associated with a given species, and the codes without brackets are habitats (from text and specified in captions). All photographs are by the author unless credited otherwise .

Discussion

The high snake diversity of Trinidad likely reflects vicariance from the time when Trinidad was attached to the mainland prior to its pre-4 Ma pull-apart from Venezuela and during subsequent low sea level stands. Many colonization events could have originated from the Paria Peninsula, which once extended into Trinidad as a more extensive peninsula. Peninsulas are known to have the capacity to harbor a significant number of reptile and amphibian taxa as they often represent distributional limits for species dispersal through changing conditions such as alternating glacial and interglacial periods (Hewitt 2000; Borzee *et al.* 2017).

Biogeography

Phylogenetic studies in the region suggest that the frequently changing topography facilitated connections and colonization events between the mainland (Venezuela

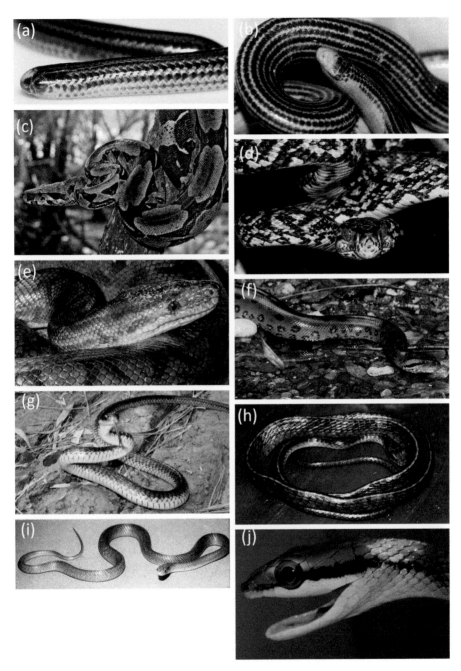

Figure 5.4 Leptotyphlopidae: (a) *Epictia tenella*. Trinidad, unknown from Tobago, TTMF and AB. [LGE]. Typhlopidae: *Amerotyphlops brongersmianus* Trinidad and Tobago [W]. Not illustrated. [A]. TTM. (b) A*merotyphlops trinitatus*. Trinidad and Tobago. TTMF. [TTE]. Anomalepididae: *Helminthophis* sp. Trinidad. TTMF. [TrE]. It has been called *Helminthophis flavoterminatus*. Not illustrated. Aniliidae: *Anilius scytale phelpsorum*.

Figure 5.4 Continued

Reported from Trinidad by Boos (2001) based on a single now lost specimen. [OB]. Not illustrated. Boidae (c) *Boa constrictor* common on Trinidad but seems much more abundant on Tobago where it is the largest predator on the island. TTMF and AB. [W]. (d) *Corallus ruschenbergerii*. TTMF [W]. (e) *Epicrates maurus* abundant in agricultural and urban areas. TTMF. [W]. (f) *Eunectes murinus*. Trinidad. [A]. Colubridae (g) *Chironius carinatus* Trinidad. TTMF and AB. [A]. (h) *Chironius septentrionalis*. Trinidad. TTMF. [CCR]. (i) *Drymarchon corais corais*. Known from Trinidad and Tobago. TTMF. [W]. (j) *Leptophis coeruleodorsus*. Trinidad and Tobago in TTMF, TTDF, and AB. [CCR].

Photographs a– by J. C. Murphy. Photograph f by Stephen L. S. Smith; photographs g and h by Adam Fifi; photograph i by Alvin Braswell.

and the Guianas) and the islands of Trinidad and Tobago (Jowers *et al.* 2011, 2015, 2020; Murphy *et al.* 2019a, 2019b). Studies using molecular clocks to date the arrival of snake species or their populations to Trinidad and Tobago are few (Murphy *et al.* 2016, 2019b; Jowers *et al.* 2020). Nevertheless, much of the evidence thus far points to a pattern of recent colonization from the mainland rather than ancient lineages remaining isolated after Trinidad detached from northern Venezuela in the Early Pliocene. The pattern of genetic similarity between snake populations in Trinidad and Tobago and the mainland can provide additional important constraints regarding the timing of the topographic and geological events that took place in the region.

The most unexpected findings seem to be the identical or close genetic relationship between related populations from the mainland and the islands. For example, phylogenetic work has shown that despite the considerable geographical distances between Guyana and Trinidad, the coral snake *Micrurus diutius* from both localities are genetically similar and suggest a Late Pleistocene–Holocene vicariance when sea level rises separated Trinidad from the mainland (Jowers *et al.* 2019). Similarly, the other coral snake *Micrurus circinalis* from Trinidad and northern Venezuela shows very low genetic differentiation (Jowers unpublished), again evidence of recent gene flow between the mainland and Trinidad. Another interesting example is that of the fossorial threadsnake from Trinidad, *Epictia tenella*, which shows a remarkable genetic similarity with specimens from Guyana, suggesting colonization by recent dispersal rather than by ancient vicariance (Murphy *et al.* 2016).

Snake species have colonized the islands from the mainland at different times, and some of these movements can be linked to climatic events. The mean age of divergence for the three-lined snake, *Atractus trilineatus*, between the mainland Venezuela-Guyana and the islands (Trinidad-Tobago) populations date to around 410 kya (Murphy *et al.* 2019b), suggesting that the island populations of *A. trilineatus* remained isolated from the mainland during later post-MIS 11 periods of higher sea levels (*e.g.*, MIS 9, 7, 5, 1) that were associated with interglacial periods of the Pleistocene (Railsback *et al.* 2015). Similarly, the genetic divergence of the

Figure 5.5 Colubridae (cont.) (a) *Leptophis haileyi*. Known only from one specimen collected in northeast Tobago. TTMF. [ToE]. (b) *Leptophis stimsoni*. TTMF. [TrE].
(c) *Mastigodryas boddaerti*. Trinidad. TTMF and TTDF. [A]. (d) *Mastigodryas dunni*. TTDF. [ToE]. (e) *Oxybelis rutherfordi* Trinidad and Tobago. TTMF, TTDF, AB. [CCR]. (f) *Phrynonax polylepis*. Trinidad. TTMF and TTDF. [A] (g) *Spilotes pullatus*. Trinidad and Tobago. TTMF, TTDF. [W]. (h) *Spilotes sulphureus.* Trinidad. TTMF. [A]. (i) *Tantilla melanocephala*. Trinidad and Tobago. TTMF, TTDF. [W]. Dipsadidae (j) *Atractus fuliginosus*. Tobago. TTMF. [CCR].

Photographs a, c, e–j by J. C. Murphy; photographs b and f by Stephen L. S. Smith.

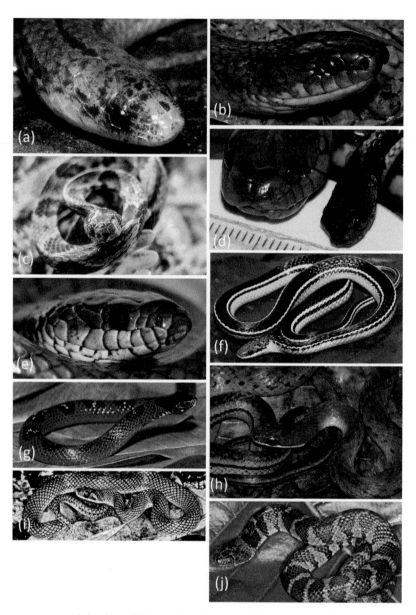

Figure 5.6 Dipsadidae (cont.) (a) *Atractus trilineatus*. Trinidad and Tobago. TMF, TTDF, AB. [LGE]. (b) *Clelia clelia*. Trinidad and Monos Island. TTMF, TTDF. [W]. (c) *Dipsas trinitatis*. TTMF. [TrE]. (d) *Dipsas variegata*. Trinidad. TTMF. [CCR]. Comparison of *D. trinitatus* (right) and *D. variegata* (left), both from Trinidad. *Erythrolamprus aesculapii*. Trinidad. Known from one specimen. Not illustrated. TTMF. [A]. *Erythrolamprus bizona*. Trinidad. TTMF. [W]. Not illustrated. (e) *Erythrolamprus cobella*. Known from Trinidad. TTMF, TTDF, MF. [LGE] Semi-aquatic. (f) *Erythrolamprus melanotus nesos*. Trinidad and Tobago. TTMF, TTDF. [TTE]. (g) *Erythrolamprus ocellatus*. TTMF. (h) *Erythrolamprus pseudoreginae*. TTMF. [ToE]. (i) *Erythrolamprus zweifeli*. Trinidad. TTMF, TTDF. [CCR]. (j) *Helicops angulatus*. Trinidad. TTMF, MF. [A]. Semi-aquatic.

Photographs b, f and g by Stephen L. S. Smith; all other photographs by J. C. Murphy.

Figure 5.7 Dipsadidae (cont.) (a) *Hydrops triangularis neglectus*. Trinidad. TTMF, TTDF, MF. [LGE]. (b) *Imantodes cenchoa cenchoa*. Trinidad and Tobago. TTMF, TTDF, MF. [W]. (c) *Leptodeira ashmeadi*. Trinidad and Tobago. TTMF, TTDF, MF. [OB]. (d) *Ninia atrata*. Trinidad and Tobago. TTMF, TTDF, AB. [W]. *Ninia franciscoi*. Trinidad. TTMF. [TrE]. Not illustrated. (e) *Oxyrhopus petolarius*. Trinidad and Tobago. TTMF, TTDF, AB. [W]. (f) *Pseudoboa neuwiedii*. Trinidad and Tobago TTMF, TTDF, AB [W]. (g) *Sibon nebulata*. Trinidad and Tobago. TTMF, TTDF, MF. [W]. (h) *Siphlophis cervinus*. Trinidad. TTMF. [W] (i) *Siphlophis compressus*. Trinidad. TTMF, TTDF, MF. [W]. (j) *Thamnodynastes ramonriveroi*. Trinidad. TTDF. [CCR].

Photograph a by William W. Lamar; photographs b–j by J. C. Murphy.

Figure 5.8 Elapidae (a) *Micrurus circinalis*. Trinidad. TTMF, TTDF, AB. [CCR]. (b) *Micrurus diutius*. Trinidad. TTMF, TTDF. [LGE]. Viperidae (c) *Bothrops atrox*. Trinidad. TTMF, TTDF, MF, AB. [A]. (d) *Lachesis muta*. Trinidad. TTMF, TTDF. [A].
Photographs by J. C. Murphy.

Black-headed Snake *Tantilla melanocephala* between Venezuela and Trinidad dates to about 200 kya. Therefore, the low genetic divergence of *T. melanocephala* recovered between Venezuela and Trinidad is somewhat similar to that found in *A. trilineatus* between the mainland and the islands (Murphy *et al.* 2019b; Jowers *et al.* 2020), and it likely results from the changing topography that facilitated connections between regions (Murphy 1997; Murphy *et al.* 2019b).

On the other hand, some populations of the above-mentioned species show deeper time genetic divergences, such as *Atractus trilineatus* from coastal Guyana and Trinidad-Tobago, which suggests a split at about 1 Ma. Murphy *et al.* (2019b) argued that different waves of colonization could reflect evolving topography on the mainland, such as discontinuities of habitat caused by evolving river systems between northern coastal Guyana and Trinidad and those flowing to the Guyana and Orinoco estuaries. In contrast, the southern Guyana populations of *A. trilineatus* might show less restricted population connectivity with northern localities. Similar landscape evolution events could have taken place to differentiate *T. melanocephala* between Venezuela-Trinidad-Tobago and Guyana-French Guiana populations circa 1.8 Ma.

To better understand the colonization of both islands, more estimates based on genetic studies of snakes are needed, but several key species that have been studied across the region are absent in Tobago (*i.e.*, *Micrurus diutius*, *M. circinalis*, *Epictia tenella*). The three published phylogenetic studies based on snakes between both islands give different timings; the *Atractus trilineatus* genetic split between Trinidad and Tobago populations was at ~180 kya, while *T. melanocephala*'s genetic difference between Trinidad-Venezuela and Tobago dates to 1.4 Ma (Jowers *et al.* 2020). Both dates are consistent with the proposed land bridge connections during the Pleistocene (Murphy 1997). Trinidad and Tobago's near endemic Oliver's Parrot Snake *Leptophis coeruleodorsus* shows very low genetic divergence between islands (0.26%), which is further evidence of a recent connection between the two main islands or a recent over-water colonization event of Tobago from Trinidad (Murphy *et al.* 2013).

Another significant remaining biogeographical enigma is the case of the fossorial-cryptozoic Hallowell's Ground Snake *Atractus fuliginosus*, a species absent from Trinidad but found in Western Venezuela and Tobago. Despite these two populations being currently more than 1,000 km apart, the genetic divergence between the Venezuela and Tobago populations dates to only 550 kya (Jowers *et al.* 2021). The authors of this phylogenetic study argue that a recent vicariance event of the Tobago population is the most probable cause. This likely occurred from a continuous population that existed along a coastal plain from Venezuela to Tobago until the sea level rose during a glacial maximum about 435 kya (MIS 12; Routledge and Hansen 2016; Railsback *et al.* 2015). Multiple periods of low sea levels (≤116 m) (at ca. 15, 140, 340, 640 kya) (Rohling *et al.* 2014; Railsback *et al.* 2015), allowed the *A. fuliginosus* population to become established on Tobago during these low-stand events (Jowers *et al.* 2021). The authors similarly argue that a possible scenario to account for the absence of *A. fuliginosus* on Trinidad is the extinction of intermediate populations in the mountain ranges of northern Venezuela and Trinidad, but this scenario is unlikely given the broad distribution of *A. trilineatus* throughout Trinidad, a species with similar natural history. The timing of the genetic divergence rules out the possibility of anthropogenic introductions.

Several authors have proposed colonization from Central to South America via island chains or through rafting (Rosen 1985 Lillegraven *et al.* 1979). Under marine or sea drift scenarios, Trinidad's colonization may have been influenced by outflow from the Orinoco Delta and the northward-flowing North Equatorial Current and offshoot west-flowing currents through the Serpent's Mouth as opposed to those in the opposite directions from northern Venezuela toward the Lesser Antilles (van Andel 1967; Iturralde-Vinent and MacPhee 1999; Murphy *et al.* 2016). In fact, the arrival of *A. fuliginosus* in Tobago from Trinidad via overwater dispersal cannot be disregarded, and this transport mechanism has been suggested for other reptiles in Trinidad and Tobago (*e.g.*, Boos 1984a, 1984b; Murphy *et al.* 2016). This dispersal route may explain the low genetic divergence between the Trinidad and Tobago populations of the parrot snake *Leptophis coeruleodorsus* and those in Tobago. Mangrove island detachment events and floating mats of vegetation following hurricanes and flooding in the Orinoco Delta could displace fossorial species as well as arboreal species and move them to southern Trinidad or even Tobago (Vidal *et al.* 2008; Murphy *et al.* 2019b). As a related example, the population of *Atractus trilineatus* from Guyana's Kaow Island, at the mouth of the Mazaruni River, is compatible with overwater dispersal during flood stages. In addition, the recovery of the same haplotypes or the extremely low genetic variability of both *Micrurus* and *Epictia tenella* from Trinidad might further corroborate such possibility.

The number of endemics in Tobago are evidence for long periods of isolation from Trinidad. Tobago has nine recognized endemic amphibian and reptile species (three frogs, two lizards, and four snakes; Murphy *et al.* 2018; Rivas *et al.* 2021), and evidence suggests that most of their closest living relatives are in the Venezuelan Central Coastal Range, not in Trinidad (Rivas *et al.* 2021). Some reptiles seem related to mainland forms (*e.g.*, the gymnophthalmid *Bachia whitei* and the mainland *B. flavescens*; Murphy *et al.* 2019c). The Tobago Stream Snake *Erythrolamprus pseudoreginae* is known only from Tobago, but its closest relative (*Erythrolamprus epinephalus*) is present in North-Central Venezuela (Murphy *et al.* 2019a). Evidence of Trinidad and Tobago connections can also be seen in the endemic Tobago snake *Erythrolamprus ocellatus*, which is the sister to the coral snake mimic *E. aesculapii* (known from both Trinidad and Venezuela) (Hodson and Lehtinen 2017). The relationship between the Amazonian *Mastigodryas boddaerti* on Trinidad and what has been considered the Tobago-endemic *M. dunni* remains to be investigated using molecular phylogenetic methods. However, *M. dunni* was long considered a subspecies of *M. boddaerti*. Overall, these phylogenetic relationships can be explained by ancient cladogenic events in the region.

The present distribution patterns of snakes in the Trinidad and Tobago archipelago are the result of both vicariance and dispersal events occurring from the Paleogene to the Holocene, which have been, in turn, affected by the complex geologic and geomorphic history of the region. Periods favorable for faunal exchange have varied greatly in duration and have not affected all parts of the archipelago equally. In addition, the ability of snake lineages to exploit opportunities for genetic and faunal exchange has been controlled by the varying ecological requirements of the lineages. Although it is difficult to date speciation and dispersal events in the absence of a robust fossil record, it appears that many of the species or lineages now restricted to a single area or a single island within the Trinidad and Tobago archipelago represent the effects of relatively recent evolutionary events that took place since the Miocene. During the Quaternary, the alternating intervals of consolidation and fragmentation of terrestrial areas apparently occurred when many of the species having wider distributions maintained gene flow among insular populations or dispersed and established their present distributions. The routes of gene exchange and dispersal utilized at that time also varied among species according to their ecological requirements. Some of our statements concerning timing of speciation and isolation and routes of dispersal are speculative in the absence of a robust fossil record and with only a few genetic studies. Additionally, many studies include incomplete sampling that could lead to incorrect conclusions about vicariance and/or dispersal dates. For example, Jadin *et al.* (2019) suggested that populations of the Brown Vine Snake of the *Oxybelis aeneus* complex from Trinidad and Tobago separated from their closest population in Panama approximately 3 Ma. An updated work by Jadin *et al.* (2020), which greatly increased sampling to include more mainland populations, found that populations from Trinidad and Tobago diverged considerably earlier from mainland French Guiana and Venezuelan populations. Consequently, we hope our preliminary hypotheses presented here are tested by additional faunal sampling and the use of additional molecular data sets.

Acknowledgments

The authors would like to thank Renoir Auguste and Mike G. Rutherford (Zoology Museum, University of the West Indies) and J. Roger Downie (University of Glasgow) for assistance with field work and logistics; Romano McFarlane (the Trinidad and Tobago Forestry Department) and Angela Ramsey (the Tobago House of Assembly, Department of Natural Resources) for collecting permits; and Gilson Rivas for providing tissues from Venezuelan specimens that could be compared to Trinidad and Tobago specimens.

References

Arkle J. C., L. A. Owen, and J. C. Weber. 2017b. Geomorphology and Quaternary landscape evolution of Trinidad and Tobago. In C. D. Allen (ed.), *Landscapes and Landforms of the Lesser Antilles*, Geomorphological Landscapes of the World. Springer, pp. 267–291.

Arkle J. C., L. A. Owen, J. C. Weber, M. Caffee, and S. Hammer. 2017a. Transient Quaternary erosion and tectonic inversion of the Northern Range, Trinidad. *Geomorphology* 295:337–353.

Arkle, J. C., Weber, J., Enkelmann, E., Owen, L. A., Govers, R., Jess, S., *et al.* 2021. Exhumation of the costal metamorphic belt above the subduction-to-transform transition, in the southeast Caribbean plate corner. *Tectonics* 40, e2020TC006414.

Boos, H. E. A. 1984a. A new snake for Trinidad. *Living World, Journal of the Trinidad and Tobago Field Naturalist's Club* 1983–84:3.

Boos, H. E. A. 1984b. A consideration of the terrestrial reptile fauna on some offshore islands north west of Trinidad. *Living World, Journal of the Trinidad and Tobago Field Naturalists' Club* 1984:12–31.

Boos, H. E. A. 2001. *The Snakes of Trinidad and Tobago*. Texas A&M University Press.

Borzee, A, J. L. Santos, S. Sánchez-Ramirez, Y. Bae, K. Heo, Y. Jang, and M. J. Jowers. 2017. Phylogeographic and population insights of the Asian common toad (*Bufo gargarizans*) in Korea and China: Population isolation and expansions as response to the ice ages. *PeerJ* 5: e4044. https://doi.org/10.7717/peerj.4044

Cope, E. D. 1879. Eleventh contribution to the herpetology of tropical America. *Proceedings of the American Philosophical Society* 18:261–277.

de Verteuil, L. A. A. G. 1858. *Trinidad: Its Geography, Natural Resources, Administration, Present Condition, and Prospects*. Ward and Lock.

Emsley, M. G. 1977. Snakes, and Trinidad and Tobago. *Bulletin of the Maryland Herpetological Society* 13(4):201–304.

Guedes, T. B., R. J. Sawaya, A. Zizka, S. Laffan, S. Faurby, R. A. Pyron, R. S. Bernils, *et al.* 2017. Patterns, biases and prospects in the distribution and diversity of Neotropical snakes. *Global Ecology and Biogeography* 27(1):14–21.

Hardy, J. D. 1982. Biogeography of Tobago, West Indies, with special reference to amphibians and reptiles, a review. *Bulletin of the Maryland Herpetological Society* 18(2):37–142.

Hewitt, G. 2000. The genetic legacy of the Quaternary ice ages. *Nature* 405:907–913.

Hodson, E. E, and R. M. Lehtinen. 2017. Diverse evidence for the decline of an adaptation in a coral snake mimic. *Evolutionary Biology* 44(3):401–410.

Iturralde-Vinent, M., and R. D. MacPhee. 1999. Paleogeography of the Caribbean region: Implications for Cenozoic biogeography. *Bulletin of the American Museum of Natural History* 238:1–95.

Jadin, R. C., C. Blair, M. J. Jowers, A. Carmona, and J. C. Murphy. 2019. Hiding in the lianas of the tree of life: Molecular phylogenetics and species delimitation reveal considerable cryptic diversity of New World Vine Snakes. *Molecular Phylogenetics and Evolution* 134:61–65.

Jadin, R. C., C. Blair, S. A. Orlofske, M. J. Jowers, G. A. Rivas, L. J. Vitt, J. M. Ray, *et al.* 2020. Not withering on the evolutionary vine: Systematic revision of the brown vine snake (Reptilia: Squamata: *Oxybelis*) from its northern distribution. *Organisms Diversity & Evolution* 20(4):723–746.

Jowers, M. J., J. L. Garcia Mudarra, S. P, Charles, and J. C. Murphy 2019. Phylogeography of West Indies Coral snakes (*Micrurus*): Island colonisation and banding patterns. *Zoologica Scripta* 48(3):263–276.

Jowers, M. J., R, M. Lethinen, J. R. Downie, A. P. Georgialis, and J. C. Murphy. 2015. Molecular phylogenetics of the glass frog *Hyalinobatrachium orientale* (Anura: Centrolenidae): Evidence for Pliocene connections between mainland Venezuela and the oceanic island of Tobago. *Mitochondrial DNA* 26:613–618.

Jowers, M. J., I. Martínez-Solano, B. L. Cohen, J. J. Manzanilla, and J. R. Downie. 2011. Genetic differentiation in the Trinidad endemic *Mannophryne trinitatis* (Anura: Aromobatidae): Miocene vicariance, in situ diversification and lack of geographical structuring across the island. *Journal of Zoological Systematics and Evolutionary Research* 49:133–140.

Jowers, M. J., G. A. Rivas, R. C. Jadin, A. L. Braswell, R. J. Auguste, A. Borzée, and J. C. Murphy. 2020. Unearthing the species diversity of a cryptozoic snake, *Tantilla melanocephala*, in its northern distribution with emphasis on the colonization of the Lesser Antilles. *Amphibian & Reptile Conservation* 14(3):206–217.

Jowers, M. J., W. E. Schargel, A. Muñoz-Mérida, S. Sánchez-Ramírez, J. C. Weber, J. F. Faria, D. J. Harris, and J. C. Murphy. 2021. The enigmatic biogeography of Tobago's marooned relics: The case study of a fossorial snake (Squamata, Dipsadidae). *Journal of Zoological Systematics and Evolutionary Research* 59(6):1382–1389.

Lillegraven J. A., M. J. Kraus, and T. M. Bown. 1979. Paleogeography of the world of the Mesozoic. In J. A. Lillegraven, Z. Kielan-Jaworowska, and W. A. Clemens (eds.), *Mesozoic Mammals: The First Two-Thirds of Mammalian History*. University of California Press, pp. 277–308.

Mertens, R. 1972. Herpetofauna tobagana. *Stungarter Beitr. zur Naturkunde* 252:1–22.

Mole, R. R. 1910. Economic zoology in relation to agriculture. Part I–Snakes. *Bulletin Department of Agriculture, Trinidad and Tobago* 9(65):140–141.

Mole, R. R. 1924. The Trinidad snakes. *Proceedings of the Zoological Society of London* (1):235–278.

Mole, R. R., and F. W. Urich. 1891. Notes on some reptiles from Trinidad. *Proceedings of the Zoological Society of London* 1891:447–449.

Murphy, J. C. 1997. *Amphibians and Reptiles of Trinidad and Tobago*. Krieger Publishing.

Murphy, J. C., A. L. Braswell, S. P. Charles, R. J. Auguste, G. A. Rivas, R. M. Lehtinen, and M. J. Jowers. 2019a. A new species of *Erythrolamprus* from the oceanic island of Tobago (Squamata, Dipsadidae). *Zookeys* 817:131.

Murphy, J. C., S. P. Charles, R. M. Lehtinen, and K. L. Koeller. 2013. A molecular and morphological characterization of Oliver's parrot snake, Leptophis coeruleodorsus (Squamata: Serpentes: Colubridae) with the description of a new species from Tobago. *Zootaxa* 3718(6):561–574.

Murphy, J. C., J. R. Downie, J. M. Smith, S. M. Livingstone, R. S. Mohammed, R. M. Lehtinen, M. Eyre, *et al.* 2018. *A Field Guide to the Amphibians and Reptiles of Trinidad & Tobago.* Trinidad & Tobago Field Naturalists' Club, Port of Spain, Republic of Trinidad and Tobago.

Murphy, J. C., M. G. Rutherford, and M. J. Jowers. 2016. The threadsnake tangle: Lack of genetic divergence in *Epictia tenella* (Squamata, Leptotyphlopidae): Evidence for introductions or recent rafting to the West Indies. *Studies on Neotropical Fauna and Environment* 51(3):197–205.

Murphy, J. C., D. Salvi, A. L. Braswell, and M. J. Jowers. 2019b. Phylogenetic position and biogeography of three-lined snakes (*Atractus trilineatus*: Squamata, Dipsadidae) in the Eastern Caribbean. *Herpetologica* 75(3):247–253.

Murphy, J. C., D. Salvi, J. L. Santos, A. L. Braswell, S. P. Charles, A. Borzée, and M. J. Jowers (2019c). The reduced limbed lizards of the genus *Bachia* (Reptilia, Squamata, Gymnophthalmidae): Biogeography, cryptic diversity and morphological convergence in the eastern Caribbean. *Organisms Diversity & Evolution* 19(2):321–340.

Railsback, L. B., P. L. Gibbard, M. J. Head, N. R. G. Voarintsoa, and S. Toucanne. 2015. An optimized scheme of lettered marine isotope substages for the last 1.0 million years, and the climatostratigraphic nature of isotope stages and substages. *Quaternary Science Reviews* 111:94–106.

Reinhardt, J., and C. F. Lütken. 1862. Bidgrad til Kundskab om Brasiliens Padder og Krybdyr. *Videnskabelige Meddelelser fra den Naturhistoriske Forening i Kjøbenhavn* 3(10–15):143–242.

Ritter, J. B., and J. C. Weber. 2007. *Geomorphology and Quaternary Geology of the Northern Range, Trinidad and Paria Peninsula, Venezuela: Recording Quaternary Subsidence and Uplift Associated with a Pull-Apart Basin.* Proceedings, Geological Society of Trinidad and Tobago, Fourth Geological Conference.

Rivas, G. A., O. M. Lasso-Alcalá, D. Rodríguez-Olarte, M. De Freitas, J. C. Murphy, C. Pizzigalli, J. C. Weber, *et al.* 2021. Biogeographical patterns of amphibians and reptiles in the northernmost coastal montane complex of South America. *PloS One* 16(3):e0246829.

Rohling, E. J., G. L. Foster, K. M. Grant, G. Marino, A. P. Roberts, M. E. Tamisiea, and F. Williams. 2014. Sea-level and deep-sea-temperature variability over the past 5.3 million years. *Nature* 508(7497):477–482.

Rosen, D. E. 1985. Geological hierarchies and biogeographic congruence in the Caribbean. *Annals of the Missouri Botanical Garden* 72:636–659.

Routledge, R, and J. Hansen. 2016. Sea level change during the last 5 million years. TeGrate's Earth-focused Modules and Courses for the Undergraduate Classroom. https://serc.carleton.edu/integrate/teaching_materials/coastlines/student_materials/901

Tuck, R. G., and J. D. Hardy. 1973. Status of the Ober Tobago Collection, Smithsonian Institution, and the proper allocation of *Amieva suranamensis tobaganus* Cope (Sauria: Teiidae). *Biological Society of Washington* 86(19):231–242.

van Andel, T. H. 1967. The Orinoco Delta. *Journal of Sedimentary Research* 37(2):297–310.

Vidal N, A. Azvolinsky, C. Cruaud, and S. B. Hedges. 2008. Origin of tropical American burrowing reptiles by transatlantic rafting. *Biology Letters* 4(1):115–118.

Weber, J. C., D. A. Ferrill, and M. K. Roden-Tice. 2001. Calcite and quartz microstructural geothermometry of low-grade metasedimentary rocks, Northern Range, Trinidad. *Journal of Structural Geology* 23:93–112.

Weber, J. C., H. Geirsson, J. L. Latchman, K. Shaw, P. La Femina, S. Wdowinski, M. Higgins, *et al.* 2015. Tectonic inversion in the Caribbean-South American plate boundary: GPS geodesy, seismology, and tectonics of the Mw 6.7 April 22, 1997 Tobago earthquake. *Tectonics* 34(6):1181–1194.

6

Giant Snakes and Tiny Seabirds on a Small Japanese Island

Masami Hasegawa and Akira Mori

Introduction

About 15 minutes after the *Salvia Maru* departed the island of Niijima, I (Hasegawa) spotted a flock of four or five small seabirds floating on the calm sea surface. It was April 20, 1998, and I had sailed from the volcanic islands of the Izu Archipelago to the Japanese mainland countless times. Each journey was a unique natural history experience, so while the ship traveled to Tokyo approximately100 km to the north, I perched on the deck with binoculars and notebook ready to record my observations.

The birds had short bills and blue-black and gray dorsal plumage that was a striking contrast to the pure white of their throats and backs (Figure 6.1). The activity of the birds increased. One plunged headfirst into the seawater and disappeared, just like the Little Grebe, *Tachybaptus ruficollis*, that I had observed in the freshwater ponds of Japan. As if responding to a radio broadcast that only seabirds can hear, another bird of the same species appeared, then two, then 15, then 30. I stopped recording their numbers in my notebook as the flock reached at least 100 individuals flapping their short wings just above the ocean surface. Within five minutes, the flock slowly vanished from sight as the *Salvia Maru* steamed away at 18 knots. To the ship's other passengers, these five minutes may have been just a sliver of time on a monotonous journey home. If only they could share the excitement of observing the Crested Murrelet, *Synthliboramphus wumizusume*—one of the most endangered bird species in the world!

The total population of the Crested Murrelet (also called the Japanese Murrelet) is estimated to be less than 10,000 individuals, with only about 1,000 inhabiting the Izu Islands (Carter *et al.* 2002; Miller *et al.* 2019). Human disturbance of nesting habitat and predation on eggs and nestlings by introduced species (*e.g.*, *Rattus* sp.) are two of the more important factors that explain the rarity of the bird in the present day (Carter *et al.* 2002). Yet, on the rocky shores of a small, approximately10-ha island in the Izu Archipelago, Crested Murrelet predation is just one part of a fascinating story of how ecological and geological processes in a marine ecosystem produced the giant ratsnakes of Tadanae Island.

Masami Hasegawa and Akira Mori, *Giant Snakes and Tiny Seabirds on a Small Japanese Island* In: *Islands and Snakes*. Edited by: Harvey B. Lillywhite and Marcio Martins, Oxford University Press. © Oxford University Press 2023.
DOI: 10.1093/oso/9780197641521.003.0006

Figure 6.1 Seabird species of Tadanae Island, the Izu Islands, Japan. (a) The foraging Streaked Shearwaters at sea. (b) The Crested Murrelet at sea. (c) The Tristram's Storm Petrel captured for banding survey on Tadanae. (d) The colony of the Black-tailed Gull on Tadanae.

Photographs by Masami Hasegawa (a, c, d) and Yasunari Hattori (b), with permission.

Introduction to the Izu Islands System

The Izu Island Archipelago is a chain of volcanic islands off the coast of Tokyo, Japan. The islands range in size from approximately 10 to 9,900 ha, and the distance of islands from the Japanese mainland ranges from about 23 to 260 km (Figure 6.2). Most of the islands are covered with well-developed broadleaved forests and receive annual rainfall of more than 3,000 mm. Like other volcanic archipelagos such as the Galapagos or Hawaiian Islands, the Izu Islands arose de novo with no connection to a larger landmass. The Izu Islands therefore provide excellent opportunities to examine numerous aspects of biogeography and biotic formation on islands (MacArthur and Wilson 1967; Whittaker *et al.* 2008; Martins and Lillywhite 2019). For example, how do different island colonization patterns affect the ecological interactions in the past and present (Hasegawa 2003; Kuriyama *et al.* 2011; Brandley *et al.* 2014)?

In 1977, I began to investigate the ecology and life history of Okada's Skink (*Plestiodon latiscutatus*, formerly *Eumeces okadae*, and began a collaborative study of the Japanese Four-lined Ratsnake (*Elaphe quadrivirgata*) with Akira Mori in 1988. The project developed into a comprehensive program exploring the interrelationships

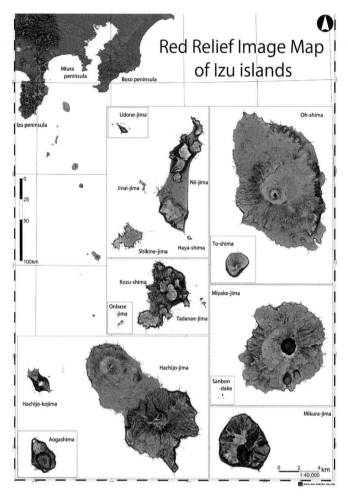

Figure 6.2 Red relief image map of the respective Izu Islands, associated with index location map showing three peninsulas of mainland Japan (Izu, Miura, and Boso), adjacent to the Izu Islands.

This red relief image map was produced by Asia Air Survey Co., Ltd with digital national land information datasets.

among phylogenetically distant but ecologically connected species on the Izu Islands. These studies have focused on topics such as predation (Hasegawa 1990, 1999; Hasegawa and Taniguchi 1996), parasitism (Hayashi and Hasegawa 1984), pollination (Kunitake *et al.* 2004; Abe and Hasegawa 2008; Mizusawa *et al.* 2014), and seed dispersal (Abe *et al.* 2006). Yet, the predator–prey relationship between *P. latiscutatus* and *E. quadrivirgata* has been the primary focus of collaborators and ourselves (Hasegawa and Moriguchi 1989; Hasegawa 1994; Mori and Hasegawa 1999, 2002; Tanaka *et al.* 2001; Kuriyama *et al.* 2006, 2016; Hasegawa and Mori 2008; Brandley *et al.* 2014; Kuriyama and Hasegawa 2017; Landry Yuan *et al.* 2021).

The dynamic geohistorical events of the formation of the Izu Islands (associated with Philippine Sea Plate activity) and the collision and accretion of proto-islands with the largest Japanese island of Honshu (*e.g.*, Hirata *et al.* 2010) produced a series of oceanic islands that present "blank slates" on which to observe insular ecological and evolutionary processes. Therefore, to understand the biological processes that shaped the Izu flora and fauna, it is first necessary to understand the geological history of the archipelago.

The Izu Islands (Figures 6.2 and 6.3) are part of a submarine mountain range on the Philippine Sea Plate called the Izu-Ogasawara Arc, extending approximately 1,200 km from north to south, that also includes the Ogasawara Islands. All of the Izu Islands, and most of the Izu Peninsula and central Japan, are volcanic, and they formed in the complex interface of the Eurasian, Pacific, and Philippine plates. The islands are also situated in a zone where blocks of the Izu–Bonin arc's upper crust

Figure 6.3 The Izu Islands, south of Tokyo and the Izu Peninsula. (a) A group of the volcanic islands north of Kozu. The nearest are Shikinejima, Niijima, Jinaijima, Udonejima, and Toshima, and the furthest Oshima. (b) The Izu Islands southeast of Kozu. The closer two small islands are Jino-Tadanae and Okino-Tadanae, where we have been studying *Elaphe quadrivirgata*. The island at the center is Miyake, the volcanically the most active Izu island, and the furthest is Mikura, which is home to the *E. quadrivirgata* population with the second largest body size. (c) A northwestern view of mainland Japan showing Mt Fuji and the Izu peninsula, with Toshima and Oshima, from the summit of Niijima.

Photographs by Masami Hasegawa.

have been successively accreted onto the Honshu arc for the past 15 million years (Amano 1991; Kitazato 1997; Hirata *et al.* 2010). Indeed, geochronology data indicate that at least Kozu and Niijima Islands are younger (0.88–0.93 mya; Kaneoka *et al.* 1970) than the Proto-Izu block (1.7–7.4 mya; Tani *et al.* 2011), suggesting that the collision and accretion of older island blocks triggered the formation of new volcanic islands behind the collision front, thereby creating a series of Izu proto-islands in a conveyor-like process. It is therefore reasonable to assume that the volcanic activity that formed the current Izu Islands was present in the past and that islands, although not necessarily the present-day islands, have been continually present in the region for millions of years.

The geologic history of the Izu Islands played a considerable role in shaping the ecology, distribution, and phylogeographical history of *E. quadrivirgata* and other plants and animals in the Izu Islands and the adjacent area of Japan's main islands (Okamoto *et al.* 2006; Kuriyama *et al.* 2011; Brandley *et al.* 2014). Understanding the historical processes of the insular ecological and evolutionary dynamics requires us to study the phylogeographies of both predatory *E. quadrivirgata* and its *P. latiscutatus* prey. A molecular phylogeographic analysis by Kuriyama *et al.* (2011) found evidence that *E. quadrivirgata* independently colonized three separate regions of the Izu Islands from mainland ancestors within the past 0.58–0.20 mya. The Izu Peninsula and Oshima and Mikura Islands were both colonized independently from lineages inhabiting eastern mainland Japan. The populations of Toshima, Niijima, Shikine, Kozu, and Tadanae Islands all derive from a single colonization from western mainland Japan.

The time-calibrated phylogeographic study of *P. latiscutatus* (Brandley *et al.* 2014) supports the conclusion that the ancestor to modern Izu Island populations dispersed from the Japan mainland to the Izu proto-islands between 3 and 7.6 mya. These lineages remained present in the area during successive formation of the islands, with one lineage recolonizing the mainland 0.24–0.7 mya. Therefore, each Izu island is an independent experiment studying how a single prey species (*P. latiscutatus*) co-evolves with varied predator assemblages. These different prey and predator colonization histories are associated with the evolution of color patterns and life histories among Izu Island *P. latiscutatus* lizard populations (Brandley *et al.* 2014).

Our research has shown that apex predators are components of trophic cascades that strongly affect the ecology and evolution of each island's vertebrate fauna (Hasegawa 1999, 2003; Brandley *et al.* 2014). For example, the island of Oshima (Figure 6.2) is the only Izu island with a native carnivorous mammal, the Japanese Weasel (*Mustela itasi*), and this species is at the top of the food chain. However, one or more species of snakes are the apex predators on the islands of Niijima, Shikine, Mikura, and Kozu, whereas birds are the dominant predators on Miyake, Hachijokojima, and Aogashima. When weasels are at the top of the food web, the abundance of lizards, snakes, and birds is low, but when both weasels and snakes are naturally absent, both birds and lizards are extremely abundant. Human introduction of *Mustela itasi* onto Miyake and Aogashima as a rat eradication program drastically reduced both lizards and avian predators, and this resulted in a strong trophic cascade since the weasel introduction (Hasegawa 1999). This indicates that the apex predators are keystone species that determine the structure of the vertebrate community of the Izu Islands (Hasegawa 2003). However, we note that the colonization histories of other Izu Island native carnivorous mammals and birds have not been revealed through time-calibrated phylogeographic analyses.

Snake Communities, Their Diets, and Body Size of the Izu Island Rat Snake

Five species of native snakes inhabit at least one Izu island (Figure 6.4), including *Elaphe quadrivirgata, E. climacophora, Euprepiophis conspicillatus, Lycodon orientalis,* and *Gloydius blomhoffii. Elaphe quadrivirgata* is endemic to the Japanese Archipelago and is a common snake feeding mainly on frogs in the rice paddies of mainland Japan (Fukada 1992; Kadowaki 1992). Although no frogs inhabit the Izu Islands, *E. quadrivirgata* is the most widespread and abundant snake species of the Izu Islands. The lizard *Plestiodon latiscutatus* is the principal prey in the snakes' diet (Hasegawa and Moriguchi 1989), with a notable exception of the Tadanae population where snakes also consume seabird eggs.

On Oshima, the nearest to the mainland and largest of the Izu Islands (Figure 6.3), *E. quadrivirgata* is sympatric with four snake species (*E. climacophora, Eu. conspicillatus, L. orientalis,* and *G. blomhoffii*), and its diet is limited to the lizards *P. latiscutatus* and *Takydromus tachydromoides.* The Niijima, Shikine, and Kozu

Figure 6.4 Native snake species of the Izu Islands. *Elaphe quadrivirgata* from Kozu (a), Niijima (b), Mikura (c), and Oshima (d) vary in color and pattern. Other species of snakes are less common on the Izu Islands, including (e) *Gloydius blomhoffii,* (f) *Lycodon orientalis,* (g) *Euprepiophis conspicillatus,* and (h) *Elaphe climacophora.*
Photographs by Masami Hasegawa.

Table 6.1 Summary of the among-island variation in snout to vent (SVL; mm) of the Izu Island *Elaphe quadrivirgata*.

Island	Largest 10 males			Total number of males	Largest 10 females			Total number of females
	Mean	SD	Maximum		Mean	SD	Maximum	
Tadanae	1716.4	16.4	1755	436	1358	36.9	1429	144
Mikura	1298.6	93.1	1500	36	923.5	91.4	1081	10
Shikine	1343.0	23.8	1392	179	1066	62.6	1220	58
Niijima	1342.1	31.6	1400	690	1072	18.4	1105	432
Kozu	1290.7	10.1	1310	936	1060	62.9	1190	342
Toshima	1034.4	41.5	1113	17	866.9	62.2	1003	13
Oshima	987.1	58.0	1085	25	755.6	98.4	846	9

populations of *E. quadrivirgata* are sympatric only with *E. climacophora*. *Elaphe quadrivirgata* feeds on a variety of vertebrates, including the Japanese Field Mouse (*Apodemus speciosus*), the Dsinezumi Shrew (*Crocidura dsinezumi*), juvenile ratsnakes (*E. climacophora*), lizards (*P. latiscutatus* and *T. tachydromoides*), and eggs and chicks of passerine birds. *Elaphe climacophora* on the Izu Islands feeds primarily on small mammals. The Tadanae and Mikura populations of *E. quadrivirgata* are solitary without other snake species, and both feed on lizards while juvenile, but on seabird eggs and chicks during adulthood. The Toshima population of *E. quadrivirgata* is solitary and its diet consists of *P. latiscutatus* (Hasegawa and Moriguchi 1989).

The striking feature of the Izu Island *E. quadrivirgata* populations is the difference in body size associated with their diets (Hasegawa and Moriguchi 1989). The mean snout to vent lengths (SVLs) of the largest 10 males and females from seven of the Izu Islands are shown in Table 6.1. Tadanae had the largest (1,716 mm), with that of Mikura the second largest. These gigantic insular populations feeding on seabird eggs and chicks are larger than snakes on Niijima and Kozu. The body size of the Oshima population is the smallest (987 mm) of the eight major Izu Islands. Female *E. quadrivirgata* body size was about 80% of male body size, and the trend of among-island female body size was the same as male body size. Adult male and female snakes from Oshima and Toshima tended to be smaller than or similar to the mainland populations. As a result, the other populations in Niijima, Shikine, Kozu, and Mikura were intermediate between the dwarf Oshima and the giant Tadanae populations.

The breeding colonies of seabirds on small uninhabited islands are the link between the terrestrial and marine Izu Island food webs. The island of Tadanae (Figures 6.5 and 6.6) is home to breeding colonies of the Crested Murrelet

Figure 6.5 The Japanese Four-lined Ratsnake, *Elaphe quadrivirgata* on Kozu (a) and Tadanae (b). The island behind the seashore of Kozu (a) is Tadanae, on which Masami Hasegawa is measuring snout-vent length of a snake with two assistants.
Photographs by Masashiro Maeda, with permission.

(*Synthliboramphus wumizusume*), Streaked Shearwater (*Calonectris leucomelas*), Tristram's Storm Petrel (*Oceandroma tristrami*), and Black-tailed Gull (*Larus crassirostris*) (Figure 6.1). These seabirds not only provide food resources to gigantic *E. quadrivirgata* but also guano as fertilizer for the terrestrial plants, which in turn support the terrestrial fauna, including herbivorous insects and the spiders and lizards that feed on them (Figure 6.6). On Tadanae, *E. quadrivirgata* is the only snake species that feeds on eggs and chicks of storm petrels, shearwaters, and murrelets. During the winter season in January, the Tristram's Storm Petrel digs holes in the ground to lay a single-egg clutch, while the Streaked Shearwater lays a single-egg clutch in June, and the Crested Murrelet lays a clutch of two eggs in deep rock crevices in April. Adults of *E. quadrivirgata* on Tadanae prey on the seabird eggs and chicks from early spring to summer, and their body size is the largest among the Japanese islands (Hasegawa and Moriguchi 1989). From mid-June to mid-August, the Mikura population of *E. quadrivirgata* also feeds on the eggs and chicks of the shearwater, the only seabird that roosts on that island. Body size of the Mikura *E. quadrivirgata* population is larger than other Izu Island populations except Tadanae. The occurrence of the most gigantic population on Tadanae is consistent with the fact that three seabird species breed from winter to summer for five months on Tadanae, whereas only a single seabird species breeds in summer on Mikura (Hasegawa and Mori 2008).

Figure 6.6 Elements and ecological events of the terrestrial food web of Tadanae.
(a) Central part of Tadanae, where the Streaked Shearwaters and Tristram's Storm
Petrels nest burrows are constructed below the pale green-colored *Miscanthus* grass
and the yellow green-colored *Carex* grass. (b) A group of newly hatched nymphs of
the short-winged *Parapodisma* grasshopper on the leaf of *Chrysanthemum pacificum*.
(c) A full adult *Parapodisma* grasshopper. (d) Extensive feeding damage caused by
Prapodisma grasshoppers on the herbaceous plant, *Elaeagnus macrophylla*. (e) The
skink *Plestiodon latiscutatus*, a predator on grasshoppers and spiders, and the only prey
item of Tadanae *E. quadrivirgata* before the snake is large enough to prey on seabird
eggs and chicks. (f) An orb-weaving spider common in the *Miscanthus* and *Carex* grasses
and often feeds on the *Parapodisma* grasshopper.
Photographs by Masami Hasegawa.

Field Studies on Tadanae

Snakes are gape-limited and they can only consume whole prey that will fit through
the snake feeding apparatus (Cundall and Greene 2000). Increasing gape size is
in large part a function of increasing body size. Our hypothesis was that selection

pressure to prey on large, nutrient-rich seabird eggs favored the evolution of large body size of Tadanae *E. quadrivirgata*, though we had no specific hypothesis to explain gigantic body size when we began field studies on Tadanae in June 1984. When visiting Tadanae, our primary mission was to collect basic natural history data of *E. quadrivirgata*. Finding the snakes was easy, but getting on the island was the difficult part. Our experience in May 1994 is typical.

Our ship, the *Mansaku Maru*, left Tako Bay on the east side of Kozu for the 10-minute trip to Tadanae. This expedition was special because it included esteemed researchers of the Pacific Seabird Group from the US West Coast to study the crested murrelet (Carter and de Forest 1994). The northeast winds and ocean swells around Tadanae often create treacherous boating conditions. On that day, the waves broke so roughly on the rocks that it seemed impossible to land (Figure 6.7). However, the local boat captains are experts at placing their (typically angler) clients on the safest spot of craggy inlets. From the elongated bow of the fishing boat, I (Hasegawa) leaped onto the rocky ledge of Tadanae, and Akira followed. Takayuki-*san*, the retired captain of the boat, threw over to us a set of camping gear, food, and drinking water that were placed in a rock crevice. The most difficult part of the route is a rounded rock that looms 10 m above the sea. After about 3 m, the rocky ledge is blocked by a rocky outcropping. We stuck to the rock like a gecko and pulled ourselves up onto it with the strength of our arms. We tried not to think about how even the slight weight of our backpacks could throw off our center of gravity and cause us to fall into the sea. After overcoming this difficult point, the rest of the climb became much easier. The quartz crystals on the surface of the rhyolite rock were rough like sandpaper, making it a perfect non-slip surface. The rest of the passengers jumped onto the rocky ledge

Figure 6.7 Approach to the landing site, Kochigahana, the northeastern corner of Tadanae by the fishing boat *Mansaku Maru* piloted by the captain Kiyomi (a) and guided by his son Manabu at the bow (b).
Photographs by Masami Hasegawa.

and made the same climb safely. Takayuki led the way, with Harry Carter right behind him, being light on his feet after climbing so many rocky seabird breeding grounds in the United States.

An hour had already passed since we landed at 10:00 AM. The sea conditions were not improving so we had to work quickly. Immediately, we saw several huge snakes with their straw-color bodies and brown stripes peeking out from the dense *Carex* grass. When I lifted one of the snakes, one or two others lifted up in an entangled mass, and all of them were put in a collecting bag. For each individual, we recorded body temperature, body mass, SVL, head length and width, reproductive condition (gravid or not), and stomach contents. We also examined certain ventral scales to confirm the identity of the snake using a unique scale-clip coding system. In 40 minutes, we captured 10 snakes. Nine of these individuals had been previously marked and captured at least once before last year. One was a newcomer who received a new identification number.

To collect the important dietary data, we had to palpate the abdomens of each snake to encourage them to disgorge their prey. On that day, only one male (#111) had prey in his stomach, and it immediately caught the attention of the murrelet researchers who gathered to see the contents. The snake had spit out four petrel eggs that were otherwise almost ready to hatch (Figure 6.8). This began a discussion about how the presence of these snake predators threaten the murrelet colony. Our companions suggested ways to reduce predation pressure on murrelet eggs, perhaps creating artificial nest burrows inaccessible to the snakes.

History has shown that concerns about the long-term stability of murrelet colonies are valid. For example, Norway rats introduced to the uninhabited Japanese islands of the Bay of Kitakyushu and Hakata caused mass mortality in the crested murrelet breeding colonies (Takeishi 1987). However, the results of our long-term study of the diets of *E. quadrivirgata* show that the predation pressure from the giant Tadanae snakes is minimal. Between 1985 and 1995, we captured a total of 86 snakes during the April and early May murrelet breeding season. Over this time period, we recorded nine snakes that consumed a total of 16 eggs, or about one egg for every five snakes. We concluded that there is a natural balance between the snake predator and murrelet prey, and it is reasonable to assume this relationship evolved since *E. quadrivirgata* first colonized Tadanae from nearby Kozu thousands of years ago (Kuriyama et al. 2011). Therefore, instead of a threat to the endangered murrelet, we view seabird egg predation as the key to understanding the evolution of gigantism in Tadanae ratsnakes—a unique evolutionary phenomenon on a tiny Japanese island (Hasegawa and Moriguchi 1989; Hasegawa and Mori 2008).

How Do Tadanae Ratsnakes Grow So Large?

Correlating egg consumption with body size in the giant snakes of Tadanae is simply the baseline information required to test more detailed evolutionary hypotheses. For

Figure 6.8 Various prey items of *Elaphe quadrivirgata* that were eaten and regurgitated. (a) A chick of the Crested Murrelet on Tadanae. (b) An egg of Tristram's Storm Petrel with a chick on the verge of hatching on Tadanae. (c) A group of baby mice on Kozu. (d) An adult male *Plestiodon latiscutatus* on Kozu. (e) An adult male *E. quadrivirgata* photographed during an attempt to swallow an adult female frog (*Pelophylax nigromaculatus*) in the paddy field of the Izu Peninsula, mainland Japan.
Photographs by Masami Hasegawa.

example, how do the snakes grow large? By what physiological mechanism did they achieve such a large size? Do they grow more rapidly as juveniles when compared to the neighboring Kozu source population, or do they simply live longer, continuing indeterminate growth?

To answer these questions, we conducted a new project in 1994, to track the growth rate of juvenile *E. quadrivirgata* on Tadanae. We captured gravid female snakes on Tadanae, hatched their eggs in captivity, individually marked them, and released these juveniles back onto Tadanae to be recaptured and measured—a procedure developed by Professor Hajime Fukada, a pioneer of snake population ecology in Japan (Fukada 1992). Over a period of three years, from 1994 to 1996, we captured a total of 20 gravid female snakes and transported them to Kyoto University to lay their eggs.

The mother snakes were returned to the island in August, and more than 100 hatchlings in total were returned in October. We tracked the growth of these age cohorts via mark-and-recapture over the next 23 years. To determine the life spans of the Kozu and Tadanae snake populations, we analyzed growth records from long-term mark and recapture studies started in 1984, thereby giving us the opportunity to test both the better juvenile growth and the longer indeterminate growth hypotheses.

In 2008, we published a paper summarizing our long-term mark and recapture study (Hasegawa and Mori 2008). Previous studies assumed that populations with larger than average body size should exhibit greater juvenile growth based on the presumption that growth generally slows after maturation even in animals exhibiting rates of indeterminate growth (Madsen and Shine 2000; Mori and Hasegawa 2002). In contrast, our data demonstrated clearly that the giant size of *E. quadrivirgata* on Tadanae is not due to rapid juvenile growth but instead is caused by prolonged, continuous, and seemingly enhanced growth throughout adulthood (Figure 6.9). The result that shocked us was that snakes on Tadanae live twice as long as snakes on all other Izu Islands. Despite the slow growth during juvenile stages, Tadanae males continued to grow substantially until nearly 20 years of age, becoming larger than individuals in the other populations at around the age of 10 years.

Figure 6.9 Facial profiles of the Tadanae Island's *Elaphe quadrivirgata* from the oldest (a) to the youngest (d) individuals. The oldest snake showed bleeding from gums, a sign of gingivitis, after biting the finger of Hasegawa.
Photographs by Masami Hasegawa.

Although undoubtedly important, the evolution of large body size is not the only adaptation to eating bird eggs. Rather, evolutionary modifications of the trophic apparatus that change the allometry of gape size (Figure 6.10) may be necessary to exploit novel (large-sized) food resources (*e.g.*, Forsman 1991; Aubret *et al.* 2004). When compared to snakes on nearby Kozu, the Tadanae *E. quadrivirgata* population has noticeably more soft, flexible tissue around the jaw (Figure 6.10) that presumably increases the maximum gape of these snakes to accommodate the size of their egg prey (*e.g.*, Jayne *et al.* 2022). Moreover, an adaptation seen in other species of egg-eating snakes (de Queiroz and Rodríguez-Robles 2006) is expanded or curved vertebral hypophyses that aid in piercing and crushing the eggshell while in the esophagus (Gans 1952; Gans and Oshima 1952). Although mainly frog-eaters, mainland populations of *E. quadrivirgata* will also consume small bird eggs. An examination of their vertebral anatomy reveals similar egg-crushing hypophyses (Goris 1963), and this may be a prerequisite trait to facilitate a local adaptation for predation on large eggs (Mullin 1996; Cundall and Greene 2000). It is not yet known if these vertebral hypophyses are more developed in the Tadanae *E. quadrivirgata* population. Such modified trophic structures and possible behavioral and physiological characteristics

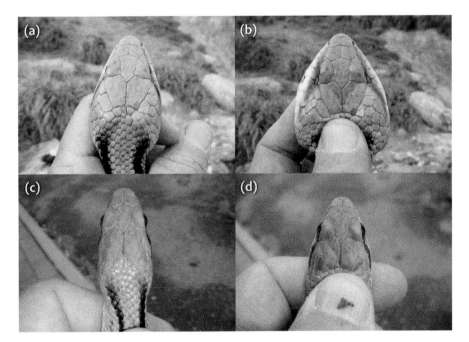

Figure 6.10 Head shapes of *Elaphe quadrivirgata* from Tadanae (above, a and b) and from Kozu (below, c and d). Left photographs (a and c) are natural head shapes, and right photographs (b and d) are the head shape differences emphasized by pressing between the thumb and the index finger.
Photographs by Masami Hasegawa.

make the giant snakes of Tadanae a unique system to study the evolution of correlated traits related to avian egg predation.

That gigantism is realized by enhanced growth at older ages rather than rapid growth in juveniles poses an interesting question: Is the gigantic size and longer life span of Tadanae *E. quadrivirgata* a phenotypically plastic response to a particularly rich food resource or an adaptive trait unique to that population? Would adult snakes from Tadanae and other Izu Island populations attain a similar gigantic body size if their diets are similarly energy-rich? Although not yet tested, a future common garden experiment could address this question by measuring the growth rates of adults from different Izu Island populations while feeding a high-energy diet (Aubret and Shine 2007; Boback and Carpenter 2007). Recall that snakes are gape-limited, and large prey such as seabird eggs and nestlings are exploitable only by large individuals. Our data show that the largest snakes in the non-gigantic, non-Tadanae populations are as large as the critical minimum body size (ca. 1,100 mm SVL) capable of swallowing eggs and nestlings in the Tadanae population. Therefore, if adult snakes on other Izu Islands had an opportunity to exploit larger prey items, such as seabird eggs or hatchlings, they could perhaps grow as large or live as long as snakes from Tadanae.

The Bank of the Pacific

Like many human families who rely on the rich marine life around the Izu Islands for their livelihoods, Kiyomi Hamakawa, the captain of the *Mansaku Maru* (Figure 6.11), is a commercial fisherman. Before he left port to catch commercially important fishes such as the Golden Alfonsino (*Beryx splendens*), flying fish (*Cheilopogon pinnatibarbatus*), and squid (*Ommastrephes bartramii*), the captain would say to us "I'm off to withdraw money from the Bank of the Pacific."

Why and how the sea around the Izu Islands is rich and sustainable are questions that were first posed by Dr. Tasaku Kitahara, a pioneer of fishery oceanographer in Japan, in 1913 (Uda 1972), and the questions have been attracting scientists of the present day (Saito 2019). The surface water of the warm Kuroshio Current, a western boundary current in the North Pacific Ocean, is nutrient-depleted and has relatively low primary productivity, yet the region around the Izu Islands supports abundant fish populations. This so-called Kuroshio Paradox (Saito 2019) is also a key to understanding the diversity and abundance of seabirds around the Izu Islands and their relationships to the gigantic *E. quadrivirgata* on Tadanae.

When nutrient-rich but light-limited upwelling subsurface water moves to shallower depths, it is subject to increased sun exposure (Uda 1972). This sun exposure, in combination with tidal action, internal waves, eddies, vortices, and wakes induced by the Kuroshio Current and submarine volcanos (Hasegawa *et al.* 2004; Isoguchi *et al.* 2009; Masunaga *et al.* 2018; Kodaira and Waseda 2019) promote high-productivity patches of phyto- and zooplankton populations around the Izu Islands (Takahashi

Figure 6.11 A dinner with the Hamakawa family after the herpetologists and ornithologists came back from the field work on Tadanae in 1994 (A). The herpetologists: (a) Mori is on the left, and M. Hasegawa is to the right of Mori. The seabird ornithologists from the US west coast: H. R. Carter is the first right and L. K. Ochikubo Chan is the second right of the last column, and J. N. Fries is the second right of front row. The Hamakawa family of the *Mansaku Maru*: the retired captain Takayuki is the first right of the front row, and his wife Yoriko is between Hasegawa and Fries. The last column are two grandchildren: Masashi is the first left, Takayoshi is the second left, and their father Norio Hamakawa is right of Masashi. (b) The current captain Kiyomi with a splendid alfonsino *Beryx splendens* he caught.

Photograph a by Akira Mori and photograph b by Masami Hasegawa.

et al. 1981; Takahashi and Kishi 1984; Atkinson *et al.* 1987; Toda 1989). This enhanced productivity ripples through the marine food web, ultimately sustaining large populations of animals at higher trophic levels such as pelagic planktivorous fishes, large carnivorous fishes, and seabirds. However, understanding the spatial and temporal dynamics of upwelling-based food webs is difficult and presents challenging tasks to follow and reveal (Uda 1972; Saito 2019; Yatsu 2019).

Our experience surveying Izu Island seabirds from 1995 to the present demonstrates that there are very few degrees of separation between the rich marine life around the Izu Islands and the evolution of giant snakes on Tadanae. In May 2021, in a known fishing ground for the common mackerel (*Scomber japonicus*) near Niijima (Kato *et al.* 2019), I (Hasegawa) witnessed a huge (>1,000) flock of Streaked Shearwaters where I once witnessed flocks of Crested Murrelets in May 1998 (Figure 6.12). Not only does the number of shearwaters counted along the sea route between Toshima and Niijima vary greatly (unpublished data), but the annual catch of common mackerels varies greatly, too (Kato et al. 2019). In northeastern Japan, Takahashi (2000) observed that fishing grounds are formed when densely distributed Japanese Anchovy (*Engraulis japonicus*) escape from Common Mackerel predation and move to the surface, allowing shearwaters to prey on them. Similarly, when attacked by large predators such as tuna or dolphins, schools of small pelagic fishes such as the sardine (*Sardinops melanostictus*) and anchovy are driven to the ocean's

Figure 6.12 Several flocks of the Streaked Shearwater, *Calonectris leucomelas*, just above schools of pelagic fishes probably fleeing predatory fish. This phenomenon is called *Nabura* in Japan.
Photographed May 2021 near Niijima, the Izu Islands by M. Hasegawa.

surface (Au and Pitman 1986; Hebshi *et al.* 2008). Flocks of raucous seabirds respond and aggregate around the rich food resource, a phenomenon called *Nabura* or *Toriyama* in Japan, which is a traditional method for locating schools of skipjack and tuna (Hotta *et al.* 1961). Consequently, the population dynamics of the giant ratsnake, the apex terrestrial predator on Tadanae, is tied to changes in marine food webs, all of which is ultimately influenced by the complex geological processes of the Izu Island (Kawasaki 1983; Ito 2012; Noguchi-Aita *et al.* 2018; Yatsu 2019).

The research by ourselves and others has demonstrated that the evolution, ecology, and ongoing health of the Izu Islands' terrestrial fauna is ultimately tied to the "Bank of the Pacific." The healthy population of fish provides food for local fishermen and prey for seabird colonies. These birds fertilize the flora of their roosting islands, and their eggs and the natural selective pressure to eat those eggs have promoted the rapid evolution of gigantic body size of *E. quadrivirgata* on the small island of Tadanae. We hope our long-term population and ecological studies of Izu Island ratsnakes and seabirds will contribute to the understanding of ecological mechanisms, functions, and dynamics of the ocean and insular terrestrial ecosystems, in general.

Acknowledgments

We sincerely appreciate the help, support, and encouragement by friends and students, especially the Hamakawa family of Kozu, Takayuki, Kiyomi, and Norio, Hajime Moriguchi, Takayo Nakayama, Takeo Kuriyama, Matthew Brandley, Yasunari Hattori, Rika Usui, and Arata Murakami. Our long-term field study of the Izu Islands was supported by funding (19H03307, 15H04426, 24570031, 21570024, 19570026) from the Japan Society for the Promotion of Science (JSPS) to M. Hasegawa. Fieldwork on Kozu and Niijima was approved by local village governments in accordance with their ordinances governing plant and animal protection. No specific permission was required for fieldwork on the other Izu Islands. Our deepest thanks goes to Matthew

Brandley for contributing to improvement of the English of the manuscript. We thank the late Shoichi Sengoku for introducing us to the fun of studying reptiles on the islands while Hasegawa was a high school student and Mori was an undergraduate student.

References

Abe, H., and M. Hasegawa. 2008. Impact of volcanic activity on a plant-pollinator module in an island ecosystem: The example of the association of *Camellia japonica* and *Zosterops japonica*. *Ecological Research* 23:141–150.

Abe, H., R. Matsuki, S. Ueno, M. Nashimoto, and M. Hasegawa. 2006. Dispersal of *Camellia japonica* seeds by *Apodemus speciosus* revealed by maternity analyses of plants and behavioral observations of animal vectors. *Ecological Research* 21:732–740.

Amano, K. 1991. Multiple collision tectonics of the Southern Fossa Magna in Central Japan. *Modern Geology* 15:315–329.

Atkinson, L., P. Blanton, J. O. McClain, T. H. C. Lee, M. Takahashi, and T. Ishimaru. 1987. Observation of upwelling around the Izu Peninsula, Japan: May 1982. *Journal of the Oceanographic Society of Japan* 43:89–103.

Au, D. W. K., and R. Pitman. 1986. Seabird interactions with dorphins and tuna in the eastern tropical pacific. *The Condor* 88:304–317.

Aubret, F., X. Bonnet, S. Maumelar, D. Bradshaw, and T. Schwaner. 2004. Diet divergence, jaw size and scale counts in two neighbouring populations of tiger snakes (*Notechis scutatus*). *Amphibia-Reptilia* 25:9–17.

Aubret, F., and R. Shine. 2007. Rapid prey-induced shift in body size in an isolated snake population (*Notechis scutatus*, Elapidae). *Austral Ecology* 32:889–899.

Boback, S. M., and D. M. Carpenter. 2007. Body size and head shape of island *Boa constrictor* in Belize: Environmental versus genetic contributions. In R. W. Henderson and R. Powell (eds.), *Biology of the Boas and Pythons*. Eagle Mountain Publishing, pp. 103–116.

Brandley, M. C., T. Kuriyama, and M. Hasegawa. 2014. Snake and bird predation drive the repeated convergent evolution of correlated life history traits and phenotype in the Izu Island scincid lizard (*Plestiodon latiscutatus*). *Plos One* 9: e9223.

Carter, H. R., and L. N. de Forest. 1994. Pacific seabird group goes to Japan: Part 2 (Izu Islands). *Pacific Seabirds* 21:17–21, 25.

Carter, H. R., K. Ono, J. N. Fries, H. Hasegawa, M. Ueta, H. Higuchi, J. T. Moyer, *et al.* 2002. Status and conservation of the Japanese murrelet (*Synthliboramphus wumizusume*) in the Izu Islands, Japan. *Journal of Yamashina Institute of Ornithology* 33:61–87.

Cundall, D., and H. W. Greene. 2000. Feeding in snakes. In K. Schwenk (ed.), *Feeding: Form, Function, and Evolution in Tetrapod Vertebrates*. Academic Press, pp. 293–333.

de Queiroz, A., and J. A. Rodríguez-Robles. 2006. Historical contingency and animal diets: The origins of egg eating in snakes. *American Naturalist* 167:684–694.

Forsman, A. 1991. Adaptive variation in head size in *Vipera berus* L. populations. *Biological Journal of the Linnean Society* 43:281–296.

Fukada, H. 1992. *Snake Life History in Kyoto*. Impact Shuppankai.

Gans, C. 1952. The functional morphology of the egg-eating adaptation in the snake genus *Dasypeltis*. *Zoologica* 37:209–244.

Gans, C., and M. Oshima. 1952. Adaptations for egg eating in the snake *Elaphe climacophora* (Boie). *American Museum Novitates* 1571:1–16.

Goris, R. C. 1963. Observations on the egg-crushing habits of the Japanese four-lined rat snake, *Elaphe quadrivirgata* (Boie). *Copeia* 1963:573–575.

Hasegawa, D., H. Yamazaki, R. G. Lueck, and L. Seuront. 2004. How islands stir and fertilize the upper ocean. *Geophysical Research Letters* 31, L16303.

Hasegawa, M. 1990. The thrush *Turdus celaenops*, as an avian predator of juvenile *Eumeces okadae* on Miyake-jima, Izu Islands. *Japanese Journal of Herpetology* 33:65–69.

Hasegawa, M. 1994. Insular radiation in life history of the lizard *Eumeces okadae* in the Izu Islands, Japan. *Copeia* 1994:732–747.

Hasegawa, M. 1999. Impacts of introduced weasel on the insular food web. In H. Ota (ed.), *Diversity of Reptiles, Amphibians and Other Terrestrial Animals on Tropical Islands: Origin, Current Status and Conservation.* Elsevier, pp. 129–154.

Hasegawa, M. 2003. Ecological diversification of insular terrestrial reptiles: A review of the studies on the lizard and snakes of the Izu Islands. *Global Environmental Research* 7:59–67.

Hasegawa, M., and A. Mori. 2008. Does a gigantic insular snake grow faster or live longer to be gigantic? Evidence from a long-term field study. *South American Journal of Herpetology* 3:145–154.

Hasegawa, M., and H. Moriguchi. 1989. Geographic variation in food habits, body size and life history traits of the snakes on the Izu Islands. In M. Matui, T. Hikida, and R. C. Goris (eds.), *Current Herpetology in East Asia.* Herpetological Society of Japan, pp. 414–432.

Hasegawa, M., and Y. Taniguchi. 1996. Behavioral discrimination of prey with various defense mechanisms by the lizard *Eumeces okadae. Journal of Ethology* 14:89–97.

Hayashi, F., and M. Hasegawa. 1984. Selective parasitism of the tick, *Ixodes asanumai* (Acarina: Ixodidae) and its influence on the host lizard *Eumeces okadae* in Miyake-jima, Izu Islands. *Applied Entomology and Zoology* 19:181–191.

Hebshi, A. J., D. C. Duffy, and K. D. Hyrenbach. 2008. Associations between seabirds and subsurface predators around Oahu, Hawaii. *Aquatic Biology* 4:89–98.

Hirata, D. H., K. Yamashita, Y. Suzuki, Y. Hirata, and T. Bing. 2010. Collision accretion tectonics of the proto-Izu–Mariana Arc: A review. *Journal of Geography* 119:1125–1160. (in Japanese with English summary)

Hotta, H., S. Fukushima, S. Odate, and Y. Aizawa. 1961. Observations of the fish schools and seabird flocks in the Tohoku sea area of Japan. *Bulletin of Tohoku Regional Fisheries Research Laboratory* 19:49–71.

Isoguchi, I., M. Shimada, F. Sakaida, and H. Kawamura. 2009. Investigation of Kuroshio-induced cold-core eddy trains in the lee of the Izu Islands using high-resolution satellite images and numerical simulations. *Remote Sensing of Environment* 113:1912–1925.

Ito, S. 2012. Studies on water mass variability related to dynamics of ecosystems in the Oyashio/Kuroshio region. *Oceanography in Japan* 21:33–50. (in Japanese with English summary)

Jayne, B. C., A. L. Bamberger, D. R. Mader, and I. A. Bartoszek. 2022. Scaling relationships of maximal gape in two species of large invasive snakes, brown treesnakes and Burmese pythons, and implications for maximal prey size. *Integrative Organismal Biology*, 1–18. https://doi.org/10.1093/iob/obac033.

Kadowaki, S. 1992. Food resource overlap between the two sympatric Japanese snakes (*Elaphe quadrivirgata* and *Rhabdophis tigrinus*) in the paddy fields. *Japanese Journal of Ecology* 42:1–7. (in Japanese with English synopsis)

Kaneoka, I., N. Ishiki, and S. Zashu. 1970. K-Ar ages of the Izu-Bonin Islands. *Geochemical Journal* 4:53–60.

Kato, M., S. Ohata, K. Takiguchi, T. Nakagawa, and K. Nakamura. 2019. Characteristics of common mackerel *Scomber japonicas* fishing conditions around Izu Islands in 2018. *Fisheries Biology and Oceanography in the Kuroshio* 20:63–68. (in Japanese with English abstract)

Kawasaki, T. 1983. Why do some pelagic fishes have wide fluctuations in their numbers? Biological basis of fluctuation from the viewpoint of evolutionary ecology. Proceedings of the Expert Consultation to Examine Changes in Abundance and Species Composition of Neritic Fish Resources, San Jose, Costa Rica, FAO Fisheries Report, 1065–1080.

Kitazato, H. 1997. Paleogeographic changes in central Honshu, Japan, during the late Cenozoic in relation to the collision of the Izu–Ogasawara Arc with the Honshu Arc. *Island Arc* 6:144–157.

Kodaira, T., and T. Waseda. 2019. Tidally generated island wakes and surface water cooling over Izu ridge. *Ocean Dynamics* 69:1373–1385.

Kunitake, Y., M. Hasegawa, T. Miyashita, and H. Higuchi. 2004. Role of a seasonally specialist bird *Zosterops japonica* on pollen transfer and reproductive success of *Camellia japonica* in a temperate area. *Plant Species Biology* 19:197–201.

Kuriyama, T., M. C. Brandley, A. Katayama, A. Mori, M. Honda, and M. Hasegawa. 2011. A time-calibrated phylogenetic approach to assessing the phylogeography, colonization history and phenotypic evolution of snakes in the Japanese Izu Islands. *Journal of Biogeography* 38:259–271.

Kuriyama, T., and M. Hasegawa. 2017. Embryonic developmental process governing the conspicuousness of body stripes and blue tail coloration in the lizard *Plestiodon latiscutatus*. *Evolution and Development* 19:29–39.

Kuriyama, T., K. Miyaji, M. Sugimoto, and M. Hasegawa. 2006. Ultrastructure of the dermal chromatophores in a lizard with conspicuous body and tail coloration (Scincidae: *Plestiodon latiscutatus*). *Zoological Science* 23:793–799.

Kuriyama, T., G. Morimoto, K. Miyaji, and M. Hasegawa. 2016. Cellular basis of anti-predator adaptation in a lizard with autotomizable blue tail against specific predators with different colour vision. *Journal of Zoology* 300:89–98.

Landry Yuan, F., S. Ito, T. P. N. Tsang, T. Kuriyama, T. Yamakazi, T. C. Bonebrake, and M. Hasegawa. 2021. Predator presence and recent climatic warming raise body temperatures of island lizards. *Ecology Letters* 24:533–524.

MacArthur, R., and E. O. Wilson. 1967. *The Theory of Island Biogeography*. Princeton University Press.

Madsen, T., and R. Shine. 2000. Silver spoons and snake body sizes: Prey availability early in life influences long-term growth rates of free-ranging pythons. *Journal of Animal Ecology* 69:952–958.

Martins, M., and H. B. Lillywhite. 2019. Ecology of snakes on islands. In H. B. Lillywhite and M. Martins (eds.), *Islands and Snakes: Isolation and Adaptive Evolution*. Oxford University Press, pp. 1–44.

Masunaga, E., Y. Uchiyama, Y. Suzue, and H. Yamazaki. 2018. Dynamics of internal tides over a shallow ridge investigated with a high-resolution downscaling regional ocean model. *Geophysical Research Letters* 45:3550–3558.

Miller, M. G., Y. Yamamoto, M. Sato, B. Lascelles, Y. Nakamura, H. Sato, Y. Ando, *et al.* 2019. At-sea distribution and habitat of breeding Japanese murrelets *Synthliboramphus wumizusume*: Implications for conservation management. *Bird Conservation International* 29:370–385.

Mizusawa, L., G. Takimoto, M. Yamasaki, Y. Isagi, and M. Hasegawa. 2014. Comparison of pollination characteristics between the insular shrub *Clerodendrum izuinsulare* and its widespread congener *C. trichotomum*. *Plant Species Biology* 29:73–84.

Mori, A., and M. Hasegawa. 1999. Geographic differences in behavioral responses of hatchling lizards (*Eumeces okadae*) to snake-predator chemicals. *Japanese Journal of Herpetology* 18:45–56.

Mori, A., and M. Hasegawa. 2002. Early growth of *Elaphe quadrivirgata* from an insular gigantic population. *Current Herpetology* 21:43–50.

Mullin, S. J. 1996. Adaptations facilitating facultative oophagy in the gray rat snake, *Elaphe obsoleta spiloides*. *Amphibia-Reptilia* 17:387–394.

Noguchi-Aita, M., S. Chiba, and K. Tadokoro. 2018. Response of lower ecosystems to decadal scale variation of climate system in the North Pacific Ocean. *Oceanography in Japan* 27:43–57. (in Japanese with English abstract)

Okamoto, T., J. Motokawa, M. Toda, and T. Hikida. 2006. Parapatric distribution of the lizards *Plestiodon* (formerly *Eumeces*) *latiscutatus* and *P. japonicus* (Reptilia: Scincidae) around the Izu Peninsula, central Japan, and its biogeographic implications. *Zoological Science* 23:419–425.

Saito, H. 2019. The Kuroshio: Its recognition, scientific activities and emerging issues. In T. Nagai, H. Saito, K. Suzuki, and M. Takahashi (eds.), *Kuroshio Current: Physical, Biogeochemical, and Ecosystem Dynamics*, Geophysical Monograph 243 (1st ed.). Wiley, pp. 1–11.

Takahashi, K. 2000. Fishery grounds with sea bird of Sendai Bay in summer and autumn. *Bulletin of Miyagi Prefecture Fisheries Research and Development Center* 16:55–59.

Takahashi, M., and M. J. Kishi. 1984. Phytoplankton growth response to wind induced regional upwelling occurring around the Izu Islands off Japan. *Journal of the Oceanographical Society of Japan* 40:221–229.

Takahashi, M., Y. Yasuoka, and M. Watanabe. 1981. Local upwelling associated with vortex motion off Oshima Island, Japan. *Coastal Upwelling* 1:119–124.

Takeishi, M. 1987. The mass mortality of Japanese murrelet *Synthliboramphus wumizusume* on the Koyashima Islet in Fukuoka. *Bulletin of Kitakyushu Museum of Natural History* 7:121–131.

Tanaka, K., A. Mori, and M. Hasegawa. 2001. Apparent decoupling of prey recognition ability with prey availability in an insular snake population. *Journal of Ethology* 19:27–32.

Tani, K., R. S. Fiske, D. J. Dunkley, O. Ishizuka, and T. Oikawa. 2011. The Izu Peninsula, Japan: Zircon geochronology reveals a record of intra-oceanic rear arc magmatism in an accreted block of Izu-Bonin upper crust. *Earth Planet Science Letter* 303:225–239.

Toda, H. 1989. Surface distributions of copepods in relation to regional upwelling around the Izu Islands in summer of 1988. *Journal of the Oceanographical Society of Japan* 45:251–257.

Uda, M. 1972. Historical development of fisheries oceanography in Japan. *Proceedings of the Royal Society of Edinburgh, Section B: Biological Sciences* 73:391–398.

Yatsu, A. 2019. Review of population dynamics and management of small pelagic fishes around the Japanese Archipelago. *Fisheries Science* 85:611–639.

Whittaker, R. J., K. A. Triantis, and R. J. Ladle. 2008. A general dynamic theory of oceanic island biogeography. *Journal of Biogeography* 35:977–994.

7

Distribution Patterns of Snakes and Conservation Importance of Islands in the Andaman and Nicobar Archipelago, Bay of Bengal, India

S. R. Chandramouli

Introduction

Islands offer interesting systems for investigating patterns of species distribution. This is particularly interesting for animal groups with reduced abilities for transoceanic dispersal and residing on island ecosystems. Reptiles—particularly snakes—are appropriate candidates for investigating these patterns. The Andaman and Nicobar archipelago of India is a fairly large island system, comprising a chain of more than 300 islands and islets and situated as a series of two separated archipelagos east of the Indian peninsula. These islands have been the subjects of several faunal explorations since the late 1800s that have cumulatively documented the insular snake fauna at a fairly fine level of detail. Since the beginning of herpetological explorations on these islands in the 1840s, totals of 22 and 23 species have been recorded for snakes from the Andaman and the Nicobar archipelagos, respectively. Various studies over the past two centuries have continued to document the snake fauna inhabiting these islands, including both additions and deletions (Table 7.1, Figure 7.1). Although these studies recorded the presence of snakes on these islands, no attempt has been made to examine and relate the distribution of these species with the existing network of protected areas. In this chapter, I examine in detail the taxonomic composition, species–area relationships, distribution patterns, and similarity of the snake fauna on these archipelagos, and I evaluate the effectiveness of the existing protected area network in conserving range-restricted snakes of the Andaman and Nicobar Islands.

Methods

A total of 26 islands (19 in the Andaman and 7 in the Nicobar archipelago) were sampled for snakes during this study. Data on snake species richness of Narcondam—a

S. R. Chandramouli, *Distribution Patterns of Snakes and Conservation Importance of Islands in the Andaman and Nicobar Archipelago, Bay of Bengal, India* In: *Islands and Snakes*. Edited by: Harvey B. Lillywhite and Marcio Martins, Oxford University Press. © Oxford University Press 2023. DOI: 10.1093/oso/9780197641521.003.0007

Table 7.1 Select literature on snakes of the Andaman and Nicobar Islands in chronological order. Species names follow Uetz *et al.* (2022).

Reference	Changes
Blyth (1846)	Described *Trimeresurus cantori*
Blyth (1859)	Reported *Ophiophagus hannah* from the Andamans
Blyth (1863)	Described *Fowlea tytleri*
Fitzinger in: Steindachner (1867)	Described *Trimeresurus labialis* from Nicobars
Theobald (1868)	Described *Trimeresurus andersoni* and *Lycodon hypsirhinoides*
Stoliczka (1870)	Described *Trimeresurus mutabilis* and *Gongylosoma nicobariense*
Stoliczka (1871)	Described *Gerrhopilus andamanensis*
Boulenger (1890)	Described *Argyrophis oatesii*
Sclater (1891)	Described *Hebius nicobarensis* and *Oligodon woodmasoni*
Wall (1909)	Described *Boiga andamanensis*
Wall (1913)	Described *Naja sagittifera*
Smith (1940)	Reported *Xenopeltis unicolor, Indotyphlops braminus, Gonyosoma oxycephalum, Coelognathus flavolineatus, Dendrelaphis andamanensis, Dendrelaphis cyanochloris, Chrysopelea paradisi* and *Xenochrophis melanzostus*
Smith (1943)	Reported *Amphiesma stolatum* and *Boiga ochracea walli* from the Andamans
Whitaker (1978)	Reported *Acrochordus granulatus*
Biswas and Sanyal (1965)	Described *Lycodon tiwarii*
Biswas and Sanyal (1978)	Described *Bungarus andamanensis*
Tiwari and Biswas (1973)	Described *Dendrelaphis humayuni*
Biswas and Sanyal (1980)	Reported *Laticauda laticaudata* from the Andamans
Das and Chandra (1994)	Reported *Microcephalophis cantoris*
Das (1998)	Described *Boiga wallachi* from Great Nicobar
Ghodke and Andrews (2001)	Reported *Hypsiscopus plumbea* from Great Nicobar
Vijayakumar and David (2006)	Rediscovered *Oligodon woodmasoni*
Harikrishnan *et al.* (2010)	Reported *Coelognathus* sp., and *Lycodon subcinctus* from Great Nicobar
Vogel *et al.* (2014)	Revalidated *Trimeresurus mutabilis*
Chandramouli (2017)	Deleted *Boiga cyanea* from Nicobars
Chandramouli *et al.* (2020)	Described *Trimeresurus davidi* from Car Nicobar
Gokulakrishnan *et al.* (2021)	Reported *Boiga cyanea* from Narcondam, Andaman

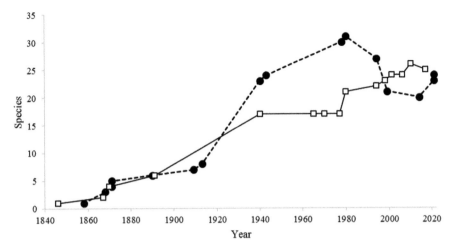

Figure 7.1 Trend-line showing the rate of discovery/reports of snakes in the Andaman (black dots, dashed line) and Nicobar Islands (white squares, solid line).

distant, isolated, volcanic island of the Andaman archipelago—were gathered from the literature (Raman *et al.* 2013; Chandramouli and Adhikari 2022; Gokulakrishnan *et al.* 2021). Snake species richness on these islands was assessed by a combination of methods ranging from visual encounter surveys to opportunistic photographic records with precisely known localities that were shared with the author. Surveys were conducted both by day and night in order to record nocturnal and diurnal species.

Habitat types surveyed for detecting snake species in these islands include primary evergreen, secondary evergreen, littoral forests, mangroves, and anthropogenic habitats such as gardens, fields, and roadside vegetation. Potential microhabitats such as rocks, fallen logs, leaf litter, streams, and ponds were inspected carefully to detect the presence of snakes. Roads with vehicular traffic passing through forested areas were visited periodically to observe cases of mortality, if any. Records based on reported instances of stray snakes into human habitation and those recorded as roadkills were included in the database.

The distribution of each species in the Andaman and Nicobar Islands was mapped based on observed distributions during the study period in addition to confirmed, precise literature records (most recently Gokulakrishnan *et al.* 2021; also see Table 7.1). Regression analyses to test the effect of island area (log-transformed) on snake species richness of the islands in each archipelago (Andaman and Nicobar) were performed. Hierarchical unweighted pair group method with arithmetic mean (UPGMA) cluster analysis with Euclidean distance was performed to identify clusters of islands with similar or unique snake species or assemblages. The protected area network in the Andaman and Nicobar archipelago was overlaid on the species

Table 7.2 Island-wise snake species occurrence records for the Andaman Islands

Island	InBr	GeAn	ArOa	LyHy	CrPa	DeAn	BoAn	Bocy	PtMc	CoFl	GoOx	FoTy	CeRy	AcGr	CaVi	GePr	BuAn	NaSa	OpHa	TrAn	LaCo	LaLa
NA	0	0	0	1	0	1	0	0	1	1	1	1	1	0	1	0	1	0	0	1	0	0
PGT	0	0	0	1	0	1	0	0	0	0	0	1	1	0	0	0	0	0	0	0	1	0
NR	0	0	0	0	0	1	0	0	0	0	0	0	1	1	1	1	0	0	0	0	1	0
INT	0	0	0	1	0	1	0	0	0	0	1	1	1	0	0	0	0	0	0	1	0	0
MA	0	0	1	1	0	1	1	0	0	0	1	1	1	0	0	1	1	1	0	1	1	1
LO	0	0	0	1	0	1	1	0	1	0	0	1	1	0	0	0	1	0	0	1	1	0
HVL	1	0	1	1	0	1	0	0	1	1	0	1	1	0	0	0	1	1	0	1	0	0
NIL	0	0	0	0	0	1	0	0	0	1	0	0	1	0	0	0	0	0	0	0	0	0
JLR	0	0	0	0	0	0	0	0	0	0	0	1	0	0	0	0	0	0	0	0	0	0
HLR	0	0	0	0	0	0	0	0	0	0	0	0	1	0	0	0	0	0	0	0	0	0
SA	1	1	1	1	0	1	1	0	1	1	1	0	1	1	1	1	1	1	1	1	1	1
LA	0	0	0	1	0	1	1	0	0	0	0	1	0	0	0	0	1	0	0	1	0	0
KYD	0	0	0	0	0	0	0	0	0	0	0	0	1	0	0	0	0	0	0	1	0	0
TMG	0	0	1	0	0	0	0	0	0	0	0	0	1	0	0	0	0	0	0	1	0	0
RSK	0	0	0	0	0	0	0	0	0	0	0	0	1	0	0	0	0	0	0	1	1	0
RUT	0	0	1	1	0	0	1	0	0	0	0	1	1	0	0	0	0	0	0	0	0	0
ALX	0	0	0	0	0	0	0	0	0	0	0	0	0	0	0	0	0	0	0	1	0	0
HOB	0	0	0	0	0	0	0	0	0	0	0	0	0	0	0	0	0	0	0	1	0	0
NAR	0	0	0	0	1	0	0	1	0	0	0	0	0	0	0	0	0	0	0	0	1	0

Abbreviations (species): GeAn, *Gerrhopilus andamanensis*; CrPa, *Chrysopelea paradisi*; HeNi, *Hebius nicobarensis*; GePr, *Gerarda prevostiana*; CaVi, *Cantoria violacea*; LaLa, *Laticauda laticaudata*; AcGr, *Acrochordus granulatus*; ArOa, *Argyrophis oatesii*; GoOx, *Gonyosoma oxycephalum*; BoCy, *Boiga cyanea*; CoFl, *Coelognathus flavolineatus*; NaSa, *Naja sagittifera*; OpHa, *Ophiophagus hannah*; BoAn, *Boiga andamanensis*; PtMc, *Ptyas mucosa*; InBr, *Indotyphlops braminus*; BuAn, *Bungarus andamanensis*; LyHy, *Lycodon hypsirhinoides*; CeRy, *Cerberus rynchops*; LaCo, *Laticauda colubrina*; DeAn, *Dendrelaphis andamanensis*; TrAn, *Trimeresurus andersoni*; FoTy, *Fowlea tytleri*; LyTi, *Lycodon tiwarii*; LySu, *Lycodon subcinctus*; DeHu, *Dendrelaphis humayuni*; BoWa, *Boiga wallachi*; CoSp, *Coelognathus sp*.; XeTr, *Xenochrophis trianguligerus*; FoLe, *Fordonia leucobalia*; XeUn, *Xenopeltis unicolor*; MaRe, *Malayopython reticulatus*; GoNi, *Gongylosoma nicobariense*; OlWo, *Oligodon woodmasoni*; HyPl, *Hypsicopus plumbea*; TrLa, *Trimeresurus labialis*; TrDa, *Trimeresurus davidi*; TrCa, *Trimeresurus cantori*; TrMu, *Trimeresurus mutabilis*.

Islands: ALX, Alexandra; JLR, John Lawrance; HLR, Henry Lawrance; RSK, Redskin; NR, North Reef; TMG, Tarmugli, KYD, Kyd, HOB, Hobday, NAR, Narcondam, INT, Interview, RUT, Rutland, PGT, Paget, NIL, Neil (Shaheed dweep), LO, Long, LA, Little Andaman, HVL, Havelock (Swarajdweep), NA, North Andaman, MA, Middle Andaman, CAR, Car Nicobar, TER, Teressa, CAM, Camorta, NAN, Nancowry, KAT, Katchall, LNI, Little Nicobar, GNI, Great Nicobar.

Table 7.3 Island-wise snake species occurrence records for the Nicobar Islands (see Table 7.2 for abbreviations)

Island	InBr	LyTi	LySu	DeHu	BoWa	CoSp	XeTr	HeNi	AcGr	CaVi	FoLe	XeUn	MaRe	CeRy	GoNi	OlWo	HyPl	LaCo	TrLa	TrAn	TrDa	TrCA	TrMu
CAR	1	1	0	0	0	0	0	0	0	0	0	0	1	1	0	0	0	0	1	1	1	0	0
TER	0	1	0	1	0	0	0	0	0	0	0	0	1	1	0	1	0	0	0	0	0	1	1
CAM	1	1	0	1	0	0	0	1	1	1	0	0	1	1	1	1	0	0	0	0	0	1	1
NAN	0	1	0	1	0	0	0	0	0	0	0	0	1	1	0	1	0	1	0	0	0	1	1
KAT	0	1	0	1	0	0	0	0	0	0	0	0	1	1	0	1	0	1	0	0	0	1	1
LNI	0	0	0	1	1	0	1	0	0	0	0	0	1	1	0	0	0	0	0	0	0	0	0
GNI	1	0	1	1	1	1	1	0	0	0	1	1	1	1	0	0	1	0	0	0	0	0	0

distribution maps to identify conservation gaps (*i.e.*, species whose distribution is not adequately covered by protected areas; Rodrigues *et al.* 2004).

Results and Discussion

Species Composition and Species–Area Relationships

The snake fauna of the Andaman and Nicobar archipelagos was found to be nearly equally rich, with both of these island groups harboring 22 and 23 species, respectively (22 genera and seven families in the Andaman Islands; 18 genera and eight families in the Nicobar Islands; Figure 7.2). Island-wise species richness ranged from 1 to 19 species in the Andaman Islands and 5 to 12 species in the Nicobar Islands. The proportion of endemic species was higher in the Nicobar Islands (47%) than the Andamans (40%). Island area explained 42% and 21% of the variation in species richness in the Andaman and Nicobar archipelagos, respectively (Figure 7.3).

Snake Species Distribution

Cerberus rynchops, being a brackish-water snake with better dispersal abilities than terrestrial or freshwater snakes, was found to be the most widely distributed species,

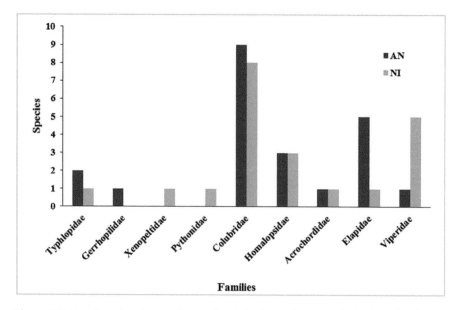

Figure 7.2 Number of snake species per family in the Andaman and Nicobar Islands.

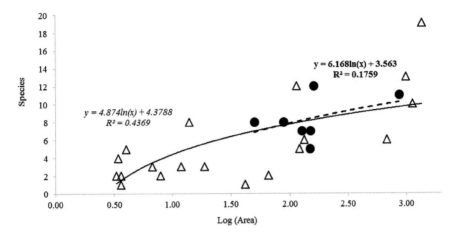

Figure 7.3 Species–area relationships for the snake fauna of the Andaman and Nicobar Islands. Solid line, white triangles, and italics: Andaman archipelago. Dotted line, black dots, and bold: Nicobar archipelago. Note that the curve for the Nicobar archipelago has a steeper slope than that of the Andamans.

occurring on 15 of the islands surveyed in both the Andaman and Nicobar archipelagos. This was closely followed by the tree snake *Dendrelaphis andamanensis*, which was recorded from 13 islands; *Trimeresurus andersoni*, which was recorded from 12 islands; *Lycodon hypsirhinoides*, which was recorded from 11 islands; and *Fowlea tytleri*, which was recorded from 9 islands within the Andaman Archipelago. The only snake found to be widespread in the Nicobar Archipelago was *Malayopython reticulatus*, which occurs on all islands in the Nicobar Archipelago. It is worth mentioning that most of the widespread species in the Andaman Archipelago are endemic to this group of islands (with the exception of *Cerberus rynchops*). Although certain species such as *Laticauda colubrina*, *Bungarus andamanensis*, *Indotyphlops braminus*, *Cantoria violacea*, *Boiga andamanensis*, *Ptyas mucosa*, *Argyrophis oatesii*, *Coelognathus flavolineatus*, *Gonyosoma oxyxcephalum*, *Naja sagittifera*, *Ophiophagus hannah*, and *Laticauda laticaudata* from the Andaman Archipelago have been recorded only from a few (2–6) islands, they could potentially occur on a larger number of islands within the archipelago. Four of the endemic species within the Andaman Archipelago have widespread distributions, while in the Nicobar Archipelago, the most widespread snake was the *M. reticulatus*, which is distributed widely throughout Sundaland (Murray-Dickson *et al.* 2017). Unlike in the Andaman Archipelago, several species recorded from the Nicobar Islands (*e.g.*, *Dendrelaphis humayuni*, *Oligodon woodmasoni*, *Trimeresurus cantori*, *T. davidi*, and *T. labialis*) have truly restricted distributions, limited to one to six islands within the archipelago (see Figure 7.4).

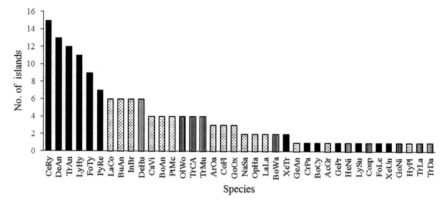

Figure 7.4 Number of islands occupied by each snake species recorded during the present study. Bars with stipple denote species that could potentially be found on more islands in the archipelago, striped bars denote species restricted to one or a few islands within the Andaman and Nicobar archipelago that would need increased conservation attention. For species abbreviations, see Table 7.2.

Similarity of the Islands' Snake Fauna

As for the islands of the Andaman Archipelago, the similarity of their snake assemblages was primarily determined by their size, with the larger islands showing a similar snake fauna to each other than the smaller ones (Figure 7.5). One exception to this pattern was the relatively large (675 km²) but somewhat isolated Little Andaman Island, from which only five species of snakes were recorded. This could possibly be due to the incomplete inventory on this island and not represent its true species richness. On the other hand, though relatively smaller (113.93 km²) than the former, Havelock Island was found to harbor more than twice the number (12) of snakes. On the other hand, the snake fauna of the Nicobar Islands was more discrete, with two separate assemblages of snakes on two groups of islands forming two well-defined clusters (Figure 7.5). The northern and central Nicobar Islands formed a cluster with shared species, while the Little and Great Nicobar islands, situated farther south, across the Sombrero Channel, had a different assemblage of snakes. Several species of snakes found in the central and northern group of Nicobar Islands are endemic (*e.g.,* *Trimeresurus labialis* and *T. davidi* for Car Nicobar; *T. mutabilis*, *T. cantori*, and *Oligodon woodmasoni* for central Nicobars; and species like *Lycodon tiwarii* being common to both northern and central Nicobar Islands).

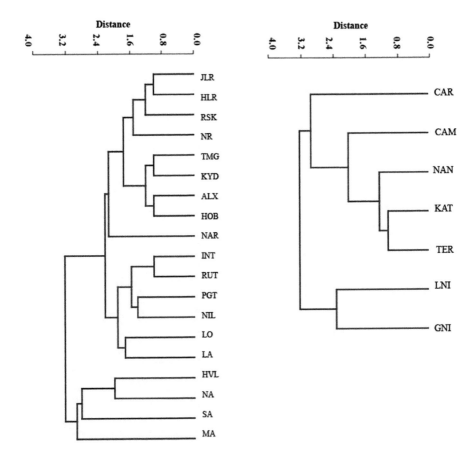

Figure 7.5 Clusters of islands with similar snake species assemblages in Andaman Islands (left) and Nicobar Islands (right). For island abbreviations see Table 7.2.

Conservation Status of Snakes: Protected Areas Versus Endemic Species

Most of the snake species occurring in the Andaman and Nicobar Archipelago were found to be categorized as Least Concern ($n = 23$) in the International Union for Conservation of Nature (IUCN) Red List (IUCN 2022), with five species being Data Deficient, four being Endangered, two species each being Near Threatened and Not Assessed, and one species each being vulnerable and Critically Endangered (Figure 7.6).

In the Andaman Archipelago, protected areas span across the length of the Andaman Islands and cover a fair proportion of the islands. This would represent a sufficiently vast network of protected areas within the Andaman Islands because

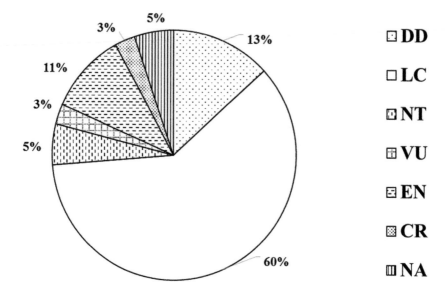

Figure 7.6 Conservation status of snakes of the Andaman and Nicobar Archipelago in the International Union for Conservation of Nature (IUCN) Red List (2022). DD, Data deficient; LC, Least concern; NT, Near-threatened; VU, Vulnerable; EN, Endangered; CR, Critically endangered; NA, Not assessed.

their snake fauna is more or less uniform throughout; only the isolated Narcondam Island Wildlife Sanctuary has two species of snakes (*Boiga cyanea* and *Chrysopelea paradisi*) restricted to it within the archipelago (widespread elsewhere). On the other hand, in the Nicobar Archipelago, the northern and central group of islands, despite each of them being home to 4–6 endemic species, are nearly devoid of protected areas with only two relatively small islands being protected (Batti Malv Wildlife Sanctuary in the Northern Nicobars and Tillanchong Island Wildlife Sanctuary in the central Nicobars). In the southern Nicobars, the Campbell Bay National Park and Galathea National Park cover the majority of Great Nicobar's area and would thus enjoy a better legal protection status than the northern and central Nicobar Islands. However, these islands harbor only two endemic snake species when compared to 4–6 species being endemic to each island of the northern and central Nicobars (Figure 7.7).

This points to a mismatch between the existing protected area network and endemic snake species distributions. Although the northern and central Nicobar Islands have been designated as restricted tribal reserves, snakes are killed in fairly

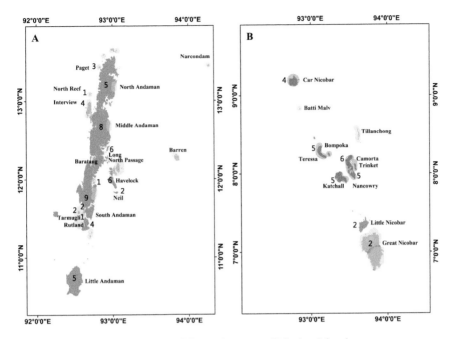

Figure 7.7 Protected areas (PAs) of the Andaman and Nicobar Islands versus distribution of endemic snakes (yellow-shaded regions denote protected areas; red-shaded areas denote regions with endemic snakes but without PAs; numbers beside the islands represent the number of endemic snakes).

large numbers locally (pers. obs.). Therefore, I propose that for the long-term survival of range-restricted snake species, more areas within the northern and central Nicobar Islands need to be protected via both gazettement and enforcement of relevant legislation within the archipelago (Figure 7.8).

Acknowledgments

I thank the Department of Environment and Forests, Andaman and Nicobar Islands, for permission to conduct the study and for the infrastructure provided. I am thankful to the Mohamed bin Zayed Species Conservation fund for grants (#14058387 and #160514249) which partly facilitated this study. I also thank the faculty of the Department of Ecology and Environmental Sciences and the Department of Ocean Studies and Marine Biology, Pondicherry University, for the laboratory space and support extended. Finally, I am grateful to Romulus Whitaker for suggesting that I contribute this chapter on island snakes.

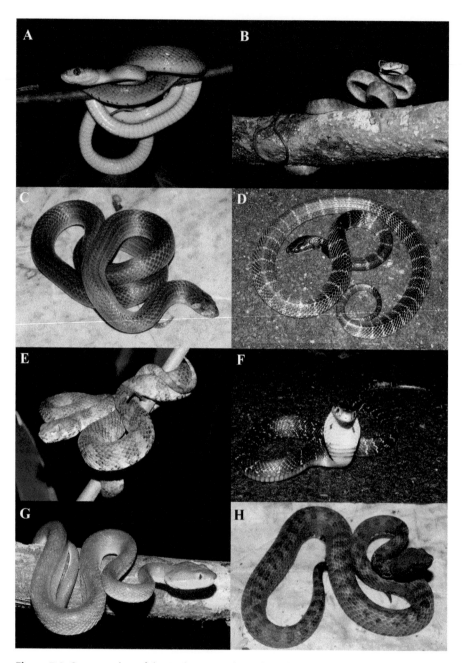

Figure 7.8 Some snakes of the Andaman and Nicobar Islands. (A) *Boiga wallachi*, (B) *Boiga andamanensis*, (C) *Oligodon woodmasoni*, (D) *Bungarus andamanensis*, (E) *Trimeresurus andersoni*, (F) *Naja sagittifera*, (G) *Trimeresurus cantori*, (H) *Trimeresurus labialis*.
Photos by S. R. Chandramouli.

References

Biswas, S., and D. P. Sanyal. 1965. A new species of wolf-snake of the genus *Lycodon* Boie (Reptilia: Serpentes: Colubridae) from the Andaman and Nicobar Islands. *Proceedings of the Zoological Society of Calcutta* 18 (2):137–141.

Biswas, S., and D. P. Sanyal. 1978. A new species of krait of the genus *Bungarus* Daudin, 1803 (Serpents: Elapidae) from the Andaman Island. *Journal of the Bombay Natural History Society* 75(1):179–183.

Biswas, S., and D. P. Sanyal. 1980. A report on the reptilian fauna of Andaman and Nicobars Islands in the collection of Zoological Survey of India. *Records of the Zoological Survey of India* 77:255–292.

Blyth, E. 1846. Notes on the fauna of Nicobar Islands: Reptilia. *Journal of the Asiatic Society of Bengal* 15:367–379.

Blyth, E. 1859. Report of the curator, Zoological department. *Journal of the Asiatic Society of Bengal* 28:411–419.

Blyth, E. 1863. The zoology of the Andaman Islands. In: F. J. Mouat. Adventures and researches among the Andaman Islanders. *Journal of the Asiatic Society of Bengal* 32:345–367.

Boulenger, G. A. 1890. *The Fauna of British India, Including Ceylon and Burma. Reptilia and Batrachia*. Taylor & Francis, i–xviii + 541 pp.

Chandramouli, S. R. 2017. A critical appraisal of the record of the green cat snake, *Boiga cyanea*, from Great Nicobar Island and its deletion from the Nicobar Islands faunal list. *Sauria* 39(3):47–50.

Chandramouli, S. R., and O. D. Adhikari. 2022. Recent sighting record of the paradise flying snake *Chrysopelea paradisi* Boie, 1827 (Reptilia: Colubridae) from Narcondam, Andaman Islands. *Sauria* 44(3):50–52.

Chandramouli, S. R., P. D. Campbell, and G. Vogel. 2020. A new species of green pit viper of the genus *Trimeresurus* Lacpde, 1804 (Reptilia: Serpentes: Viperidae) from the Nicobar Archipelago, Indian Ocean. *Amphibian & Reptile Conservation* 14(3) [Taxonomy Section]:169–176.

Das, I. 1998 A new species of *Boiga* (Serpents: Colubridae) from Nicobar Archipelago. *Journal of South Asian Natural History* 31(1):59–67.

Das, I., and K. Chandra. 1994. Two snakes new to Andaman and Nicobar Islands. *Journal of the Andaman Science Association* 10(1 &2):114–115.Ghodke, S., and H. V. Andrews. 2001. *Enhydris plumbea* (Boie, 1827) (Serpents: Colubridae: Homalopsinae): A new record for India. *Hamadryad* 26(2):373–375.

Gokulakrishnan, G., A. K. Das, C. Sivaperuman, and S. R. Chandramouli. 2021. First record of the green catsnake, *Boiga cyanea* (Duméril, Bibron, and Duméril 1854) (Reptilia: Colubridae), from Narcondam Island, India. *Reptiles & Amphibians* 28(2):262–263.

Harikrishnan, S., B. C. Choudhary, and K. Vasudevan. 2010. Recent records of snakes (Squamata: Serpentes) from Nicobar Islands, India. *Journal of Threatened Taxa* 2 (11):1297–1300.

IUCN. 2022. The IUCN Red List of Threatened Species. Version 2022-1. Accessed August 20, 2022. https://www.iucnredlist.org.

Murray-Dickson, G., M. Ghazali, R. Ogden, R. Brown, and M. Auliya. 2017. Phylogeography of the reticulated python (*Malayopython reticulatus* ssp.): Conservation implications for the worlds' most traded snake species. *PLoS ONE* 12(8):e0182049. https://doi.org/10.1371/journal.pone.0182049

Raman, T. S., D. Mudappa, T. Khan, U. Mistry, A. Saxena, K. Varma, N. Ekka, *et al.* 2013. An expedition to Narcondam: Observations of marine and terrestrial fauna including the island-endemic hornbill. *Current Science* 105(3):346–360.

Rodrigues, A. S., H. R. Akcakaya, S. J. Andelman, M. I. Bakarr, L. Boitani, T. M. Brooks, S. J. Chanson, *et al.* 2004. Global gap analysis: Priority regions for expanding the global protected-area network. *BioScience* 54(12):1092–1100.

Sclater, L. 1891. Notes on the collection of snakes in the Indian museum with descriptions of several new species. *Journal of the Asiatic Society of Bengal* 60(2):230–250.

Smith, M. A. 1940. The herpetology of the Andaman and Nicobar Islands. *Proceedings of the Linnaean Society of London* 3:150–158.

Smith, M. A. 1943. *The Fauna of British India, Ceylon and Burma, Including the Whole of Indochinese Sub-Region. Reptilia and Amphibia, Vol. 3, Serpentes.* Taylor and Francis.

Steindachner, F. 1867. Reptilien. In Anonymous (1865–1869) (ed.), *Reise der Österreichischen Fregatte Novara um die Erde in den Jahren 1857, 1858, 1859. B. von Wüllerstorf-Urbair. Zoologischer Theil. Erster Band (Wirbelthiere), KaiserlichKöniglichen Hof- und Staatsdrückerei* 1(3): 1–98.

Stoliczka, F. 1870. Observations on some Indian and Malayan Amphibia and Reptilia. *Journal of the Asiatic Society of Bengal* 39:134–228.

Stoliczka, F. 1871. Notes on some Indian and Burmese ophidians. *Journal of Asiatic Society of Bengal* 40:421–455.

Tiwari, K. K., and S. Biswas. 1973. Two new reptiles from the Great Nicobar Island. *Journal of the Zoological Society of India* 25(1 & 2):57–63.

Theobald, W. 1868. Catalogue of reptiles in the Museum of the Asiatic Society of Bengal. *Journal of the Asiatic Society of Bengal* 37 (Suppl. 146/2):vi, 7–88.

Uetz, P., P. Freed, R. Aguilar, and J. Hošek (eds.). 2022. The reptile database. Accessed August 20, 2022. http://www.reptile-database.org.

Vijayakumar, S. P., and P. David. 2006. Taxonomy, natural history, and distribution of the snakes of the Nicobar Islands (India), based on new materials and with an emphasis on endemic species. *Russian Journal of Herpetology* 13(1):11–40.

Vogel, G., P. David, and S. R. Chandramouli. 2014. On the systematics of *Trimeresurus labialis* Fitzinger in Steindachner, 1867, a pitviper from the Nicobar Islands (India), with revalidation of *Trimeresurus mutabilis* Stoliczka, 1870 (Squamata, Viperidae, Crotalinae). *Zootaxa* 3786 (5):557–573.

Wall, F. 1909. Remarks on some forms of *Dipsadomorphus*. *Records of the Indian Museum* 3:151–155.

Wall, F. 1913. A popular treatise on the common Indian snakes (part 2). *Journal of the Bombay Natural History Society* 22:243–259.

Whitaker, R. 1978. Herpetological survey in the Andamans. *Hamadryad* 3:9–16.

8

Diversity, Endemism, and Biogeography of Island Snakes of the Gulf of California, Mexico

Gustavo Arnaud, Ricardo J. Sawaya, and Marcio Martins

Introduction

The Gulf of California, also known as Mar Bermejo or Mar de Cortés, is located in northwestern Mexico, between the Baja California peninsula (states of Baja California and Baja California Sur) and the coast of mainland Mexico (states of Sonora and Sinaloa). It has a length of approximately 1,600 km, with a width varying from 205 to 85 km, and a surface of 283,000 km². There are 898 islands and islets in the Gulf of California, which together occupy an area of 3,000 km². The Gulf of California has one of the highest productivity levels of the planet, caused mainly by upwelling of deeper, cooler, nutrient-rich water. The islands of the Gulf of California are internationally recognized as one of the more ecologically intact island ecosystems in the world.

The Gulf of California has changed since its formation in the mid-Cretaceous, with tectonic shifts causing uplift and subsidence of land masses (Durham and Allison 1960). During the Pleistocene glaciations, the sea level fluctuated in association with the advance and retreat of the continental ice sheets. It is estimated that during the last glaciation the sea level decreased approximately 110 m, rising gradually in the last 17,000 years until its current level, approximately 6,000 years ago (Fairbridge 1960; Godwin *et al.* 1958). These sea level oscillations caused the formation of many land-bridge islands. It is assumed, therefore, that the islands with adjacent waters less than 110 m deep relative to the peninsula, mainland Mexico, or much larger islands are less than 17,000 years old. Among the islands found in this situation are El Muerto, Tiburón, San Marcos, Coronados, San José, San Francisco, and Espiritu Santo, as well as many small offshore islands (Soulé and Sloan 1966; Figure 8.1). Detailed information regarding the age and geology of the islands is provided by Carreño and Helenes (2002).

In general, the islands in the Gulf of California can be classified into three categories: (1) land-bridge islands (cf. continental shelf of Ali 2017), which were connected to land but became isolated with the rise in sea level during the Pleistocene glaciations (Grismer 2002); (2) continental (see different types in Ali 2017), which were once

Gustavo Arnaud, Ricardo J. Sawaya, and Marcio Martins, *Diversity, Endemism, and Biogeography of Island Snakes of the Gulf of California, Mexico* In: *Islands and Snakes*. Edited by: Harvey B. Lillywhite and Marcio Martins, Oxford University Press.
© Oxford University Press 2023. DOI: 10.1093/oso/9780197641521.003.0008

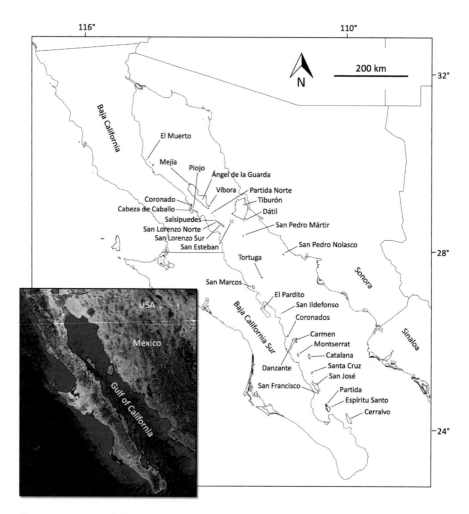

Figure 8.1 The Gulf of California between the peninsula of Baja California to the west and mainland Mexico (Sonora and Sinaloa states) to the east. The islands inhabited by snakes are indicated by their names (see Table 8.1). The inset shows a satellite image of the same region.

connected to the peninsula or mainland but later became separated by the displacement of tectonic plates and are now surrounded by deep water (see discussion on the origin of Ángel de la Guarda by Murphy and Aguirre-León 2002); and (3) oceanic (see different types in Ali 2017), those that have not had contact with the Baja California peninsula or mainland Mexico, having emerged from the sea, and are thus surrounded by deep water. Because the exact origin of some islands with no recent connection to the peninsula or mainland is not known (see examples in Murphy and Aguirre-León 2002), we use in some analyses a simpler classification with two island types: "oceanic" (oceanic + continental) and land-bridge islands (Table 8.1).

Table 8.1 Area (km²); distance to peninsula, mainland Mexico, or much larger island (km); elevation (m); island type (see text); species richness; and number of endemic species for the islands of the Gulf of California in which snakes occur

Island	Area (km²)	Distance (km)	Distance from	Elevation (m)	Island type	Species richness	Number of endemic species
Ángel de la Guarda	936.04	12.12	Peninsula	1316	"oceanic"	6	1
Cabeza de Caballo	0.77	2.05	Peninsula	140	land-bridge	1	1
Carmen	143.03	6.03	Peninsula	479	land-bridge	7	0
Catalana (= Santa Catalina)	40.99	25.15	Peninsula	470	"oceanic"	4	3
Cerralvo	140.46	8.73	Peninsula	767	"oceanic"	13	2
Coronado (= Smith)	9.13	2.18	Peninsula	465	land-bridge	2	0
Coronados	7.59	2.61	Peninsula	283	land-bridge	6	0
Danzante	4.64	2.61	Peninsula	106	land-bridge	8	0
Dátil (= Turner)	1.25	1.94	Tiburón	170	land-bridge	2	0
El Muerto (= Miramar, = Link)	1.33	3.39	Peninsula	192	land-bridge	4	0
El Pardito (= Coyote)	0.30	9.41	Peninsula	12	land-bridge	1	0
Espíritu Santo	87.55	6.15	Peninsula	576	land-bridge	7	0
Mejía	2.26	0.75	Ángel de le Guarda	262	land-bridge	2	0
Monserrat	19.86	13.70	Peninsula	223	"oceanic"	7	0
Partida	19.29	0.17	Espíritu Santo	335	land-bridge	6	0
Partida Norte (= Cardonosa)	1.36	12.26	Ángel de la Guarda	122	land-bridge	1	0

(continued)

Table 8.1 Continued

Island	Area (km²)	Distance (km)	Distance from	Elevation (m)	Island type	Species richness	Number of endemic species
Piojo	0.57	6.30	Peninsula	38	land-bridge	1	1
Salsipuedes	1.16	1.49	San Lorenzo Norte	114	land-bridge	3	0
San Esteban	40.72	34.50	Peninsula	431	"oceanic"	4	1
San Francisco	4.49	2.49	San José	210	land-bridge	2	0
San Idelfonso	1.33	9.94	Peninsula	117	land-bridge	1	0
San José	187.16	4.16	Peninsula	633	land-bridge	11	0
San Lorenzo Norte	4.26	0.12	San Lorenzo	200	land-bridge	1	0
San Lorenzo Sur	33.03	16.36	Peninsula	485	"oceanic"	3	1
San Marcos	30.07	4.91	Peninsula	271	land-bridge	12	0
San Pedro Mártir	2.90	50.15	Peninsula	320	"oceanic"	2	0
San Pedro Nolasco	3.45	14.61	Mainland Mexico	326	"oceanic"	1	0
Santa Cruz	13.06	19.82	Peninsula	457	"oceanic"	4	0
Tiburón	1223.53	1.70	Mainland Mexico	1218	land-bridge	13	0
Tortuga	11.36	36.30	Peninsula	309	"oceanic"	3	0
Víbora (= Pond, Estanque)	1.03	0.61	Ángel de le Guarda	122	land-bridge	1	0

Area and elevation are from Murphy et al. (2002). Distance from Murphy et al. (2002) and obtained in this study. The remaining data were obtained in this study.

Characteristics of the Islands Inhabited by Snakes

The area of the islands inhabited by snakes in the Gulf of California varies from 0.3 to 1,223.5 km^2 (El Pardito and Tiburón, respectively; Table 8.1). Their distances from the coast vary from 0.17 to 39.09 km (Partida and San Pedro Mártir, respectively; Table 8.1), and the depth of the sea surrounding islands varies from 3 to 1,200 m (Víbora and Tortuga, respectively). The islands off the Baja California peninsula are similar in climate and vegetation, forming part of the Central Gulf Coast ecoregion, except for the islands at the northern end of the gulf, which belong to the San Felipe Desert ecoregion (González *et al.* 2010). The physiography of the islands is variable, with plains and ravines, hills of various altitudes—with a maximum of 1,318 m in Ángel de la Guarda—steep slopes, and steep coasts and cliffs (Figure 8.2). Most of the beaches are rocky, with gravel and pebbles, and to a lesser extent they are sandy (Figure 8.2). The soils are represented by regosols (thick eutric and calcaric), arenosols, and podzoluvisols (DOF 2001). The prevailing vegetation is a xeric scrub of the sarcocaulous type, characteristic of the Sonoran Desert (Shreve and Wiggins 1964; Figure 8.2). As in desert areas in general, the vegetation is higher and denser along streams and at the bottoms of the ravines, while on some islands the slopes are devoid of vegetation (Soulé and Sloan 1966; Figure 8.2). The fauna is adapted to low productivity due to the low rainfall, which is why they may use subsidies from the sea as a strategy (Stapp and Polis 2003). The climate during the summer reaches temperatures of 45–50°C, and winter is mild. The rainfall pattern is variable between the north and the south of the Gulf; to the north of Bahía de los Ángeles (latitude 29°), it is concentrated mainly in winter (Aschmann 1959), while to the south, it is concentrated in the summer months, mainly due to the influence of tropical storms or hurricanes (Salinas *et al.* 1990). Precipitation is greater in the extreme south of the Gulf, decreasing latitudinally.

Probably because of the arid conditions and absence of water (see Pliego-Sánchez *et al.* 2021), the islands inhabited by snakes in the Gulf of California are uninhabited by humans, with only four exceptions. San Marcos (30 km^2) has a small town with about 370 inhabitants. San José (187 km^2) is inhabited by two families at the Las Palmas ranch. Although very small (0.3 km^2), 13 people live in El Pardito (see Dewitt *et al.* 2019). Finally, approximately five people live on Carmen (143 km^2). In general, the islands are refuges for fishermen who commonly establish temporary fishing camps. The conservation status of the islands inhabited by snakes is generally good, despite the human impact of the introduction of exotic species such as cats (*Felis catus*), goats (*Capra hircus*), and rodents (*Mus musculus* and *Rattus rattus*) on several islands (Aguirre-Muñoz *et al.* 2011; but see the section "Threats to Island Snakes" below). Gypsum is extracted on San Marcos Island where the natural vegetation was removed from more than one-third of the island. In a study on the insular herpetofauna of Mexico, Pliego-Sánchez *et al.* (2021) indicate that the islands Carmen and San Marcos are among those with the greatest major threats.

Figure 8.2 Examples of island habitats in selected islands of the Gulf of California, Mexico, in which snakes occur. Top left, San Estéban; top right, Cabeza de Caballo; middle left, Tortuga; middle right, Coronado (Smith); bottom left, Espíritu Santo; bottom right, Catalana (Santa Catalina).
Photographs by Marcio Martins (top left) and Gustavo Arnaud (all the remaining).

The Snake Assemblages on Islands

Inventories of snakes on the islands of the Gulf of California report a total of 38 species present on 31 islands (see examples of snakes in Figure 8.3): 1 species of the family Charinidae, 19 Colubridae, 4 Dipsadidae, 1 Elapidae, 1 Leptotyphlopidae, and 14 Viperidae (Table 8.2; Soulé and Sloan 1966; Grismer 1999, 2002; Murphy and

Figure 8.3 Examples of island snakes from the Gulf of California, Mexico. Top left, *Lichanura trivirgata* from Coronados; top right, the endemic *Hypsiglena catalinae* from Catalana (Santa Catalina); Bottom left, the endemic *Crotalus angelensis* from Ángel de la Guarda; bottom right, the endemic *Crotalus estebanensis* from San Estéban.
Photographs by Marcio Martins (top right) and Gustavo Arnaud (all the remaining).

Aguirre-León 2002; Devitt *et al.* 2008; Lovich *et al.* 2009; Arnaud *et al.* 2014; Wallach *et al.* 2014; Frick *et al.* 2016; Arnaud and Blázquez 2018; Meik *et al.* 2018; Uetz *et al.* 2022). Eleven species are endemic to the island where they occur: *Hypsiglena catalinae, Lampropeltis catalinensis,* and *Crotalus catalinensis* on Catalana; *Masticophis slevini* and *C. estebanensis* on San Estéban; *Rhinocheilus etheridgei* and *Sonora savagei* on Cerralvo; *C. angelensis* on Ángel de la Guarda; *C. polisi* on Cabeza de Caballo; *C. thalassoporus* on Piojo; and *C. lorenzoensis* on San Lorenzo Sur (Table 8.2). The genus *Crotalus* is the most widely represented among the islands, including 14 species on 26 islands (Table 8.2). The islands with the highest number of snake species are Tiburón (13 species), Cerralvo and San Marcos (12 species each), and San José (11 species), all of them relatively large islands (30.1 to 1,223.5 km^2; Tables 8.1 and 8.2).

Snake species vary in the number of islands on which they occur and can be grouped into four categories of island occurrence: (1) widespread for those species occurring on 10 or more islands, including three species (7.9%), *Lampropeltis californiae, Masticophis fuliginosus,* and *Hypsiglena ochrorhynchus*; (2) intermediate frequency, for those species occurring on five to nine islands, represented by nine

Table 8.2 Species of island snakes with the islands where they occur in the Gulf of California, Mexico, and the non-island region where the species occur (Baja California peninsula or mainland Mexico).

Species/Islands	Ángel de la Guarda	Cabeza de caballo	Carmen	Catalana (= Santa Catalina)	Cerralvo	Coronados	Coronado (= Smith)	Danzante	Dátil (= Turners)	El Muerto	El Pardito	Espíritu Santo	Mejía	Monserrat	Partida Norte (= Cardonosa)	Partida	Piojo	Salsipuedes	San Esteban	San Francisco	San Ildefonso	San José	San Lorenzo Norte	San Lorenzo Sur	San Marcos	San Pedro Mártir	San Pedro Nolasco	Santa Cruz	Tiburón	Tortuga	Víbora (= Pond, Estanque)	Baja California Peninsula	Mainland Mexico
Charinidae																																	
Lichanura trivirgata	x		x		x	x						x	x												x				x			x	x
Colubridae																																	
Bogertophis rosaliae								x																	x							x	
Lampropeltis californiae	x				x									x				x					x	x	x		x	x	x			x	x
Lampropeltis catalinensis (E)				x																													
Masticophis barbouri												x				x																x	
Masticophis bilineatus																													x				x
Masticophis fuliginosus			x		x	x		x				x		x		x					x	x			x				x			x	x
Masticophis slevini (E)																			x														
Phyllorhynchus decurtatus	x				x									x								x			x							x	x
Pituophis catenifer																												x	x			x	x
Pituophis vertebralis																					x	x										x	
Rhinocheilus etheridgei (E)					x																												
Salvadora hexalepis												x										x							x			x	x
Sonora savagei (E)					x																												
Sonora semiannulata																					x			x	x								x
Sonora straminea								x				x		x		x					x	x		x	x				x			x	x

Species																		
Tantilla planiceps							x											x
Trimorphodon lyrophanes		x	x		X			x				x					x	x
Dipsadidae																		
Hypsiglena catalinae (E)				x														
Hypsiglena chlorophaea	x	x		x		x	x				x	x	x		x		x	x
Hypsiglena ochrorhynchus		x	x	x		x	x	x	x		x	x	x	x			x	
Hypsiglena slevini		x	x		x						x						x	
Elapidae																		
Micruroides euryxanthus														x				x
Leptotyphlopidae																		
Rena humilis	x	x	x		x		x				x	x	x		x		x	x
Viperidae																		
Crotalus angelensis (E)	x																	
Crotalus atrox			x		x			x	x						x	x x x		x
Crotalus catalinensis (E)		x																
Crotalus cerastes													x		x		x	x
Crotalus enyo	x	x		x x		x	x				x	x			x		x	x
Crotalus estebanensis (E)								x										
Crotalus lorenzoensis (E)									x									
Crotalus mitchellii	x	x		x x		x	x				x	x			x		x	x
Crotalus molossus															x			x
Crotalus polisi (E)	x																	
Crotalus pyrrhus		x	x		x												x	x
Crotalus ruber	x	x	x	x x		x	x				x	x			x	x	x	x
Crotalus thalassoporus (E)										x								
Crotalus tigris														x			x	x

Occurrence data from Soulé and Sloan (1966), Grismer (1999, 2002), Murphy and Aguirre-Leon (2002), Devitt et al. (2008), Lovich et al. (2009), Arnaud et al. (2014), Wallach et al. (2014), Frick et al. (2016), Arnaud and Blazquez (2018), Meik et al. (2018), and Uetz et al. (2022).

E, endemic species.

species (23.7%), *Lichanura trivirgata, Sonora straminea, Phyllorhynchus decurtatus, Trimorphodon lyrophanes, Hypsiglena slevini, Rena humilis, Crotalus enyo, Crotalus mitchellii,* and *Crotalus ruber*; (3) low frequency, for those species occurring on three or four islands, represented by two species (5.3%), *Salvadora hexalepis* and *Crotalus atrox*; and (4) very low frequency, for those species occurring on one or two islands only, represented by 24 species (63.2%) (Table 8.2).

The Origin of Island Populations

The snake fauna of the Gulf of California islands includes 11 endemic species and 27 non-endemics. Among the non-endemics, 22 (81.5%) occur also in the Baja California Peninsula and 16 (59.3%) occur also in mainland Mexico, with 10 species occurring on both sides of the Gulf (Table 8.2). Both endemics and non-endemics may have originated from ancestral populations from these regions either (1) by the isolation in land-bridge islands of once widespread coastal populations during Pleistocene glaciations; (2) by immigration through over-water dispersal from the peninsula or mainland to islands; or (3) from populations that remained in a piece of the peninsula or mainland that became separated by the displacement of tectonic plates (continental islands; Murphy 1983; Murphy and Aguirre-León 2002; see also a recent review on the origin of island snake populations in Martins and Lillywhite 2019). It is worth noting that the second and the third processes may be involved in the origin of the snake fauna of a given island. For instance, the snake fauna of a continental island may have originated both from populations that remained on the island when it became separated from a continent and from later immigration through over-water dispersal (*e.g.*, the snake fauna of Madagascar originated by both processes; see review in Martins and Lillywhite 2019).

The first process above may be synthetized by a few steps (see a detailed explanation of this process in Martins and Lillywhite 2019): (i) when coastal lowlands become exposed due to falling sea level during Pleistocene glaciations, coastal populations spread through these areas; (ii) when sea level rises during interglacial periods, portions of these populations become trapped on mountains that become land-bridge islands in the new landscape; (iii) in general, a series of extinctions occur after these populations become isolated, while some populations may persist (the proportion of extinctions tend to be higher in small islands and lower in large islands). This may be the origin, for example, of snake populations on the islands Danzante, Espiritu Santo, and San José. In some cases, these isolated populations on land-bridge islands diverge enough from the parent population to be considered a distinct species. As an example, there is strong molecular evidence indicating that *C. polisi*, from Cabeza de Caballo, and *C. thalassoporus*, from Piojo, originated from an ancestral *C. pyrrhus* population from the peninsula (Meik *et al.* 2018). A similar process might be involved in the occurrence of populations on small islands close to much larger islands, like the population of *C. atrox* on Dátil Island, which may have originated from the population of this species occurring in the much larger Tiburón Island.

As for the origin of snakes from oceanic islands of the Gulf of California, immigration through over-water dispersal may be the process involved in most cases (Murphy and Aguirre-León 2002; see review in Martins and Lillywhite 2019). The heavy rainfall that occurs in the region—due to the presence of tropical storms—brings from the Baja California Peninsula plants, trunks, and roots in conglomerates in the form of "floating islands", which can transport live animals; snakes are specially able to persist for long periods in these "floating islands" (see Martins and Lillywhite 2019). This colonization hypothesis was proposed to explain the presence of *C. ruber* on Montserrat (Murphy and Aguirre-León 2002) and Coronados (Arnaud *et al.* 2014) and may also explain the origin of *C. catalinensis* from an ancestral population related to *C. ruber* on Catalana (see Ruiz-Sanchez *et al.* 2019). This process may indeed continue to occur presently because approximately 10 hurricanes per decade affect the Baja California peninsula, and several of the storms that are generated in the eastern Pacific basin, with a northward trajectory, enter the Gulf of California (Martínez-Gutierrez and Mayer 2004).

Finally, regarding continental islands, as stated above, their snake populations may have originated from populations that have remained in the piece of the peninsula or mainland that moved away through movements of tectonic plates. For instance, Murphy and Aguirre-León (2002) suggested that the snake fauna of Ángel de la Guarda has originated from populations that remained in the block when it separated from the peninsula around 1 million years ago.

Diversity Patterns

We used regression analyses and non-parametric tests to explore the effect of island area and isolation (distance from the peninsula, mainland Mexico, or much larger neighbor island; see Table 8.1) on snake species richness in the islands of the Gulf of California. We also compared snake species richness and endemism in different island types (land-bridge or "oceanic"; see above) using non-parametric Kruskal-Wallis tests. Only islands inhabited by snakes were used in such analyses (Table 8.1). We hypothesized that larger islands might present more resources (in amount and diversity) and consequently support more species. Furthermore, islands closer to the peninsula, mainland, or a larger island would have a higher chance of receiving migrants, which might also contribute to their species richness. Finally, "oceanic" islands might have been more isolated from the peninsula or mainland than land-bridge islands. Thus, our predictions are that (i) larger; (ii) nearer to the mainland, peninsula, or larger island; and (iii) land-bridge islands will show higher species richness. We also hypothesize that isolation might favor endemism by reducing gene flow between the islands and the mainland, peninsula, or larger island. Smaller islands would also favor endemism as smaller populations might evolve faster than larger

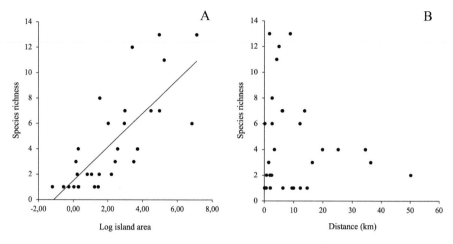

Figure 8.4 (A) The number of snake species (species richness) in the islands of the Gulf of California is significantly and positively related to island area (natural logarithm of island area; $R^2 = 0.57$, $p < 0.01$); B) The number of snake species (species richness) in the islands of the Gulf of California showed a weak, negative, and non-significant relationship with island distance from the mainland, peninsula, or larger island ($R^2 = 0.03$, $p = 0.95$); however, closer islands showed a higher variation in species richness than farther islands, and islands farther than 14 km showed a maximum of four species.

populations. We then expect that smaller, farther, and oceanic islands must present more endemic species.

Species richness was significantly and positively related to island area (Figure 8.4A). Island area explained 57% of the variation in species richness. The area of the eight islands harboring only a single species varied between 0.30 and 4.26 km² (Table 8.1). The highest richness (13) was recorded in Cerralvo (140.5 km²), San Marcos (30.1 km²), and Tiburón (1,223.5 km²). Species richness was not significantly related to island distance from the mainland, peninsula, or larger island (*i.e.*, degree of isolation; Figure 8.4B). Only 3% of the variation in species richness was explained by island isolation. Similar results were found for reptiles generally in islands of the Gulf of California by Case (2002) and for the insular herpetofauna of Mexico by Pliego-Sánchez *et al.* (2021) (*i.e.*, a relatively high proportion of the variance of richness is explained by area and a low proportion is explained by distance). Even though there is no relationship between richness and island distance, the closest islands (0–14 km from the mainland, peninsula, or larger island) showed a higher variation in species richness (1–13) than farther islands (>14 km from the mainland, peninsula, or larger island), which harbor only one to four species (Figure 8.4B).

Species richness varied from 1 to 13 species in both land-bridge and oceanic islands (Figure 8.5; Table 8.1). Although oceanic islands showed a larger median

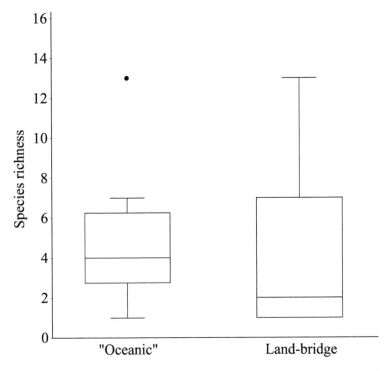

Figure 8.5 The number of snake species (species richness) is not significantly different between oceanic and land-bridge islands of the Gulf of California (Kruskal-Wallis chi-square = 0.63, p = 0.43). Horizontal line inside boxes = median; boxes = 25% to 75% of values; vertical lines = range; dot = outlier.

(4) than land-bridge islands, there was no significant difference in species richness regarding island type (Figure 8.5; Kruskal-Wallis chi-square = 0.63, p = 0.43).

According to our predictions, the degree of isolation—indicated by island distance from the mainland, peninsula, or larger island—seemed to favor endemism because islands with one to three endemic species (see Table 8.1) are generally more isolated than those without endemics (Figure 8.6A), but the difference is marginally non-significant (Kruskal-Wallis chi-square = 3.22, p = 0.07). Contrary to our predictions, we found no difference between the area of islands with endemics and those without endemics (Figure 8.6B; Kruskal-Wallis chi-square = 1.08, p = 0.30).

According to our predictions, more "oceanic" islands (N = 5) harbor endemic species than do land-bridge islands (N = 2). From 21 land-bridge islands sampled, only 2 (9.5%) harbor endemic species, whereas 5 out of 10 (50%) oceanic islands harbor endemics. Indeed, the number of oceanic islands harboring endemic species is significantly higher (Fisher's Exact Test, p = 0.02). Both land-bridge islands harboring endemics have only one endemic species, whereas three oceanic islands harboring endemics have one endemic, one has two endemics, and one has three endemic species.

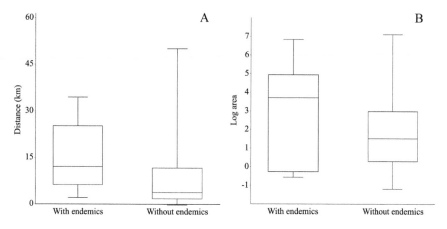

Figure 8.6 (A) Degree of isolation (indicated by island distance from the mainland, peninsula, or larger island, in km) of islands in the Gulf of California with one to three endemic species is generally larger than that of islands without endemics, but this difference is marginally non-significant (Kruskal-Wallis chi-square = 3.22, p = 0.07); (B) The area (natural logarithm of area in km^2) of islands with one to three endemic species is generally larger than that of islands without endemics, but this difference is not significant (Kruskal-Wallis chi-square = 1.08 p = 0.30). Horizontal line inside boxes = median; boxes = 25% to 75% of values; vertical lines = range; and dots = outlier.

Conservation Status of Island Snakes

Of the 38 snake species recorded on the islands of the Gulf of California, six were not evaluated (NE) by the Red List of the International Union for Conservation of Nature (IUCN 2022; Table 8.3). Of the remaining 32 species evaluated, only the Santa Catalina Rattlesnake *C. catalinensis* is listed as threatened (Critically Endangered, CR; Ávila Villegas *et al.* 2007; Table 8.3). An additional three species are listed as Data Deficient, indicating that there is no adequate information to assess their conservation status based on their distribution and/or population (IUCN 2012). Thus, the gathering of additional data on these species is urgent. The remaining 28 species are listed as Least Concern (LC; *i.e.*, not threatened; Table 8.3).

The only threatened island snake from the Gulf of California (*C. catalinensis*) needs to be re-evaluated for the IUCN Red List given that its population is stable and the main threatening factor that acted on its population (presence of introduced domestic cats) has disappeared (Aguirre-Muñoz *et al.* 2011). Furthermore, landing on the island is forbidden for tourists and fishermen. Visits are allowed only for research purposes, coupled with environmental education and surveillance activities that take place in the area of influence of the island (Arnaud and Martins 2019; DOF 2019a; Arnaud and Popoca 2022).

Table 8.3 Category in the International Union for Conservation of Nature Red List (IUCN 2022), category in the Norma Oficial Mexicana 059 (NOM-059; DOF 2019b), value and category of the Environmental Vulnerability Index (EVS; from Wilson and McCranie, 2013, except for *Crotalus polisi*, *Crotalus pyrrhus*, *Crotalus thalassoporus*, and *Hypsiglena catalinae*, which we calculated) of the snakes from the islands of the Gulf of California

Species	IUCN	NOM-059	EVS value	EVS category
Lichanura trivirgata	LC	A	10	Medium
Bogertophis rosaliae	LC	NE	10	Medium
Lampropeltis californiae	LC	NE	10	Medium
Lampropeltis catalinensis (E)	DD	NE	17	High
Masticophis barbouri	DD	NE	17	High
Masticophis bilineatus	LC	NE	11	Medium
Masticophis fuliginosus	NE	NE	9	Low
Masticophis slevini (E)	LC	NE	17	High
Phyllorhynchus decurtatus	LC	NE	11	Medium
Pituophis catenifer	LC	NE	9	Low
Pituophis vertebralis	LC	NE	12	Medium
Rhinocheilus etheridgei (E)	DD	A	16	High
Salvadora hexalepis	LC	NE	10	Medium
Sonora savagei (E)	LC*	NE	15	High
Sonora semiannulata	LC	NE	5	Low
Sonora straminea	LC*	NE	8	Low
Tantilla planiceps	LC	NE	9	Low
Trimorphodon lyrophanes	LC	NE	10	Medium
Hypsiglena catalinae (E)	NE	Pr	10	Medium
Hypsiglena chlorophaea	LC	Pr	8	Low
Hypsiglena ochrorhynchus	NE	Pr	8	Low
Hypsiglena slevini	LC	A	11	Medium
Micruroides euryxanthus	LC	A	15	High
Rena humilis	LC	NE	8	Low
Crotalus angelensis (E)	LC	NE	18	High
Crotalus atrox	LC	Pr	9	Low
Crotalus catalinensis (E)	CR	A	19	High
Crotalus cerastes	LC	Pr	16	High
Crotalus enyo	LC	A	13	Medium
Crotalus estebanensis (E)	LC	NE	19	High
Crotalus lorenzoensis (E)	LC	NE	19	High
Crotalus mitchellii	LC	Pr	12	Medium
Crotalus molossus	LC	Pr	8	Low

(continued)

Table 8.3 Continued

Species	IUCN	NOM-059	EVS value	EVS category
Crotalus polisi (E)	NE	NE	12	Medium
Crotalus pyrrhus	NE	NE	11	Medium
Crotalus ruber	LC	Pr	9	Low
Crotalus thalassoporus (E)	NE	NE	12	Medium
Crotalus tigris	LC	Pr	16	High

LC, Least Concern; DD, Data Deficient; CR, Critically Endangered; A, *Amenazada* (Threatened); Pr, Protección Especial (Special Protection); NE, Not evaluated.

E: Endemic species.

* In the IUCN red list, these species are listed in the genus *Chilomeniscus*.

The government of Mexico, through the Norma Oficial Mexicana NOM-059-SEMARNAT-2010, lists Mexican species that are at risk of extinction (DOF 2019b). A total of 15 island snakes from the Gulf of California are listed in the present version of the list, six in the Amenazada category (threatened) and nine in the Protección Especial category (special protection; DOF 2019b; Table 8.3).

We also provide evaluations of all the island snakes from the Gulf of California using the Environmental Vulnerability Index (EVS; Wilson and McCranie 2002; Table 8.3). This index was created to calculate the susceptibility of a species to future environmental threats without requiring information on their population status (see Wilson and McCranie 2002). The scores of 34 species were obtained in Wilson *et al.* (2013), and we calculated the scores for an additional four species: *C. polisi*, *C. pyrrhus*, *C. thalassoporus*, and *H. catalinae*. Among the 38 island snakes from the Gulf of California, 12 have a high vulnerability, 15 had a medium vulnerability, and 11 had a low vulnerability (Table 8.3).

Comparing the global and Mexican red lists and the EVS, four species deserve special attention among the island snakes from the Gulf of California. *Crotalus catalinensis* is Critically Endangered in the IUCN Red List (but see above), Amenazada (threatened) in Mexico, and has a high EVS score (19; Table 8.3). *Rhinocheilus etheridgei* and *Masticophis barbouri* are Data Deficient in the IUCN Red List, Amenazada (threatened) in Mexico, and have high EVS scores (16 and 17, respectively; Table 8.3). Finally, *Lampropeltis catalinensis* is Data Deficient in the IUCN Red List, was not evaluated in Mexico, and has a high EVS score (17; Table 8.3).

Finally, no island snake from the Gulf of California is listed in the appendices of the Convention on International Trade in Endangered Species of Wild Fauna and Flora (CITES 2022), an international agreement between governments whose main objective is to avoid species being threatened by the international trade.

Threats to Island Snakes

Most snakes from the islands of the Gulf of California occur also in the Baja California Peninsula and/or in mainland Mexico (11 species only in the peninsula, 6 species only in mainland Mexico, and 10 species in both; Table 8.2). Thus, although some of the island populations of these species may face local threats (see below), most species are safeguarded by additional populations from non-insular areas. But if the intent is to preserve populations instead of species, special attention should be devoted to these insular populations. Eleven species of island snakes from the Gulf of California are endemic to a single island, and, curiously, no species is endemic to more than one island in the Gulf of California (Table 8.2).

Snakes are threatened mainly by habitat loss, fragmentation and/or disturbance, climate change, invasive species, disease, environmental stochasticity, and overcollection (Gibbons *et al.* 2000; Pliego-Sánchez *et al.* 2021). At the Gulf of California, the main threats to the island endemics and to the populations of non-endemics are invasive species (especially cats, goats, and rodents; Aguirre-Muñoz *et al.* 2011), killing and illegal collection (*e.g.*, Mellink 1995; Auliya *et al.* 2016), human disturbances in their habitats, and natural disasters like hurricanes (see also Blázquez *et al.* 2018). Besides these threats, Pliego-Sánchez *et al.* (2021) also include climate change as an important threat for the insular herpetofauna of Baja California, especially related to severe climatic events.

Among invasive species, feral cats are indicated as the main threat for most species that occur in islands with this threat (see, *e.g.*, Frost 2007). However, programs have been carried out to eradicate exotic animals (cats, goats, and rodents) on 15 islands of the Gulf of California (Aguirre-Muñoz *et al.* 2011). These included six islands where snakes occur, namely Coronados, Danzante, Mejía, Montserrat, San Pedro Mártir, and Catalana; the latter is inhabited by three endemic species (Table 8.2). However, cats, goats, and rodents are still present on some of the islands where snakes occur, as in Ángel de la Guarda, Carmen, Cerralvo, Espiritu Santo, San José, and San Marcos, which are large islands (areas of 30–936 km^2); hence, the logistics for their eradication are more complicated than in small islands. Ángel de la Guarda and Cerralvo are inhabited by endemic species (Table 8.2); thus, the effects of invasive animals on these populations should be studied.

Regarding illegal collection, there are verifiable reports of people going to Catalana Island to illegally collect *C. catalinensis* (Ávila Villegas *et al.* 2007), and this may occur with the other endemics (particularly rattlesnakes), as they tend to have higher prices on the illegal pet market (Mellink 1995; Auliya *et al.* 2016). The effect of illegal collection may be greater in the populations of small islands, such as El Piojo and Cabeza de Caballo (areas of 0.6–0.8 km^2). Finally, the presence of a gypsum mine on San Marcos Island already has resulted in the removal of more than one-third of the natural vegetation of the island, with obvious negative consequences for snake populations.

Finally, in the future, tourism may become an important threat to island snakes in the Gulf, especially at the more visited islands (Carmen, Cerralvo, Coronados, Espíritu Santo, La Partida, San José, San Pedro Nolasco, and Tiburón; Blázquez *et al.* 2018). For example, more than 20,000 people visited Espíritu Santo in 2009 (Hernández-Trejo *et al.* 2012), and there is no study on the possible impact of this activity on insular snakes.

Conservation Actions That Benefit Island Snakes

Red lists and other prioritization strategies (like the EVS) indicate which species should receive special attention for conservation. Conservation actions that benefit the island snakes of the Gulf of California include laws to protect biodiversity and regulate collection (see Blázquez *et al.* 2018), protected areas (see Ezcurra *et al.* 2002; Blázquez *et al.* 2018), education (Tershy *et al.* 2002), and *ex situ* populations (see Lovich *et al.* 2009). The islands are within the Islas del Golfo de California Flora and Fauna Protection Area, whose area of influence includes coastal areas of the states of Baja California, Baja California Sur, Sonora, Sinaloa, and Nayarit (DOF 2001). The Gulf also harbors other protected areas that benefit insular snakes: two national parks (Archipiélago de San Lorenzo and Bahía de Loreto), which protect the islands San Lorenzo, Coronados, Danzante, Montserrat, Catalana, and Carmen, and two Biosphere Reserves (Bahía de los Ángeles, Canales de Ballenas y Salsipuedes, and Isla San Pedro Mártir), whose area of influence includes the islands of Ángel de la Guarda, Coronado (Smith), Cabeza de Caballo, Piojo, Salsipuedes, and San Pedro Mártir.

As an important conservation action directed to this threat, environmental education is carried out to avoid introductions of exotic animals to the islands, coordinated by the protected areas encompassed by the Islas del Golfo de California Flora and Fauna Protection Area (Tershy *et al.* 2002; G. Arnaud, pers. obs.). Additional educational programs focused on the island snakes are carried out in various communities on the coast of the peninsula facing the Gulf of California, where their inhabitants are trained through information, talks, and workshops so that they recognize the importance of snakes and become integrated into conservation processes (G. Arnaud, pers. obs.).

The maintenance of *ex situ*, captive populations in zoos and other institutions has the potential to provide individuals for population reinforcement if there is a population decline or loss of genetic variability and for reintroduction when species become extinct in the wild (*e.g.*, Odum and Reinert 2015; Ziegler 2015). According to Lovich *et al.* (2009), there are in zoos captive individuals of about 15 snake species that occur in the islands of the Gulf of California, although only one (*C. catalinensis*) is endemic to the island where it occurs. Thus, efforts should be made to maintain *ex situ* populations of the remaining ten endemics.

Conclusion

Beyond the matters of species richness and composition, studies including phylogenetic approaches should analyze the evolutionary distinctiveness of each snake species inhabiting the islands in the Gulf of California. Mapping this evolutionary distinctiveness (see Oliveira-Dalland *et al.* 2022) would make it possible to prioritize the conservation of unique evolutionary processes in addition to species under some degree of threat. It would be also interesting to analyze the functional diversity of the island snakes in order to better conserve the ecosystem functions of unique habitats.

Regarding conservation actions, the prioritization of restricted-range species and actions against habitat, invasives, and overcollection-related threats (*e.g.*, establishment of new protected areas, education about invasive species, and enforcement and control of trade, respectively) would benefit several island snakes from the Gulf of California (see Blázquez *et al.* 2018).

Acknowledgments

We thank Fundação de Amparo à Pesquisa do Estado de São Paulo (FAPESP) for a grant to MM and RJS (#2020/12658-4). MM and RJS are grateful to Conselho Nacional de Desenvolvimento Científico e Tecnológico (CNPq) for research fellowships #309772/2021-4 and #312795/2018-1, respectively. We also thank Alejandro Arnaud for helping with statistical analyses.

References

Aguirre-Muñoz, A., A. Samaniego-Herrera, L. Luna-Mendoza, A. Ortiz-Alcaraz, M. Rodríguez-Malagón, F. Méndez-Sánchez, M. Félix-Lizárraga, *et al.* 2011. Island restoration in Mexico: Ecological outcomes after systematic eradications of invasive mammals. In C. R. Veitch, M. N. Clout, and D. R. Towns (eds.), *Island Invasives: Eradication and Management.* IUCN, pp. 250–258.

Ali, J. R. 2017. Islands as biological substrates: Classification of the biological assemblage components and the physical island types. *Journal of Biogeography* 44:984–994.

Auliya, M., S. Altherr, D. Ariano-Sanchez, E. H. Baard, C. Brown, R. M. Brown, J. Cantu, *et al.* 2016. Trade in live reptiles, its impact on wild populations, and the role of the European market. *Biological Conservation* 204:103–119.

Arnaud, G., and M. C. Blázquez. 2018. First record of *Lichanura trivirgata* Cope, 1868 (Squamata: Boide) from Coronados Island, Gulf of California, México. *Herpetology Notes* 11:1025–1026.

Arnaud, G., R. Carbajal-Márquez, J. Rodríguez-Canseco, and E. Ferreyra. 2014. Primeros registros de la cascabel roja (*Crotalus ruber*) en la isla Coronados, golfo de California, México. *Revista Mexicana de Biodiversidad* 85:322–324.

Arnaud, G., and M. Martins. 2019. Living without a rattle. The biology and conservation of the rattlesnake, *Crotalus catalinensis*, from Santa Catalina Island, México. In H. B. Lillywhite and

M. Martins (eds.), *Islands and Snakes: Isolation and Adaptive Evolution.* Oxford University Press, pp. 241–257.

Arnaud, G., and E. Popoca. 2022. Parque Nacional Bahía de Loreto. In A. Cazal-Ferreira, L. López-Levi, and C. McCoy-Cador (eds.). *Áreas naturales protegidas: Entre sociedades y naturalezas. Parques nacionales* (vol. 1). Editorial Itaca, pp. 361–380. Aschmann, H. 1959. *The Central Desert of Baja California: Demography and Ecology.* University of California Press.

Avila Villegas, H., D. R. Frost, and G. Arnaud. 2007. *Crotalus catalinensis. The IUCN Red List of Threatened Species* e.T64314A12764544. https://www.iucnredlist.org/

Blázquez, M. C., P. Vázquez, and A. Ortega-Rubio. 2018. Status of the phylogeography, taxonomy and conservation of the reptiles of the Gulf of California Islands. In Ortega–Rubio, A. (ed.), *Mexican Natural Resources Management and Biodiversity Conservation.* Springer, pp. 285–304.

Carreño, A. L., and J. Helenes. 2002. Geology and ages of islands. In T. Case, M. Cody, and E. Ezcurra (eds.), *A New Island Biogeography of the Sea Cortés.* Oxford University Press, pp. 14–40.

Case, T. J. 2002. Reptiles: Ecology. In T. Case, M. Cody, and E. Ezcurra (eds.), *A New Island Biogeography of the Sea Cortés.* Oxford University Press, pp. 221–270.

CITES, Convention on International Trade in Endangered Species of Wild Fauna and Flora. 2022. Appendices I, II, and III. Accessed June 11, 2023. https://cites.org/sites/default/files/eng/app/2023/E-Appendices-2023-05-21.pdf

Dewitt, T., T. LaDuc, and J. McGuire. 2008. The *Trimorphodon biscutatus* (Squamata: Colubridae) species complex revisited: A multivariate statistical analysis of geographic variation. *Copeia* 2008:370–387.

DOF, Diario Oficial de la Federación. 2001, April 17. Aviso mediante el cual se informa al público en general que la Secretaría de Medio Ambiente y Recursos Naturales ha concluido la elaboración del Programa de Manejo del Área de Protección de Flora y Fauna Islas del Golfo de California, asimismo se da a conocer el Resumen del Programa de Manejo respectivo, el plano de localización y zonificación de dicha área. https://dof.gob.mx/nota_detalle.php?codigo=767360&fecha=17/04/2001#gsc.tab=0.

DOF, Diario Oficial de la Federación. 2019a, April 23. Programa de Manejo del Parque Nacional Bahía de Loreto. https://www.dof.gob.mx/nota_detalle.php?codigo=5558313&fecha=23/04/2019#gsc.tab=0.

DOF, Diario Oficial de la Federación. 2019b, November 14. Modificación del Anexo Normativo III, Lista de especies en riesgo de la Norma Oficial Mexicana NOM–059–SEMARNAT–2010, Protección ambiental–Especies nativas de México de flora y fauna silvestres–Categorías de riesgo y especificaciones para su inclusión, exclusión o cambio–Lista de especies en riesgo, publicada en 30 de diciembre de 2010. https://www.dof.gob.mx/nota_detalle.php?codigo=5578808&fecha=14/11/2019#gsc.tab=0.

Durham, J. W., and E. C. Allison. 1960. The geologic history of Baja California and its marine faunas. *Systematic Zoology* 9:47–91.

Ezcurra E., L. Bourillón, A. Cantú, M. E. Martínez, and A. Robles. 2002. Ecological conservation. In T. Case, M. Cody, and E. Ezcurra (eds.), *A New Island Biogeography of the Sea Cortés.* Oxford University Press, pp. 417–444.

Fairbridge, R. W. 1960. The changing level of the sea. *Scientific American* 202:70–79.

Frick, W. F., P. A. Heady, and B. Hollingwirth. 2016. *Lichanura trivirgata* (Rosy Boa). *Herpetological Review* 47:1–2.

Frost, D. R. 2007. *Masticophis slevini. The IUCN Red List of Threatened Species 2007* e.T63849A12721842. https://www.iucnredlist.org/

Gibbons, J. W., D. E. Scott, T. J. Ryan, K. A. Buhlmann, T. D. Tuberville, B. S. Metts, J. L. Greene, *et al.* 2000. The global decline of reptiles, déjà vu amphibians: Reptile species are declining on a global scale. *BioScience* 50:653–666.

Godwin, H., R. P. Suggate, and E. H. Willis. 1958. Radiocarbon dating of the eustatic rise in ocean-level. *Nature* 181:1518–1519.

González, A. C., P. P. Garcillán, and E. Ezcurra. 2010. Ecorregiones de la Península de Baja California: Una Síntesis. *Boletín de la Sociedad Botánica de México* 87:69–82.

Grismer, L. L. 1999. An evolutionary classification of reptiles on islands in the Gulf of California, Mexico. *Herpetologica* 55:446–469.

Grismer, L. L. 2002. *Amphibians and Reptiles of Baja California: Including Its Pacific Islands and the Islands in the Sea of Cortés.* University of California Press.

Hernández-Trejo, V., G. Avilés-Polanco, and M. A. Almendarez–Hernández. 2012. Beneficios económicos de los servicios recreativos provistos por la biodiversidad acuática del Parque Nacional Archipiélago Espíritu Santo. *Estudios Sociales* 20:157–177.

IUCN. 2012. *IUCN Red List Categories and Criteria*, Version 3.1 (2nd ed.). IUCN.

IUCN. 2022. *The IUCN Red List of Threatened Species*, Version 2022-1. Accessed November 28, 2022. https://www.iucnredlist.org/

Lovich, R. E., L. L. Grismer, and G. Danemann. 2009. Conservation status of the herpetofauna of Baja California, México and associated islands in the Sea of Cortez and Pacific Ocean. *Herpetological Conservation and Biology* 4:358–378.

Martínez-Gutiérrez, G., and L. Mayer. 2004. Huricanes en Baja California, México y sus implicaciones en la sedimentación en el golfo de California. *Geos* 24:57–64.

Martins, M., and H. L. Lillywhite. 2019. Ecology of snakes on islands. In Lillywhite, H. L. and M. Martins (eds.), *Islands and Snakes: Isolation and Adaptive Evolution.* Oxford University Press, pp. 1–44.

Meik, J., S. Schaack, O. Flores-Villela, and J. Streicher. 2018. Integrative taxonomy at the nexus of population divergence and speciation in insular speckled rattlesnakes. *Journal of Natural History* 52:989–1016.

Mellink, E. 1995. The potential effect of commercialization of reptiles from Mexico's Baja California Peninsula and its associated islands. *Herpetological Natural History* 3:95–99.

Murphy, R. W. 1983. The reptiles: Origins and evolution. In Case T. J., and M. L. Cody (eds.), *Island Biogeography of the Sea of Cortez.* University of California Press, pp. 130–158.

Murphy, R. W., and G. Aguirre–León. 2002. Nonavian reptiles: Origins and evolution. In T. Case, M. Cody, and E. Ezcurra (eds.), *A New Island Biogeography of the Sea Cortés.* Oxford University Press, pp. 181–220.

Murphy, R. W., F. Sánchez–Piñero, G. Polis, and R. Aalbu. 2002. New measurements of area and distance for islands in the Sea of Cortés. In T. Case, M. Cody, and E. Ezcurra (eds.), *A New Island Biogeography of the Sea Cortés.* Oxford University Press, pp. 447–464.

Odum, R. A., and H. K. Reinert, H. K. 2015. The Aruba Island rattlesnake *Crotalus unicolor.* Species survival plan: A case history in ex situ and in situ conservation. *International Zoo Yearbook* 49:104–112.

Oliveira-Dalland, L. G., L. R. V. Alencar, L. R. Tambosi, P. A. Carrasco, R. M. Rautsaw, J. Sigala-Rodriguez, G. Scrocchi, and M. Martins. 2022. Conservation gaps for Neotropical vipers: Mismatches between protected areas, species richness and evolutionary distinctiveness. *Biological Conservation* 275:1–8.

Pliego-Sánchez, J. V., C. Blair, A. H. Díaz de la Vega-Pérez, and V. H. Jiménez-Arcos 2021. The insular herpetofauna of Mexico: Composition, conservation, and biogeographic patterns. *Ecology and Evolution* 11:6579–6592.

Ruiz-Sanchez, E., G. Arnaud, O. R. Cruz–Andrés, G. D. León, and F. Javier. 2019. Phylogenetic relationships and origin of the rattlesnakes of the Gulf of California islands (Viperidae: Crotalinae: *Crotalus*). *Herpetological Journal* 29:162–172.

Salinas, Z. C., A. C. Leyva, D. B. Llunch, and E. R. Díaz. 1990. Distribución geográfica y variabilidad climática de los regímenes pluviómetros de Baja California Sur, México. *Atmósfera* 3:217–237.

Shreve, F., and I. R. Wiggins. 1964. *Vegetation and Flora of the Sonoran Desert* (vol. 591). Stanford University Press.

Soulé, M., and A. Sloan. 1966. Biogeography and distribution of the reptiles and amphibians on islands in the Gulf of California, Mexico. *Transactions of the San Diego Society of Natural History* 14:137–156.

Stapp P., and G. A. Polis. 2003. Influence of pulsed resources and marine subsidies on insular rodent populations. *Oikos* 102:111–123,

Tershy, B. R., C. J. Donlan, B. S. Keitt, D. A. Croll, J. A. Sanchez, B. Wood, M. A. Hemosillo, G. R. Howald, and N. Biavaschi. 2002. Island conservation in north-west Mexico: A conservation model integrating research, education and exotic mammal eradication. In C. R. Veitch and M. N. Clout (eds.), *Turning the Tide: The Eradication of Invasive Species.* IUCN, pp. 293–300.

Uetz, P., P. Freed, R. Aguilar, and J. Hošek. 2022. The reptile database. Accessed July 15, 2022. http://www.reptile–database.org.

Wallach, V., K. L. Williams, and J. Boundy. 2014. *Snakes of the World: A Catalogue of Living and Extinct Species.* CRC Press.

Wilson, L. D., and J. R. McCranie. 2004. The conservation status of the herpetofauna of Honduras. *Amphibian and Reptile Conservation* 3:6–33.

Wilson, L. D., V. Mata-Silva, and J. D. Johnson, 2013. A conservation reassessment of the reptiles of Mexico based on the EVS measure. *Amphibian & Reptile Conservation* 7:1–47.

Ziegler, T. 2015. Cologne Zoo: Reptile research and conservation. *International Zoo Yearbook* 49:8–21.

9

Natural History and Conservation of the Galapagos Snake Radiation

Diego F. Cisneros-Heredia and Carolina Reyes-Puig

Another [iguana] hatchling has its first glimpse of a dangerous world ... snakes' eyes aren't very good, but they can detect movement, so if the hatchling keeps its nerve, it may just avoid detection ... [dozens of snakes spill out of the volcanic rocks at Fernandina island, chasing a little iguana, managing at first to catch and constrict it, but the iguana manages to release itself and, making a dramatic leap, reaches safety up the rocks]. A near-miraculous escape.

—David Attenborough (Narrator, BBC Planet Earth II)[1]

In 2016, more than three centuries after the first known reference to snakes in the Galapagos Islands, a short 4-minute video put Galapagos snakes at the center of worldwide attention.[1] A BBC Planet Earth documentary showed a thrilling scene of dozens of Western Galapagos Racers chasing a young marine iguana. This scene remains one of the most popular snake videos ever shared on the internet, with more than 200 million views and 730 thousand likes on YouTube alone,[1] and it was the first time Galapagos snakes ever garnered so much attention.

Terrestrial non-avian reptiles are particularly rich in the Galapagos archipelago, with 9 recognized species of snakes of the genus *Pseudalsophis*, commonly called Galapagos Racers, 16 giant tortoises of the genus *Chelonoidis*, 11 geckos of the genus *Phyllodactylus*, 10 lava lizards of the genus *Microlophus*, 3 terrestrial iguanas of the genus *Conolophus*, and 1 Marine Iguana *Amblyrhynchus cristatus* (Benavides *et al.* 2009; Gentile and Snell 2009; Miralles *et al.* 2017; Zaher *et al.* 2018, Arteaga *et al.* 2019; Jensen *et al.* 2022). The genus *Pseudalsophis* (family Colubridae, subfamily Dipsadinae) includes all snake species inhabiting the Galapagos Islands and one species from the western coast of South America[2] (Zaher *et al.* 2009). In this chapter, we present an overview of the diversity, natural history, and conservation of the *Pseudalsophis* snakes from the Galapagos archipelago, one of the least studied groups of colubrids on oceanic islands.

Diego F. Cisneros-Heredia and Carolina Reyes-Puig, *Natural History and Conservation of the Galapagos Snake Radiation*
In: *Islands and Snakes.* Edited by: Harvey B. Lillywhite and Marcio Martins, Oxford University Press.
© Oxford University Press 2023. DOI: 10.1093/oso/9780197641521.003.0009

The Galapagos

> The Galapagos Islands are a great number of uninhabited islands lying
> under and on both sides of the Equator.... They are of a good height,
> most of them flat and even on the top; 4 or 5 of the easternmost are rocky,
> barren and hilly, producing neither tree, herb, nor grass, but a few dildoe-
> trees, except by the seaside.... The Spaniards when they first discovered
> these islands found multitudes of iguanas, and land-turtle or tortoise,
> and named them the Galapagos Islands.... There are some green snakes
> on these islands, but no other land animal that I did ever see.
>
> **—William Dampier[3]**

In 1684, the English pirate, privateer, and naturalist William Dampier visited the
Galapagos Islands on board the pirate ship *Batchelor's Delight*. He published the first
observations about the Galapagos flora and fauna that reached Europe in his 1697
book, *A New Voyage Round the World*. A young Charles Darwin took a copy of this
book on board the *HMS Beagle* (Mason 2015; Civallero 2022). The Galapagos archipelago became world-famous for its role as a natural laboratory where Darwin
gathered data to formulate his ideas about biological evolution and for its natural diversity, being home to many endemic species of flora and fauna that evolved in isolation not just from their continental relatives but also from each other (Quiroga and
Sevilla 2017; Kelley *et al.* 2019; Ali and Fritz 2021).

Galapagos is an oceanic archipelago formed by 19 main volcanic islands and more
than 100 islets and rocks, spread out over more than 430 km from north to south. The
archipelago is in the eastern Pacific Ocean, 930 km off the coast of Ecuador, South
America. On the lowlands, all islands are arid, warm, and covered by xerophytic vegetation; as elevation increases on the largest islands, a moister climate influenced by orogenic rainfall opens the way for semideciduous and evergreen vegetation (Wiggins
and Porter 1971; Jackson 1999). The larger islands of the Galapagos are Isabela
($4,588 \text{ km}^2$), Santa Cruz (986 km^2), Fernandina (642 km^2), Santiago (585 km^2), San
Cristóbal (558 km^2), Floreana (173 km^2), and Marchena (130 km^2) (Snell *et al.* 1996).
Five islands (Santa Cruz, San Cristóbal, Isabela, Floreana, and Baltra) have human
settlements and infrastructure, covering in total more than 260 km^2, with urban areas
mostly distributed in the lowlands and agricultural areas in the highlands (Wiggins
and Porter 1971; Guézou *et al.* 2010; López Andrade and Quiroga Ferri 2019; Laso
et al. 2019).

The modern names of most Galapagos Islands are in Spanish and correspond
to the names used in this chapter. However, older literature, including many doc-
uments dealing with Galapagos snakes, used names coined by English sailors.[4] To
avoid confusion, we present equivalents between the official names in Spanish and
the older English names (in parentheses) of the islands mentioned in this chapter
(Woram 1989; Snell *et al.* 1996): Isabela (Albermale), Santa Fe (Barrington), Floreana

(Charles), San Cristóbal (Chatham), Pinzón (Duncan), Española (Hood), Santa Cruz (Indefatigable), Santiago (James), Rábida (Jervis), Fernandina (Narborough), and Baltra (South Seymour).

The Taxonomic History of Galapagos Snakes: 188 Years and Still Ongoing

> There is one snake which is numerous; it is identical, as I am informed by
> M. Bibron, with the *Psammophis Temminckii* from Chile.
> —Charles Darwin[5]

The discovery of the diversity and evolution of Galapagos snakes has been a complex process, slowly unfolding over a long time. In 1835, *HMS Beagle* arrived at the Galapagos carrying an eager young naturalist, Charles Darwin, who collected a small snake while exploring Floreana Island, one of the southernmost islands of the archipelago. That specimen would travel back to England, to the British Museum, where Albert Günther used it to formally describe the first species of snake from the Galapagos: *Herpetodryas biserialis* Günther, 1860, marking the beginning of the taxonomic history of the Galapagos snake radiation. In 1876, Franz Steindachner named two taxa of Galapagos snakes, but as variations of a mainland species (*Dromicus chamissonis* var. *dorsalis* Steindachner, 1876 and *D. c.* var. *habelii* Steindachner, 1876), citing four islands as the localities for his specimens: Santa Cruz, Española, Floreana, and Rábida. During several decades, few specimens would arrive in scientific collections, these being identified either as conspecific with mainland species or with the same name applied to populations from different islands under the assumption that widespread species were involved and with their generic assignment constantly changing—having been included at one point or another in *Dromicus*, *Alsophis*, *Leimadophis*, *Opheomorphus*, and *Orophis* (Peters 1869; Steindachner 1876; Cope 1890; Garman 1892; Heller 1903; Amaral 1929; Dunn 1932; Parker 1932) (Figure 9.1).

It was not until the 20th century that three studies provided a much more in-depth review of the Galapagos snake radiation (Van Denburgh 1912; Mertens 1960; Thomas 1997). Van Denburgh (1912) proposed the recognition of seven taxa of Galapagos snakes under the genus *Dromicus* (six species, one with two subspecies) based on the specimens collected by Joseph R. Slevin and others from the expedition of the California Academy of Sciences to the Galapagos Islands between 1905 and 1906[6]: *Dromicus biserialis* from Floreana Island and a nearby islet; *D. dorsalis* from Rábida, Santa Cruz, Baltra, Santa Fe, and Santiago Islands; *D. hoodensis* Van Denburgh, 1912 from Española Island and a nearby islet; *D. o. occidentalis* Van Denburgh, 1912 from Fernandina Island; *D. o. helleri* Van Denburgh, 1912 from Isabela and Tortuga Islands; *D. slevini* Van Denburgh, 1912 from Pinzón, Fernandina, and Isabela Islands; and *D. steindachneri* Van Denburgh, 1912 from Santa Cruz, Baltra, and Rábida. Van Denburgh (1912) put Steindachner's *habelii* as a synonym of

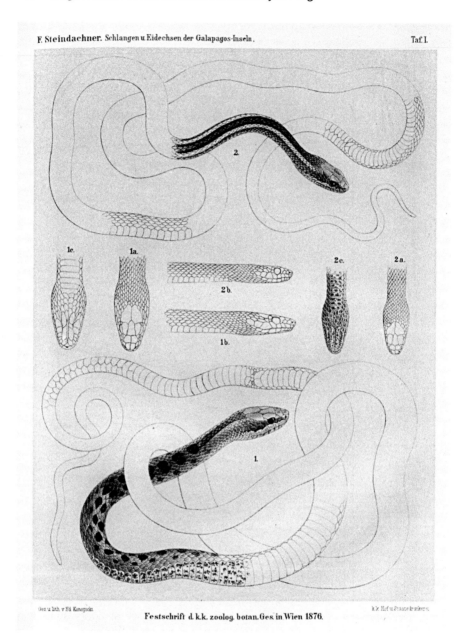

F. Steindachner. Schlangen u. Eidechsen der Galapagos-Inseln.   Taf. I.

Gez. u. lith. v. Ed. Konopicki.   k.k. Hof u. Staatsdruckerei.

Festschrift d. k.k. zoolog. botan. Ges. in Wien 1876.

Figure 9.1 Plate from Steindachner (1876) showing *Dromicus chamissonis* var. *dorsalis* and *D. c.* var. *habelii*.

dorsalis and considered that the five specimens examined by Steindachner were probably from Santa Cruz or Rábida Islands. In his "Suggestions to Future Students," Van Denburgh (1912) wrote, "Future collectors in these islands should strive to secure specimens of the snake of Chatam Island, if such there be. Doubtless, it will prove to be a most interesting new species." Unknown to all scientists at the time, a specimen from San Cristóbal had been collected and reached Paris; its locality became confused and it ended up being used to describe a mainland snake.[7] Eventually, Irenäus Eibl-Eibesfeldt collected specimens of the San Cristóbal snake, later recognized by Mertens (1960) as a distinct subspecies. Mertens (1960) accepted eight taxa (three species, eight subspecies), considering Van Derburgh's *occidentalis* and *helleri* as subspecies of *D. dorsalis*, *steindachneri* subspecies of *D. slevini*, and *hoodensis* subspecies of *D. biserialis*, and proposing a new subspecies, *D. b. eibli* Mertens, 1960 for the population of San Cristóbal Island. Thomas (1997) proposed a new taxonomic arrangement, recognizing six taxa but this time divided into three different genera: *Philodryas hoodensis*, *Antillophis slevini*, *A. steindachneri*, *Alsophis*[8] *b. biserialis*, *A. b. dorsalis*, and *A. b. occidentalis*; with *eibli* synonymized under *A. b. biserialis*, *habelii* under *A. b. dorsalis*, and *helleri* under *A. b. occidentalis* (Figure 9.2).

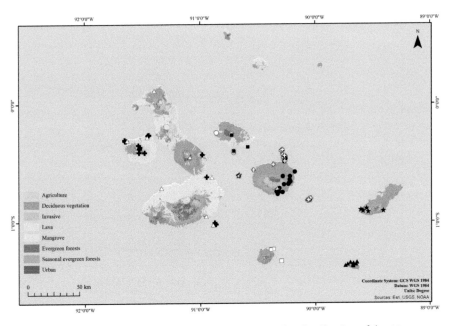

Figure 9.2 Map of the Galapagos Archipelago showing the distribution of the 10 species of Galapagos snakes of the genus *Pseudalsophis* referred to in this chapter and the vegetation types. Symbols: Open square, *P. biserialis*; open triangle, *P. darwini*; open crosses, *P. dorsalis*; black stars, *P. eibli*; black squares, *P. hephaestus*; black triangles, *P. hoodensis*; black crosses, *P. occidentalis*; white stars, *P. slevini*; black circles, *P. steindachneri*; white circles, *P. thomasi*.

It was not until Zaher *et al.* (2009) that a phylogenetic classification of the Galapagos radiation was proposed and the genus *Pseudalsophis* was described to include all species from the archipelago and *P. elegans* (Tschudi 1845) from the western coast of South America.[9] Zaher *et al.* (2018) conducted a major sampling across the Galapagos, providing a phylogenetic analysis with a time-calibrated tree for *Pseudalsophis* and recognizing nine species: *Pseudalsophis biserialis* from San Cristóbal and Floreana Islands and adjacent islets; *P. darwini* Zaher *et al.* 2018 from Isabela, Fernandina, and Tortuga Islands; *P. dorsalis* from Santa Cruz, Baltra and Santa Fe Islands and adjacent islets; *P. hephaestus* Zaher *et al.* 2018 from Santiago and Rábida Islands; *P. hoodensis* from Española Island and adjacent islets; *P. occidentalis* from Fernandina, Isabela, and Tortuga Islands; *P. slevini* from Pinzón Island; *P. steindachneri* from Baltra and Santa Cruz Islands and adjacent islets; and *P. thomasi* Zaher *et al.* 2018 from Santiago and Rábida Islands (Table 9.1).

While Zaher *et al.* (2018) is the most modern treatment of the Galapagos snake radiation, there are still pending issues. Zaher *et al.* (2018) kept the populations of San Cristóbal and Floreana Islands together as *P. biserialis*, but also reported several morphometric and meristic differences between them. Compared with the populations from Floreana and surrounding islets, the snakes from San Cristóbal Island have a lower head length to snout-vent length ratio, a higher tail length to snout-vent length ratio, and a higher male subcaudal scale count. Mertens (1960) and Thomas (1997) already pointed out differences in colorations, with San Cristóbal snakes having a striped pattern and Floreana populations a spotted pattern. Since 2017, we have been studying the snakes of San Cristóbal, and we have data supporting these consistent differences. Based on this information and the biogeographic separation between San Cristóbal and Floreana Islands, we recognized the San Cristóbal population as a valid, different species: *Pseudalsophis eibli* (Mertens 1960) new combination (Figure 9.3).[10] The taxonomy of the populations of *P. occidentalis* and *P. dorsalis* also requires further exploration. Zaher *et al.* (2018) reported differences between the populations from Fernandina Island from those of Isabela and Tortuga Islands. If integrative taxonomic studies result in these populations being different evolutionary units, the name *helleri* would be available for the populations of Isabela, as proposed by Van Denburgh (1912).

The Natural History: Food, Predators, Habits, and Habitats

> Just above the cliff dwellings of the iguanas . . . I found many interesting things. One of the first was a harmless serpent. . . . Like all of its less lowly neighbours it was quite tame, and made no objection to my picking it up, curling contentedly at the bottom of a snake bag. . . . I found later that it had been feeding on beetles, two grasshoppers and a small moth.
>
> —Beebe (1988)

Table 9.1 Taxonomic history of the Galapagos snake radiation

Current name	Species author, year	Original name	Van Denburgh (1912)	Mertens (1960)	Thomas (1997)	Zaher et al. (2018)
Pseudalsophis biserialis	Günther 1860	Herpetodryas biserialis	Dromicus biserialis	Dromicus biserialis biserialis	Alsophis biserialis biserialis	Pseudalsophis biserialis
Pseudalsophis darwini	Zaher et al. 2018	Pseudalsophis darwini	Dromicus slevini	Dromicus slevini slevini	Antillophis slevini	Pseudalsophis darwini
Pseudalsophis dorsalis	Steindachner 1876	Dromicus chamissonis var. dorsalis	Dromicus dorsalis	Dromicus dorsalis dorsalis	Alsophis biserialis dorsalis	Pseudalsophis dorsalis
Pseudalsophis dorsalis	Steindachner 1876	Dromicus chamissonis var. habelii	Dromicus dorsalis	Dromicus dorsalis dorsalis	Alsophis biserialis dorsalis	Pseudalsophis dorsalis
Pseudalsophis eibli	Mertens 1960	Dromicus biserialis eibli	–	Dromicus biserialis eibli	Alsophis biserialis biserialis	Pseudalsophis biserialis
Pseudalsophis hephaestus	Zaher et al. 2018	Pseudalsophis Hephaestus	Dromicus steindachneri	Dromicus slevini steindachneri	Antillophis steindachneri	Pseudalsophis hephaestus
Pseudalsophis hoodensis	Van Denburgh 1912	Dromicus hoodensis	Dromicus hoodensis	Dromicus biserialis hoodensis	Philodryas hoodensis	Pseudalsophis hoodensis
Pseudalsophis occidentalis	Van Denburgh 1912	Dromicus occidentalis	Dromicus occidentalis	Dromicus dorsalis occidentalis	Alsophis biserialis occidentalis	Pseudalsophis occidentalis
Pseudalsophis occidentalis	Van Denburgh 1912	Dromicus occidentalis helleri	Dromicus occidentalis helleri	Dromicus dorsalis helleri	Alsophis biserialis occidentalis	Pseudalsophis occidentalis
Pseudalsophis slevini	Van Denburgh 1912	Dromicus slevini	Dromicus slevini	Dromicus slevini slevini	Antillophis slevini	Pseudalsophis slevini
Pseudalsophis steindachneri	Van Denburgh 1912	Dromicus steindachneri	Dromicus steindachneri	Dromicus slevini steindachneri	Antillophis steindachneri	Pseudalsophis steindachneri
Pseudalsophis thomasi	Zaher et al. 2018	Pseudalsophis thomasi	Dromicus dorsalis	Dromicus dorsalis dorsalis	Alsophis biserialis dorsalis	Pseudalsophis thomasi

Figure 9.3 *Pseudalsophis eibli* (Mertens 1960) from San Cristóbal Island, Galapagos, Ecuador.
Photographs by the authors (archive Institute IBIOTROP, Universidad San Francisco de Quito USFQ).

John Van Denburgh, William Beebe, and Robert I. Bowman were the first to provide some insights into the diet of the Galapagos snakes.[11] These constrictor opisthoglyph snakes were initially thought to be mainly insectivorous, but more data have shown that squamate reptiles, like lava lizards, leaf-toed geckos, and Marine Iguanas, are their main prey (Van Denburgh 1912; Townsend 1930; Slevin 1935; Strijbos 1977; Fritts and Fritts 1982; Werner 1983; Beebe 1988; Laurie and Brown 1990; Merlen and Thomas 2013; Ortiz-Catedral *et al.* 2019). It would seem that they are opportunistic predators, preying upon a variety of vertebrates (including lava lizards, iguanas, geckos, snakes, birds,[12] bird eggs, fish, and carrion) and invertebrates, and even probably eating plants[13] (Van Denburgh 1912; Townsend 1930; Bowman 1960; Boersma 1974; Fritts and Fritts 1982; Werner 1983; Beebe 1988; Laurie and Brown 1990; Márquez *et al.* 2010; Merlen and Thomas 2013; Cadena-Ortiz *et al.* 2017; Christian 2017; Olesen *et al.* 2018; Zaher *et al.* 2018; Arteaga *et al.* 2019; Ortiz-Catedral *et al.* 2019, 2021b; Sollis 2020; this work, Table 9.2). Cannibalism seems rather frequent among Galapagos snakes (Sollis 2020; Ortiz-Catedral *et al.* 2021b) (Table 9.2).

Among all other Galapagos snakes, the population of *P. occidentalis* from Fernandina Island shows some of the more distinctive diets. On the west coast of Fernandina, snakes actively hunt for small marine fish in tidal pools, burying a third of their body into the sand and rock holes at the bottom of pools to search for fish (Merlen and Thomas 2013). Further research on this topic seems necessary to understand the hunting mechanisms and the prey–predator dynamics involved. Hunting of Marine Iguanas, made famous by the BBC Planet Earth II documentary, also occurs in Fernandina. Dozens of snakes migrate long distances to the coasts and wait for hatchling iguanas to come out of the sand. Although great numbers of snakes are involved, they do not behave as pack hunters; instead, they heavily compete with each other to grab prey (Werner 1983; Merlen and Thomas 2013).

Most information on the diet of Galapagos snakes comes from casual observations of predatory events; however, studies inspecting fecal samples have proved useful as a noninvasive method to gather data on the dietary niche of Galapagos snakes (Ortiz-Catedral *et al.* 2019, 2021b). Dietary items have been published for most species of Galapagos snakes, except for *P. darwini*, *P. eibli*, and *P. hephaestus*. During our ongoing studies at San Cristóbal Island since 2017 (Williams *et al.* 2019), we have recorded data demonstrating that the diet of *P. eibli* is similar to other congeners. An observation reported in the citizen science platform iNaturalist shows that *P. hephaestus* feeds on lava lizards (Table 9.2). The diet of the recently described *P. darwini* remains unknown, although it may probably also feed on lava lizards and geckos.

Information on the predators and parasites of Galapagos snakes is limited. Before local extinctions, Galapagos Hawk *Buteo galapagoensis*, Galapagos Barn Owl *Tyto punctatissima*, and Galapagos Short-eared Owl *Asio galapagoensis* were apparently the natural predators of *Pseudalsophis* on most islands (Wood 1939; Steadman 1986; Steadman and Burke DeLeon 1999; Jaramillo *et al.* 2016) (Table 9.3). Nowadays, nonnative mammals are probably the main predators on islands invaded by Black Rats *Rattus rattus*, domestic and feral dogs *Canis familiaris*, and domestic and feral cats *Felis catus* (Van Denburgh 1912; Clark 1981; Fritts and Fritts 1982; Sollis 2020) (Table 9.3). Other predators include native and introduced birds (domestic chicken *Gallus gallus*, Yellow-crowned Night-heron *Nyctanassa violacea*, Striated Heron *Butorides striata*, Smooth-billed Ani *Crotophaga ani*, Mockingbirds *Mimus* spp.), and Galapagos Centipede *Scolopendra galapagoensis* (Cisneros-Heredia and Márquez 2017a; Cisneros-Heredia 2018; Cooke *et al.* 2019; Ortiz-Catedral *et al.* 2021a; this work, Table 9.3). *Infidum luckeri* (Trematoda, Dicrocoeliidae) is the only parasite so

Table 9.2   List of known prey of the Galapagos snakes of the genus *Pseudalsophis*.

Snake species	Locality	Prey species	Sources
Pseudalsophis biserialis	Gardner	Aves, eggshells	Christian 2017; Ortiz-Catedral *et al.* 2019
Pseudalsophis biserialis	Gardner	Squamata, *Microlophus grayii*	Christian 2017; Ortiz-Catedral *et al.* 2019
Pseudalsophis biserialis	Gardner	Squamata, *Phyllodactylus baurii*	Christian 2017; Ortiz-Catedral *et al.* 2019
Pseudalsophis biserialis	Gardner	Insecta	Christian 2017; Ortiz-Catedral *et al.* 2019
Pseudalsophis biserialis	Champion	Squamata, *Microlophus grayii*	Christian 2017; Ortiz-Catedral *et al.* 2019; Sollis 2020
Pseudalsophis biserialis	Champion	Squamata, *Phyllodactylus baurii*	Christian 2017; Ortiz-Catedral *et al.* 2019
Pseudalsophis biserialis	Champion	Squamata, *Pseudalsophis biserialis*	Sollis 2020
Pseudalsophis biserialis	Champion	Insecta	Christian 2017; Ortiz-Catedral *et al.* 2019
Pseudalsophis dorsalis	Santa Cruz	Insecta, Coleoptera	Beebe 1988
Pseudalsophis dorsalis	Santa Cruz	Insecta, Orthoptera, cf. *Schistocerca melanocera*	Beebe 1988
Pseudalsophis dorsalis	Santa Cruz	Insecta, Lepidoptera	Beebe 1988
Pseudalsophis dorsalis	Santa Cruz	*Momordica charantia*	Olesen *et al.* 2018
Pseudalsophis dorsalis	El Edén	Squamata, *Pseudalsophis dorsalis*	Sollis 2020
Pseudalsophis dorsalis	El Edén	Squamata, *Microlophus indefatigabilis*	Sollis 2020
Pseudalsophis dorsalis	Venecia	Squamata, *Phyllodactylus galapagensis*	Sollis 2020
Pseudalsophis dorsalis	Santa Fe	Mammalis, Rodentia	Sollis 2020
Pseudalsophis dorsalis	Santa Fe	Aves, *Zenaida galapagoensis* eggs	Ortiz-Catedral *et al.* 2019
Pseudalsophis dorsalis	Santa Fe	Aves, eggshells	Ortiz-Catedral *et al.* 2019
Pseudalsophis dorsalis	Santa Fe	Squamata, *Amblyrhynchus cristatus*	Laurie and Brown 1990
Pseudalsophis dorsalis	Santa Fe	Squamata, *Conolophus pallidus*	Márquez *et al.* 2010
Pseudalsophis dorsalis	Santa Fe	Squamata, *Microlophus barringtonensis*	Ortiz-Catedral *et al.* 2019; Sollis 2020
Pseudalsophis dorsalis	Santa Fe	Squamata, *Phyllodactylus barringtonensis*	Arteaga *et al.* 2019

Table 9.2 Continued

Snake species	Locality	Prey species	Sources
Pseudalsophis dorsalis	Santa Fe	Squamata, *Pseudalsophis dorsalis*	Sollis 2020
Pseudalsophis dorsalis	Seymour Norte	Aves, *Puffinus subalaris*	Sollis 2020
Pseudalsophis dorsalis	Seymour Norte	Squamata, *Microlophus indefatigabilis*	Ortiz-Catedral *et al.* 2019; Sollis 2020
Pseudalsophis eibli	San Cristóbal	Aves, *Geospiza fuliginosa* hatchlings	This work
Pseudalsophis eibli	San Cristóbal	Squamata, *Amblyrhynchus cristatus*	This work
Pseudalsophis eibli	San Cristóbal	Squamata, *Microlophus bivittatus*	This work
Pseudalsophis eibli	San Cristóbal	Squamata, *Phyllodactylus leei*	Arteaga *et al.* 2019; this work
Pseudalsophis eibli	San Cristóbal	Squamata, *Hemidactylus frenatus*	This work
Pseudalsophis eibli	San Cristóbal	Squamata, *Gonatodes caudiscutatus*	This work
Pseudalsophis eibli	San Cristóbal	Insecta, *Schistocerca melanocera*	This work
Pseudalsophis eibli	San Cristóbal	Insecta, *Erinnys ello*	This work
Pseudalsophis hephaestus	Santiago	Squamata, *Microlophus jacobi*	This work based on iNaturalist observation
Pseudalsophis hoodensis	Española	Aves	Sollis 2020
Pseudalsophis hoodensis	Española	Squamata, *Microlophus delanonis*	Van Denburgh 1912; Werner 1978; Sollis 2020; this work
Pseudalsophis hoodensis	Española	Squamata, *Pseudalsophis hoodensis*	Sollis 2020
Pseudalsophis hoodensis	Española	Squamata, *Phyllodactylus gorii*	This work
Pseudalsophis occidentalis	Fernandina	Mammalia, *Nesoryzomys* sp.	Ortiz-Catedral *et al.* 2021; Sollis 2020
Pseudalsophis occidentalis	Fernandina	Aves	Sollis 2020
Pseudalsophis occidentalis	Fernandina/ Isabela?	Aves, *Spheniscus mendiculus* young and eggs	Boersma 1974
Pseudalsophis occidentalis	Fernandina	Squamata, *Amblyrhynchus cristatus*	Werner 1983; Merlen and Thomas 2013; Ortiz-Catedral *et al.* 2019; Sollis 2020

(*continued*)

Table 9.2 Continued

Snake species	Locality	Prey species	Sources
Pseudalsophis occidentalis	Fernandina	Squamata, *Amblyrhynchus cristatus* carrion	Ortiz-Catedral *et al.* 2019, 2021
Pseudalsophis occidentalis	Fernandina	Squamata, *Conolophus subcristatus*	Werner 1983
Pseudalsophis occidentalis	Fernandina	Squamata, *Microlophus albemarlensis*	Merlen and Thomas 2013; Ortiz-Catedral *et al.* 2019, 2021; Sollis 2020
Pseudalsophis occidentalis	Fernandina	Squamata, *Phyllodactylus simpsoni*	Ortiz-Catedral *et al.* 2019, 2021; Sollis 2020
Pseudalsophis occidentalis	Fernandina	Squamata, *Pseudalsophis occidentalis*	Ortiz-Catedral *et al.* 2021; Sollis 2020
Pseudalsophis occidentalis	Fernandina	Actinopterygii, *Dialommus fuscus*	Merlen and Thomas 2013
Pseudalsophis occidentalis	Fernandina	Actinopterygii, *Bolinichthys* sp. cf. *longipes*	Merlen and Thomas 2013
Pseudalsophis occidentalis	Fernandina	Actinopterygii, *Labrisomus dendriticus*	Bowman 1960, Merlen and Thomas 2013
Pseudalsophis occidentalis	Tortuga	Aves, eggshells	Ortiz-Catedral *et al.* 2019
Pseudalsophis occidentalis	Tortuga	Squamata, *Microlophus albemarlensis*	Ortiz-Catedral *et al.* 2019
Pseudalsophis occidentalis	Tortuga	Squamata, *Phyllodactylus simpsoni*	Ortiz-Catedral *et al.* 2019
Pseudalsophis occidentalis	Tortuga	Insecta	Ortiz-Catedral *et al.* 2019
Pseudalsophis slevini	Pinzón	Squamata, *Phyllodactylus duncanensis*	Van Denburgh 1912; Ortiz-Catedral *et al.* 2019
Pseudalsophis slevini	Pinzón	Squamata, *Phyllodactylus duncanensis* eggs	Sollis 2020
Pseudalsophis slevini	Pinzón	Squamata, *Microlophus duncanensis*	Ortiz-Catedral *et al.* 2019; Sollis 2020
Pseudalsophis slevini	Pinzón	Squamata, *Pseudalsophis slevini*	Sollis 2020
Pseudalsophis slevini	Pinzón	Insecta	Ortiz-Catedral *et al.* 2019
Pseudalsophis steindachneri	Santa Cruz	Squamata, *Microlophus indefatigabilis*	Townsend 1930, Sollis 2020; this work
Pseudalsophis steindachneri	Santa Cruz	Squamata, *Phyllodactylus galapagensis*	Sollis 2020
Pseudalsophis steindachneri	Santa Cruz?	Squamata, *Phyllodactylus galapagensis* eggs	Sollis 2020
Pseudalsophis steindachneri	Santa Cruz	Squamata, *Hemidactylus frenatus*	This work

Table 9.2 Continued

Snake species	Locality	Prey species	Sources
Pseudalsophis steindachneri	Santa Cruz	Squamata, *Pseudalsophis*	Sollis 2020
Pseudalsophis steindachneri	Baltra	Insecta: Orthoptera	Van Denburgh 1912
Pseudalsophis thomasi	Santiago	Aves	Sollis 2020
Pseudalsophis thomasi	Santiago	Squamata, *Microlophus jacobi*	Slevin in Fritts and Fritts 1982; Cadena-Ortiz *et al.* 2016; Zaher *et al.* 2018; Ortiz-Catedral *et al.* 2019; Sollis 2020
Pseudalsophis thomasi	Santiago	Squamata, *Phyllodactylus maresi*	Ortiz-Catedral *et al.* 2019; Sollis 2020
Pseudalsophis thomasi	Santiago	Squamata, *Pseudalsophis*	Sollis 2020
Pseudalsophis thomasi	Santiago	Insecta	Ortiz-Catedral *et al.* 2019
Pseudalsophis thomasi	Rábida	Squamata, *Microlophus jacobi*	Ortiz-Catedral *et al.* 2019

far reported in Galapagos snakes, inhabiting the gall bladder of *Pseudalsophis hoodensis* (McIntosh 1939; Hughes *et al.* 1942; Fernandes and Kohn 2014). The few studies that have directed efforts to understand and identify prey-predator-parasite dynamics show complex interactions between endemic Galapagos fauna and important cascades of consumption among secondary consumers on the islands, positioning Galapagos snakes as important predators of small to medium-sized organisms in the archipelago (Christian 2017; Ortiz-Catedral *et al.* 2019, 2021a; Sollis 2020).

Spotting Galapagos snakes is not always an easy task. On islands free of invasive species, snakes are rather calm, but their coloration usually provides wonderful camouflage that keeps them hidden even when in plain sight. However, on islands with invasive species, they are shy and elusive, and the quest to find them may become more difficult due to the fact that they are mostly active at dawn and dusk. Some species tend to be locally frequent, but the general pattern suggests that many species exhibit low densities and occupy habitats with low accessibility (Christian 2017; Arteaga *et al.* 2019; Sollis 2020; Ortiz-Catedral *et al.* 2021b; D. F. Cisneros-Heredia, C. Reyes-Puig *et al.* unpublished data). Galapagos snakes have activity levels with a bimodal pattern (morning and evening), mainly between 05:30 and 11:00 in the morning and between 16:00 and 19:00 in the afternoon (Christian 2017; Zaher *et al.* 2018; D. F. Cisneros-Heredia and C. Reyes-Puig *et al.* unpublished data). Although some species have been observed to be active at night (*P. eibli* and *P. dorsalis*), it does not seem to be a regular behavior (Arteaga *et al.* 2019; D. F. Cisneros-Heredia, C. Reyes-Puig *et al.* unpublished data). In many islands, snakes are detectable to the human eye when they are

Table 9.3 List of known predators of the Galapagos snakes of the genus *Pseudalsophis*

Snake species	Locality	Predator species	Sources
Pseudalsophis biserialis	Floreana	Aves, *Tyto punctatissima*	Steadman 1986; Steadman and Burke DeLeon 1999
Pseudalsophis biserialis	Gardner	Aves, *Asio galapagoensis*	Christian 2017
Pseudalsophis biserialis	Gardner	Myriapoda, *Scolopendra galapagoensis*	Ortiz-Catedral *et al.* 2021
Pseudalsophis biserialis	Champion	Squamata, *Pseudalsophis biserialis*	Sollis 2020
Pseudalsophis dorsalis	El Edén	Squamata, *Pseudalsophis dorsalis*	Sollis 2020
Pseudalsophis dorsalis	Santa Fe	Squamata, *Pseudalsophis dorsalis*	Sollis 2020
Pseudalsophis eibli	San Cristóbal	Mammalia, *Felis catus*	This work
Pseudalsophis eibli	San Cristóbal	Mammalia, *Canis familiaris*	This work
Pseudalsophis eibli	San Cristóbal	Mammalia, *Rattus rattus*	This work
Pseudalsophis eibli	San Cristóbal	Aves, *Asio galapagoensis*	This work
Pseudalsophis eibli	San Cristóbal	Aves, *Gallus gallus*	Cisneros-Heredia 2018
Pseudalsophis eibli	San Cristóbal	Aves, *Mimus melanotis*	This work
Pseudalsophis eibli	San Cristóbal	Aves, *Nyctanassa violacea*	This work
Pseudalsophis eibli	San Cristóbal	Aves, *Crotophaga ani*	Cooke *et al.* 2020; this work
Pseudalsophis eibli	San Cristóbal	Myriapoda, *Scolopendra galapagoensis*	This work
Pseudalsophis hephaestus	Santiago	Mammalia, *Rattus rattus*	Slevin in Fritts and Fritts 1982
Pseudalsophis hephaestus	Santiago	Aves, *Buteo galapagoensis*	Jaramillo *et al.* 2016
Pseudalsophis hoodensis	Española	Aves, *Buteo galapagoensis*	Cisneros-Heredia and Márquez 2017
Pseudalsophis hoodensis	Española	Aves, *Butorides striata*	This work
Pseudalsophis hoodensis	Española	Squamata, *Pseudalsophis hoodensis*	Sollis 2020

Table 9.3 Continued

Snake species	Locality	Predator species	Sources
Pseudalsophis occidentalis	Fernandina/Isabela?	Aves, *Buteo galapagoensis*	Arteaga *et al.* 2019
Pseudalsophis occidentalis	Fernandina	Squamata, *Pseudalsophis occidentalis*	Ortiz-Catedral *et al.* 2021
Pseudalsophis slevini	Pinzón	Aves, *Buteo galapagoensis*	Wood 1939
Pseudalsophis slevini	Pinzón	Squamata, *Pseudalsophis slevini*	Sollis 2020
Pseudalsophis steindachneri	Santa Cruz	Aves, *Mimus parvulus*	This work
Pseudalsophis thomasi	Santiago	Mammalis, *Sus scrofa*	Coblentz and Baber 1987
Pseudalsophis thomasi	Santiago	Mammalia, *Rattus rattus*	Slevin in Fritts and Fritts 1982
Pseudalsophis thomasi	Santiago	Aves, *Buteo galapagoensis*	Jaramillo *et al.* 2016

thermoregulating, foraging, and hunting, while they usually rest in shady hideouts during the hot sunny hours around midday (Merlen and Thomas 2013; Christian 2017; Ortiz-Catedral *et al.* 2019; Sollis 2020; D. F. Cisneros-Heredia, C. Reyes-Puig *et al.* unpublished data).

Galapagos snakes are distributed across 13 islands and 9 islets (Table 9.4). They are terrestrial, usually foraging amid volcanic rocks, fallen cacti, leaf litter, and along sandy and volcanic beaches. When resting, they hide under logs, cacti, and volcanic rocks (Van Denburgh 1912; Bowman 1960; Fritts and Fritts 1982; Beebe 1988; Merlen and Thomas 2013; Christian 2017; Arteaga *et al.* 2019; Sollis 2020; D. F. Cisneros-Heredia and C. Reyes-Puig *et al.* unpublished data). Most species inhabit the coastal lowlands in deciduous forest,[14] deciduous shrubland, and deciduous tall-grass, but *P. darwini*, *P. dorsalis*, *P. eibli*, and *P. occidentalis* can be found higher up in semideciduous and evergreen forests in the highlands of Isabela and San Cristóbal Islands (Table 9.4). *Pseudalsophis dorsalis*, *P. eibli*, *P. occidentalis*, and *P. steindachneri* (Figure 9.4) have been recorded in urban green areas and periurban areas on the lowlands, and *P. dorsalis*, and *P. eibli* in agricultural lands on the highlands (Sollis 2020; this work). *Pseudalsophis occidentalis* holds the highest elevation record among all Galapagos snakes, having been recorded in the highlands of Fernandina eating Isabela Lava Lizard, *Microlophus albermalensis*, at 1,400 m elevation (Ortiz-Catedral *et al.* 2019; Sollis 2020). We have recorded *P. eibli* feeding on the introduced gecko *Gonatodes caudiscutatus*, the only squamate reptile regularly recorded in the highlands of San Cristóbal, up to 450 m (San Cristóbal lava lizard, *Microlophus bivittatus*, does not occur above 250 m elevation). The presence of high densities of this gecko

Table 9.4 Geographic and ecological distribution of the Galapagos snakes of the genus *Pseudalsophis*

Species	P. biserialis	P. darwini	P. dorsalis	P. eibli	P. hephaestus	P. hoodensis	P. occidentalis	P. slevini	P. steindachneri	P. thomasi
Elevation range	0–70 m	0–800 m	0–600 m	0–400	0–190 m	0–135 m	0–1,400 m	0–330 m	0–40 m	0–190 m
Main islands										
Baltra			P							
Bartolomé										P
Española						P				
Fernandina		P					P			
Floreana	Ex									
Isabela		P					P			
Pinzón								P		
Rábida					P					P
San Cristóbal				P						
Santa Cruz			P						P	
Santa Fe			P							
Santiago					P					P
Seymour Norte			P						P	
Islets										
Champion	P									
Cowley							P			
Gardner por Española						P				
Gardner por Floreana	P									
El Edén			P							
Enderby	P									

Tortuga	P				P	P
Sombrero Chino			P			
Venecia		P		P		
Vegetation types						
Deciduous Forest		P	P	P	P	P
Deciduous Shrubland	P	P	P		P	P
Deciduous Tallgrass	P	P	P			P
Semideciduous Evergreen Forest		P	P			
Evergreen Forest		P	P	P		
Highland Deciduous Tallgrass		P		P		
Coastal Humid Forest and Shrubland		P				
Urban and periurban areas		P		P	P	
Agricultural areas		P	P			

P, Present; Ex, local extinction.

Figure 9.4 *Pseudalsophis steindachneri* (Van Denburgh 1912) from Santa Cruz Island, Galapagos, Ecuador.
Photograph by Joshua Addesi.

might have facilitated the expansion of *P. eibli* toward the highlands of San Cristóbal, where otherwise it was previously unknown (Lundh 2011).

Information is still greatly needed to better understand natural processes and biological interactions in the Galapagos snakes. Monitoring surveys are being conducted for some populations of *P. biserialis, P. dorsalis,* and *P. occidentalis* (Ortiz-Catedral *et al.* 2019, 2021a, 2021b; Jiménez-Uzcátegui and Ortiz-Catedral 2020) and for *P. eibli* and *P. hoodensis* (D. F. Cisneros-Heredia, C. Reyes-Puig *et al.* unpublished data), but additional continuous studies are required for all populations, especially those on islands with human and invasive wildlife populations.

The Future: Conservation

> Snakes must be very rare on Charles Island.... It is probable that the ravages of the smaller kinds of mammals that have been introduced there—particularly rats and cats—have pushed them to the verge of extinction.

—Van Denburgh (1912)

Invasive species, especially mammals, negatively impact reptile species in island ecosystems across the world (Böhm *et al.* 2013; Cox *et al.* 2022). Species like *P. biserialis* in Floreana and *P. dorsalis* in Santa Cruz[15] have seen local extinctions or severe population declines, and *P. eibli* is becoming rarer in San Cristóbal (Christian 2017; Arteaga *et al.* 2019; Jiménez-Uzcátegui and Ortiz-Catedral 2020; Sollis 2020; D. F. Cisneros-Heredia, C. Reyes-Puig *et al.* unpublished data). On these islands, Galapagos snake populations suffer predation from invasive predators, such as cats, rats, dogs, and pigs (Van Denburgh 1912; Clark 1981; Fritts and Fritts 1982; Coblentz and Baber 1987; Cisneros-Heredia 2018). Although some Galapagos snakes may be able to adapt to modified habitats in urban and agricultural matrices, they are still impacted heavily by invasive species, roadkills, and human–fauna conflicts. In contrast, it would seem that populations of *P. hoodensis* and *P. slevini* have seen an increase in population thanks to habitat protection and elimination of invasive species (Arteaga *et al.* 2019; Sollis 2020; D. F. Cisneros-Heredia, C. Reyes-Puig *et al.* unpublished data). Unfortunately, for many species, data about their population size and trends are lacking, thus limiting conservation efforts by the Galapagos National Park.

At the global level, *P. occidentalis* and *P. dorsalis* are currently classified under the International Union for Conservation of Nature (IUCN) category of Least Concern (Márquez and Yánez Muñoz 2022; Márquez *et al.* 2022); *P. biserialis*, *P. hoodensis*, and *P. steindachneri* are considered as Near Threatened (Cisneros-Heredia and Márquez 2017a, 2017b; Márquez *et al.* 2017); and *P. slevini* as Vulnerable (Márquez and Cisneros-Heredia 2019). At the national level, *P. slevini* was classified as Critically Endangered; *P. biserialis* and *P. steindachneri* as Endangered; and *P. hoodensis* as Vulnerable (Carrillo *et al.* 2005). However, all these evaluations were made before the taxonomic revision by Zaher *et al.* (2018) and with very little information available at the time; thus, they require urgent review and updating. *Pseudalsophis hoodensis* is known from an uninhabited island, and although its populations seemingly have no immediate threats, historical declines on Española have been reported, and the population remains restricted to a single island and surrounding islets. *Pseudalsophis sleveni* is also restricted to a single 18 km^2 island, but rats were eradicated thanks to the efforts of the Galapagos National Park. The populations currently assigned to *P. occidentalis* from Fernandina are apparently stable, and there are no known direct threats to them; however, the situation is different on Isabela, where invasive species might impact snake populations. *Pseudalsophis dorsalis* has suffered vast population declines on Santa Cruz, the most populated island in the Galapagos and one with many invasive species. In fact, the species was thought to be locally extinct on the island (Arteaga *et al.* 2019), but a few recent reports suggest the survival of a small population.[15] *Pseudalsophis steindachneri* also appears to be declining in Santa Cruz, and the conservation status and extinction risk of both species should be reassessed because they might be threatened. The taxonomic separation of *P. biserialis* and *P. eibli* calls for an urgent review of the conservation assessment of *P. biserialis*, which currently survives on just three islets near Floreana, being extinct on the main island. *Pseudalsophis darwini*, *P. eibli*, *P. haephastus*, and *P. thomasi* have not been assessed by the IUCN. *Pseudalsophis darwini* might not be at risk since it is widely distributed and apparently its populations are not declining, but nevertheless it remains a rare species with few records. *Pseudalsophis haephastus* and *P. thomasi* are known from a few localities, and although their populations might not be undergoing declines, predation by rats could be a problem. *Pseudalsophis eibli* is entirely restricted to San Cristóbal Island, the second most populated island of the Galapagos. Its abundance has declined in recent years, with predation by invasive species and road mortality being serious problems (Figure 9.5).

Conclusion

After more than 4 million years of evolution, the Galapagos terrestrial snake radiation includes at least 10 species and counting, distributed across one of the most famous archipelagos in the world. Despite the remarkable opportunities to investigate the evolution, ecology, biogeography, and conservation of a diverse and complex squamate clade, the Galapagos snakes remain among the least studied terrestrial vertebrates of South America. Coordinated actions developed by different organizations should be secured to conduct more research across the different populations of Galapagos snakes to support long-term conservation strategies developed by the Galapagos National Park. Top priorities for the future should include the assessment and reassessment of the conservation status and extinction risk of all species of Galapagos snakes, increasing the knowledge of their biology and ecology, eradicating invasive species, recovering natural ecosystems, and promoting the active involvement and participation of local people in the conservation of the Galapagos snakes and its ecosystems.

Acknowledgments

We thank the Galapagos National Park Directorate for their collaboration and constant support, especially Christian Sevilla, Galo Quezada, Daniel Lara, Jorge Carrión, Carlos Vera, and all authorities and park rangers of the Galapagos National Park for their valuable comments during project proposal reviews and their support with logistics and transportation to the sampling sites. Research permits for our study of snakes of the San Cristóbal and Española Islands were issued by the Galapagos National Park. All applicable international and national guidelines for the care and use of animals were followed. We are grateful to Cornelio Williams, Mateo Dávila, Paula Oléas, Sebastián Ramos, Nicole Acosta, Andrés Mena, Izan Chalen, Emilia Peñaherrera, and Giovanny Sarigu for their help during field and laboratory work. Our work was possible thanks to grants provided by the Universidad San Francisco de Quito (USFQ), Galapagos Science Center, USFQ Galápagos, Colegio de Ciencias Biológicas y Ambientales (COCIBA–USFQ), Oficina de Vinculación USFQ ("Celebrando la

Figure 9.5 Galapagos snakes in populated islands are frequently killed on the roads. This photograph was taken on the road between Puerto Baquerizo Moreno and Puerto Chino, San Cristóbal Island, Galapagos, Ecuador.

Photograph by the authors (archive Institute IBIOTROP, Universidad San Francisco de Quito USFQ).

Naturaleza Urbana" and "Conectándose con la Naturaleza de Galápagos" commu-nity outreach programmes), and Programa "Becas de Excelencia," Secretaría de Educación Superior, Ciencia, Tecnología e Innovación (SENESCYT), Ecuador.

Notes

1. The documentary was produced by the BBC Natural History Unit as part of Planet Earth II. The video is now available everywhere on the internet. On YouTube, it was originally posted in BBC ("Iguana Chased by Killer Snakes") and BBC Earth ("Iguana vs. Snakes") channels, but the version with the largest internet reach is in Zapping Sauvage channel ("Iguana vs serpentes"), with more than 200 million views, 730,000 comments and 29,000 comments.
2. *Pseudalsophis elegans* (Tschudi 1845) is the first described species of the genus and inhab-its the dry Pacific coastland and western Andean foothills and slopes of southern Ecuador, Peru, and extreme northern Chile (Guedes *et al.* 2018).
3. William Dampier (b. 1651, East Coker, Somerset, England; d. 1715, London). His detailed notes on the biodiversity he met during his travels led him to be nicknamed "The Naturalist Pirate" (Mason 2015; Civallero 2022). His first mention of snakes of Galapagos was pub-lished in his book "A New Voyage Round the World," published in 1697 (Dampier 1937).
4. English names used for islands in the Galapagos archipelago were mainly coined by William Ambrosia Conley, an English buccaneer who visited the islands in 1684, and other sailors during the 17th and 18th centuries (Woram 1989).
5. Darwin mentioned his record of a snake at Floreana Island in his "Journal of Researches into the Natural History and Geology of the Countries Visited During the Voyage of H.M.S. Beagle Round the World, Under the Command of Capt. Fitz Roy, R.N." (Darwin 1845).
6. Joseph R. Slevin published his observations on the diversity and natural history of the rep-tiles of the Galapagos in 1935 (Slevin 1935), and his notes taken during the expedition were edited and published by Fritts and Fritts (1982).
7. A second specimen of a snake from Galapagos reached European museums in the 1830s. It was apparently collected by the crew of the frigate *La Vénus*, probably at San Cristóbal when the ship visited Galapagos in 1838 (du Petit-Thouars 1841). This specimen became a syntype of *Dryophylax freminvilleii* (Duméril, Bibron, and Duméril 1854), a name now in the synonymy of *Pseudalsophis elegans* (Thomas and Ineich 1999).
8. Maglio (1970) was the first to propose the placement of *biseralis*, *dorsalis*, and *slevini* in a non-monophyletic *Alsophis*, based on a review of their skull characters. However, Thomas (1997) proposed that *slevini* and *steindachneri* were *Antillophis*, while *biserialis*, *dorsalis*, and *occidentalis* were *Alsophis*.
9. Zaher (1999) found shared derived hemipenial characters relating Galapagos snakes with *Alsophis elegans* and species of the genus *Saphenophis* from South America, disagreeing with Thomas's (1997) hypothesis about the close relationship of the Galapagos snakes with Antillean snakes. Zaher (1999) also elevated Thomas's subspecies to species status.
10. This is a new combination proposed herein, being the first time that the name *eibli* (Mertens 1960) is combined with the genus *Pseudalsophis*.
11. John Van Denburgh (1912) reported the stomach contents of *Pseudalsophis hoodensis*, *P. slevini*, and *P. steindachneri* based on his examination of the specimens collected by the expedition of the California Academy of Sciences to the Galapagos Islands between 1905

and 1906 (Table 9.2). William Beebe visited the Galapagos in 1923 and provided notes on the diet in his expedition book *Galápagos: World's End*, first published in 1924. Robert I. Bowman explored the Galapagos in 1957, together with Irenäus Eibl-Eibesfeldt, being the first to report fish consumption by a *Pseudalsophis* (Bowman 1960).

12. Snakes of the genus *Pseudalsophis* have been observed to inspect mockingbird nests at Santa Cruz and active nests of Darwin finches at Champion-by-Floreana (Ortiz-Catedral *et al.* 2017) and at San Cristóbal (D. F. Cisneros-Heredia and C. Reyes-Puig, unpublished data). Curio (1965) reported that Darwin finches reacted to the presence of snakes. However, there are few records of actual predation of birds by Galapagos snakes (Boersma 1974; Christian 2017; Ortiz-Catedral *et al.* 2019; Sollis 2020; this work).

13. Olesen *et al.* (2018) reported that "Even the Galápagos snakes may disperse seeds . . . snakes consume the pulp of the invasive squash *Momordica charantia* on the Galapagos island Santa Cruz (D. Meribeth pers. comm.)." As pointed out by Ortiz-Catedral *et al.* (2019), it remains to be determined if the pulp of *Momordica* is part of the diet of the Galapagos snakes, at least occasionally, or if it was accidentally consumed.

14. Vegetation classification follows the proposal by Rivas-Torres *et al.* (2018).

15. The report that a small population of *P. dorsalis* might still survive in the highland of Santa Cruz brings hope for a population recovery program in the island (Sollis 2020).

References

Ali, J. R., and U. Fritz. 2021. Origins of Galápagos' land-locked vertebrates: What, whence, when, how? *Biological Journal of the Linnean Society* 134(2):261–284.

Amaral, A. do. 1929. Estudos sobre ophidios neotropicos. XVIII Lista remissiva dos ophidios da região neotropica. *Memorias do Instituto Butantan* 4:129–271.

Arteaga, A. F., L. Bustamante, J. Vieira, W. Tapia, and J. M. Guayasamin. 2019. *Reptiles of the Galápagos*. Tropical Herping.

Beebe, W. 1988. *Galápagos: World's End*. Dover.

Benavides, E., R. Baum, H. M. Snell, H. L. Snell, and J. W. Sites Jr. 2009. Island biogeography of Galápagos lava lizards (Tropiduridae: *Microlophus*): Species diversity and colonization of the archipelago. *Evolution* 63(6):1606–1626.

Boersma, P. D. 1974. The Galapagos Penguin: A Study of Adaptations for Life in an Unpredictable Environment. PhD thesis. Ohio State University.

Böhm, M., B. Collen, J. E. M. Baillie, P. Bowles, J. Chanson, N. Cox, G. Hammerson, *et al.* 2013. The conservation status of the world's reptiles. *Biological Conservation* 157:372–385.

Bowman, R. I. 1960. *Report on a Biological Reconnaissance of the Galápagos Islands During 1957*. United Nations Educational, Scientific and Cultural Organization (UNESCO).

Cadena-Ortiz, H., A. Barahona-V., D. Bahamonde-Vinueza, and J. Brito M. 2017. Anecdotal predation events of some snakes in Ecuador. *Herpetozoa* 30(1/2):93–96.

Carrillo, E., S. Aldás, M. Altamirano, F. Ayala, D. Cisneros-Heredia, A. Endara, C. Márquez, *et al.* 2005. *Lista Roja de los Reptiles del Ecuador*. Fundación Novum Millenium.

Christian, E. J. 2017. Demography and Conservation of the Floreana Racer (*Pseudalsophis biserialis biserialis*) on Gardner-by-Floreana and Champion Islets, Galápagos Islands, Ecuador. Master's thesis. Massey University.

Cisneros-Heredia, D. F. 2018. The hitchhiker wave: Non-native small terrestrial vertebrates in the Galapagos. In M. Torres and C. F. Mena (eds.), *Understanding Invasive Species in the Galapagos Islands*. Springer, pp. 95–139.

Cisneros-Heredia, D. F., and C. Márquez. 2017a. *Pseudalsophis hoodensis*. The IUCN Red List of Threatened Species 2017: http://dx.doi.org/10.2305/IUCN.UK.2017-2.RLTS. T190539A54447664.en.

Cisneros-Heredia, D. F., and C. Márquez. 2017b. *Pseudalsophis steindachneri*. The IUCN Red List of Threatened Species 2017: http://dx.doi.org/10.2305/IUCN.UK.2017-2.RLTS. T190542A54447669.en.

Civallero, E. 2022. *Una historia de Galápagos en 15 documentos: Del mapa de Ortelius al diario de Darwin*. Fundación Charles Darwin.

Clark, D. A. 1981. Foraging patterns of black rats across a desert-montane forest gradient in the Galapagos Islands. *Biotropica* 13(3):182.

Coblentz, B. E., and D. W. Baber. 1987. Biology and control of feral pigs on Isla Santiago, Galapagos, Ecuador. *Journal of Applied Ecology* 24(2):403–418.

Cooke, S. C., L. E. Haskell, C. B. van Rees, and B. Fessl. 2019. A review of the introduced smooth-billed ani Crotophaga ani in Galápagos. *Biological Conservation* 229:38–49.

Cope, E. D. 1890. Scientific results of explorations by the U. S. Fish Commission steamer Albatross. No. III. Report on the batrachians and reptiles collected in 1887–88. *Proceedings of the United States National Museum* 12(769):141–147.

Cox, N., B. E. Young, P. Bowles, M. Fernandez, J. Marin, G. Rapacciuolo, M. Böhm, *et al.* 2022. A global reptile assessment highlights shared conservation needs of tetrapods. *Nature* 605(7909):285–290.

Curio, E. 1965. Zur geographischen Variation des Feinderkennens einiger Darwinfinken (Geospizidae). *Verhandlungen der Deutschen Zoologischen Gesellschaft* 28:466–492.

Dampier, W. 1937. *A New Voyage Round the World*. Adam and Charles Black.

Darwin, C. 1845. *Journal of Researches into the Natural History and Geology of the Countries Visited During the Voyage of H.M.S. Beagle Round the World, Under the Command of Capt. Fitz Roy, R.N.* (2nd ed., corrected, with additions). John Murray.

Dunn, E. R. 1932. The Colubrid snakes of the Greater Antilles. *Copeia* 1932(2):89–92.

Fernandes, B. M. M., and A. Kohn. 2014. *South American Trematodes Parasites of Amphibians and Reptiles*. Instituto Oswaldo Cruz, Fiocruz.

Fritts, T. H., and P. R. Fritts (eds.). 1982. Race with extinction: Herpetological notes of J. R. Slevin's journey to the Galapagos 1905–1906. *Herpetological Monographs* 1:1–98.

Garman, S. 1892. The reptiles of the Galapagos Islands from the collections of Dr. Geo. Baur. *Bulletin of the Essex Institute* 24(4–6):73–87.

Gentile, G., and H. Snell. 2009. *Conolophus marthae* sp. nov. (Squamata, Iguanidae), a new species of land iguana from the Galápagos archipelago. *Zootaxa* 2201(1):1–10.

Guedes, T. B., R. J. Sawaya, A. Zizka, S. Laffan, S. Faurby, R. A. Pyron, R. S. Bérnils, *et al.* 2018. Patterns, biases and prospects in the distribution and diversity of Neotropical snakes. *Global Ecology and Biogeography* 27(1):14–21.

Guézou, A., M. Trueman, C. E. Buddenhagen, S. Chamorro, A. M. Guerrero, P. Pozo, and R. Atkinson. 2010. An extensive alien plant inventory from the inhabited areas of Galapagos. *Plos One* 5(4):e10276.

Günther, A. 1860. On a new snake from the Galapagos islands. *Proceedings of the Zoological Society of London* 28:97–98.

Heller, E. 1903. Papers from the Hopkins Stanford Galapagos Expedition, 1898–1899. XIV. Reptiles. *Proceedings of the Washington Academy of Sciences* 5:39–98.

Hughes, R. C., J. W. Higginbotham, and J. W. Clary. 1942. The trematodes of reptiles, part i, systematic section. *The American Midland Naturalist* 27(1):109–134.

Jackson, M. H. 1999. *Galapagos: A Natural History* (5th print). University of Calgary Press.

Jaramillo, M., M. Donaghy-Cannon, F. H. Vargas, and P. G. Parker. 2016. The diet of the Galapagos hawk (*Buteo galapagoensis*) before and after goat eradication. *Journal of Raptor Research* 50(1):33–44.

Jensen, E. L., M. C. Quinzin, J. M. Miller, M. A. Russello, R. C. Garrick, D. L. Edwards, S. Glaberman, et al. 2022. A new lineage of Galapagos giant tortoises identified from museum samples. *Heredity* 128(4):261–270.

Jiménez-Uzcátegui, G., and L. Ortiz-Catedral. 2020. Vertebrate diversity on Floreana Island, Galapagos. *Galapagos Research* 69:18–24.

Kelley, D., K. Page, D. Quiroga, and R. Salazar. 2019. *In the Footsteps of Darwin: Geoheritage, Geotourism and Conservation in the Galapagos Islands.* Springer International.

Laso, F. J., F. L. Benítez, G. Rivas-Torres, C. Sampedro, and J. Arce-Nazario. 2019. Land cover classification of complex agroecosystems in the non-protected highlands of the Galapagos Islands. *Remote Sensing* 12(1):65.

Laurie, W. A., and D. Brown. 1990. Population biology of marine iguanas (*Amblyrhynchus cristatus*). II. Changes in annual survival rates and the effects of size, sex, age and fecundity in a population crash. *Journal of Animal Ecology* 59(2):529.

López Andrade, J. E., and D. Quiroga Ferri. 2019. The Galapagos urban context. In T. Kvan. and J. Karakiewicz (eds.), *Urban Galapagos.* Springer International, pp. 9–22.

Lundh, J. P. 2011. The Galapagos: A brief history. Accessed December 28, 2022. http://www.lundh.no/jacob/galapagos/pg05.htm.

Maglio, V. J. 1970. West Indian xenodontine colubrid snakes: Their probable origin, phylogeny, and zoogeography. *Bulletin of the Museum of Comparative Zoology* 141(1):1–53.

Márquez, C., and D. F. Cisneros-Heredia. 2019. *Pseudalsophis slevini. The IUCN Red List of Threatened Species* 2019: https://dx.doi.org/10.2305/IUCN.UK.2019-3.RLTS.T190543A54447674.en

Márquez, C., D. F. Cisneros-Heredia, and M. H. Yánez Muñoz. 2017. *Pseudalsophis biserialis. The IUCN Red List of Threatened Species* 2017: http://dx.doi.org/10.2305/IUCN.UK.2017-2.RLTS.T190541A56253872.en.

Márquez, C., D. F. Cisneros-Heredia, and M. H. Yánez Muñoz. 2022. *Pseudalsophis dorsalis. The IUCN Red List of Threatened Species* 2022: https://dx.doi.org/10.2305/IUCN.UK.2022-1.RLTS.T190538A217764645.en

Márquez, C., E. A. Muñoz H., G. Gentile, W. Tapia, F. J. Zabala, S. A. Naranjo L., and A. J. Llerena. 2010. Estado poblacional de las iguanas terrestres (*Conolophus subcristatus, C. pallidus* y *C. marthae*: Squamata, Iguanidae), Islas Galápagos. *Boletín Técnico, Serie Zoológica* 9(6):19–37.

Márquez, C., and M. H. Yánez Muñoz. 2022. *Pseudalsophis occidentalis. The IUCN Red List of Threatened Species* 2022: https://dx.doi.org/10.2305/IUCN.UK.2022-1.RLTS.T190540A217764795.en

Mason, H. 2015, April 22. The pirate naturalist. *Hakai Magazine.* https://hakaimagazine.com/article-short/pirate-naturalist/

McIntosh, A. 1939. A new dicrocoeliid trematode collected on the Presidential cruise of 1938. *Smithsonian Miscellaneous Collections* 98(16):1–2.

Merlen, G., and R. A. Thomas. 2013. A Galápagos ectothermic terrestrial snake gambles a potential chilly bath for a protein-rich dish of fish. *Herpetological Review* 44(3):415–417.

Mertens, R. 1960. Über die Schlangen der Galapagos Inseln. *Senckenbergiana biologica* 41:133–141.

Miralles, A., A. Macleod, A. Rodríguez, A. Ibáñez, G. Jiménez-Uzcategui, G. Quezada, M. Vences, and S. Steinfartz. 2017. Shedding light on the imps of darkness: An integrative taxonomic revision of the Galápagos marine iguanas (genus *Amblyrhynchus*). *Zoological Journal of the Linnean Society* 181(3):678–710.

Olesen, J. M., C. F. Damgaard, F. Fuster, R. H. Heleno, M. Nogales, B. Rumeu, K. Trøjelsgaard, et al. 2018. Disclosing the double mutualist role of birds on Galápagos. *Scientific Reports* 8(1):57.

Ortiz-Catedral, L., E. Christian, W. Chimborazo, C. Sevilla, and D. Rueda. 2021a. A Galapagos centipede *Scolopendra galapagoensis* preys on a Floreana Racer *Pseudalsophis biserialis*. *Galapagos Research* 70:2–4.

Ortiz-Catedral, L., E. Christian, M. J. A. Skirrow, D. Rueda, C. Sevilla, K. Kumar, E. M. R. Reyes, and J. C. Daltry. 2019. Diet of six species of Galapagos terrestrial snakes (*Pseudalsophis* spp.) inferred from faecal samples. *Herpetology Notes* 12:701–704.

Ortiz-Catedral, L., C. Sevilla, G. Young, and D. Rueda. 2017. Historia natural y perspectivas de conservación del cucuve de Floreana (*Mimus trifasciatus*). In L. Cayot and D. Cruz (eds.), *Informe Galápagos 2015-2016*. DPNG, CGREG, FCD, GC, pp. 171–174.

Ortiz-Catedral, L., H. Sollis, E. Moncreiffe, J. Ramirez, D. Rueda, C. Sevilla, and R. Wollocombe. 2021b. Evidence of cannibalism in a population of western Galápagos racers *Pseudalsophis occidentalis* (Serpentes: Colubridae). *Herpetological Bulletin* 157:6–11.

Parker, H. W. 1932. Some new or rare reptiles and amphibians from southern Ecuador. *Annals and Magazine of Natural History* 9(49):21–26.

Peters, W. 1869. Mittheilung über eine neue Eidechsenart, *Phyllodactylus galapagensis*, von den Galapagos-Inseln. *Monatsberichte der Königlichen Preussische Akademie des Wissenschaften zu Berlin* 1869:719–720.

du Petit-Thouars, A. 1841. *Voyage autour du monde sur la frégate La Vénus pendant les anné 1836-1839 publié par ordre du Roi, sous les auspices du Ministre de la Marine*. Gide.

Quiroga, D., and A. M. Sevilla (eds.). 2017. *Darwin, Darwinism and Conservation in the Galapagos Islands*. Springer International.

Rivas-Torres, G. F., F. L. Benítez, D. Rueda, C. Sevilla, and C. F. Mena. 2018. A methodology for mapping native and invasive vegetation coverage in archipelagos: An example from the Galápagos Islands. *Progress in Physical Geography: Earth and Environment* 42(1):83–111.

Slevin, J. R. 1935. An account of the reptiles inhabiting the Galapagos Islands. *Bulletin of the New York Zoological Society* 38(1):3–24.

Snell, H. M., P. A. Stone, and H. L. Snell. 1996. A summary of geographical characteristics of the Galapagos Islands. *Journal of Biogeography* 23(5):619–624.

Sollis, H. E. 2020. Conservation of the Central Galapagos Racer (*Pseudalsophis dorsalis*) in the Galapagos Islands, Ecuador. Master's Thesis. Massey University.

Steadman, D. W. 1986. Holocene vertebrate fossils from Isla Floreana, Galapagos. *Smithsonian Contributions to Zoology* 413:1–103.

Steadman, D. W., and V. Burke DeLeon. 1999. First highly stratified prehistoric vertebrate sequence from the Galapagos Islands, Ecuador. *Pacific Science* 53(2):129–143.

Steindachner, F. 1876. Die Schlangen und Eidechsen der Galapagos-Inseln. *Herausgegeben von der K. K. Zoologisch-Botanischen Gesellschaft in Wien* 1876:303–330.

Strijbos, J. P. 1977. De reptielen van de Galapagos Eilanden. *De Levende Natuur* 80(10):233.

Thomas, R. A. 1997. Galápagos terrestrial snakes: Biogeography and systematics. *Herpetological Natural History* 5(1):19–40.

Thomas, R. A., and I. Ineich. 1999. The identity of the syntypes of *Dryophylax freminvillei* with comments on their locality data. *Journal of Herpetology* 33(1):152–155.

Townsend, C. H. 1930. The Astor expedition to the Galápagos Islands. *Bulletin of the New York Zoological Society* 33:135–155.

Tschudi, J. J. 1845. Reptilium conspectum quae in republica Peruana reperiuntur er pleraque observata vel collecta sunt in itenere. *Archiv für Naturgeschichte* 11(1):150–170.

Van Denburgh, J. 1912. Expedition of the California Academy of Sciences to the Galapagos islands, 1905-1906. IV The snakes of the Galapagos islands. *Proceedings of the California Academy of Sciences* 1:323–374.

Werner, D. I. 1983. Reproduction in the iguana *Conolophus subcristatus* on Fernandina Island, Galapagos: Clutch size and migration costs. *American Naturalist* 121(6):757–775.

Wiggins, I. L., and D. M. Porter. 1971. *Flora of the Galápagos Islands*. Stanford University Press.

Williams, C., C. Reyes-Puig, and D. F. Cisneros-Heredia. 2019. Conociendo a la serpiente corredora de Galápagos. In E. D. V. Gonzaga (ed.), *Archivos Académicos 21: Memorias del Congreso de Ciencias Biológicas y Ambientales.* USFQ Press, p. 3.

Wood, G. C. 1939. Zoological results of the George Vanderbilt South Pacific expedition of 1937. Part IV. Galapagos reptiles. *Notulae Naturae* 16:1–2.

Woram, J. M. 1989. Galapagos island names. *Noticias de Galapagos* 48:22–32.

Zaher, H. 1999. Hemipenial morphology of the South American xenodontine snakes with a proposal for a monophyletic Xenodontinae and reappraisal of colubroid hemipenes. *Bulletin of the American Museum of Natural History* 240:1–168.

Zaher, H., F. G. Grazziotin, J. E. Cadle, R. W. Murphy, J. C. de Moura-Leite, and S. L. Bonatto. 2009. Molecular phylogeny of advanced snakes (Serpentes, Caenophidia) with an emphasis on South American Xenodontines: A revised classification and descriptions of new taxa. *Papéis Avulsos de Zoologia (São Paulo)* 49(11):115–153.

Zaher, H., M. H. Yánez-Muñoz, M. T. Rodrigues, R. Graboski, F. A. Machado, M. Altamirano-Benavides, S. L. Bonatto, and F. G. Grazziotin. 2018. Origin and hidden diversity within the poorly known Galápagos snake radiation (Serpentes: Dipsadidae). *Systematics and Biodiversity* 16(7):614–642.

10

Paradise Lost

Collapse of the Unusual Cottonmouth Population on Seahorse Key

Mark R. Sandfoss

Introduction and History

The islands that comprise the Cedar Keys in the Gulf of Mexico have a history that includes use by Seminole Indians, the Spanish, and pirates (McCarthy 2007; Collins 2011). Locals dating back to pre-Columbian times have delighted in the plentiful oysters, and anglers today flock to the area in pursuit of abundant redfish, sea trout, and mullet. These islands have also been subjects of folklore and ghastly tales of pirates and ghosts. In the early to mid-1800s, the US government established a military depot, internment camp, and hospital on Seahorse Key, and a lighthouse was constructed in 1854. The lighthouse and other historical structures are preserved on the island, as well as a graveyard dating back to the Civil War.

Many islands are inhabited by a variety of animal species considered undesirable by most people (*e.g.*, spiders [*Nephila clavipes*], rats [*Rattus rattus*], and swarms of biting midges [*Culicoides* spp.] and mosquitoes). This is in addition to raucous colonies of seasonally nesting waterbirds that splatter the vegetation white with voided urates and litter the ground with regurgitated fish carcasses in all states of decomposition. And I have yet to mention the main character of this chapter, the venomous Florida Cottonmouth (*Agkistrodon conanti* Gloyd 1969; Serpentes, Viperidae)—a venomous snake that inhabits some of these islands at very high densities (Wharton 1958; Lillywhite and Brischoux 2012). Therefore, depending on one's perspective, the Cedar Keys may inspire awe or terror.

The first scientific description of the cottonmouths inhabiting islands in the Cedar Keys was made by Archie Carr, Jr. in 1936 (Carr 1936), followed by Charles Wharton in the 1950s (Wharton 1958, 1966, 1969). Since then, contemporary research has been led by Harvey Lillywhite (*e.g.*, Lillywhite *et al.* 2002, 2008). The rich history of scientific exploration in the Cedar Keys, and research on the cottonmouths of Seahorse Key in particular, was summarized in Chapter 9 of the first volume of *Islands and Snakes* (Lillywhite and Sheehy 2019).

Mark Robert Sandfoss, *Paradise Lost* In: *Islands and Snakes*. Edited by: Harvey B. Lillywhite and Marcio Martins, Oxford University Press. © Oxford University Press 2023. DOI: 10.1093/oso/9780197641521.003.0010

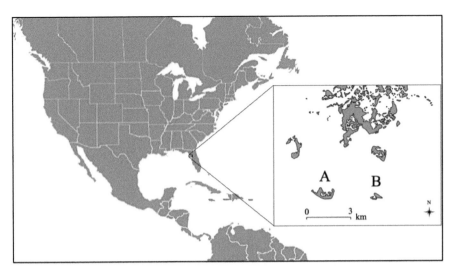

Figure 10.1 An aerial view of the study area including Seahorse Key (A) and Snake Key (B), coastal northwestern Florida, USA. Map of North America not to scale.

Briefly, Seahorse Key is a 67-ha island on the Gulf coast of Florida (Figure 10.1), which historically supported a large population of Florida Cottonmouths. The cottonmouth population on Seahorse Key was characterized by a trophic association with colonially nesting waterbirds that provided food resources to the snakes via fish carrion, largely during March to November each year (Wharton 1969; Lillywhite *et al.* 2002; Lillywhite and McCleary 2008; Lillywhite *et al.* 2008). This trophic relationship is different from expectation, based on the ordinary dietary habits of cottonmouths elsewhere, because of (1) the reliance of snakes on fish carrion scavenged after being dropped or regurgitated by waterbirds (Wharton 1958, 1966, 1969) and (2) the seemingly complete avoidance of eggs and chicks that occasionally fall from nests or are found at ground level, despite their nutritional value (Wharton 1969; Lillywhite *et al.* 2008).

These trophic features contrast with other islands around the world having similarly dense populations of venomous snake that feed directly on chicks or migratory birds (Shine *et al.* 2002; Aubret *et al.* 2006; Marques *et al.* 2012; Lillywhite and Martins 2019). In addition, the presence of the cottonmouths appears to benefit nesting waterbirds by deterring potential nest predators such as rats, raccoons, and other snakes (Lillywhite and Zaidan 2004; Lillywhite and Sheehy 2019). This unique relationship between nesting waterbirds and cottonmouth snakes on Seahorse Key has persisted for at least the past 80 years (Carr 1936; Wharton 1969; McCue *et al.* 2012). As a result of this long-standing trophic relationship and ephemeral "manna from heaven," the population of cottonmouths on Seahorse Key was unusually dense (Lillywhite and Sheehy 2019). In essence, the cottonmouth snakes of

Seahorse Key appeared to be living without any major predators in a relatively iso-lated trophic "paradise."

Paradise Lost

Unexpectedly, between April 21 and 23, 2015, the entire colony of waterbirds com-pletely abandoned Seahorse Key, and the nesting birds have not returned as of the writing of this chapter. The abandonment of nesting activities for all species simul-taneously, combined with the fact that many individuals abandoned nests contain-ing energetically expensive eggs, suggests that a dramatic disturbance must have occurred. Investigations have found no evidence of widespread adult mortality, and the few carcasses that were recovered tested negative for disease. The exact cause for the mass abandonment remains unknown.

The sudden abandonment of Seahorse Key by waterbirds after so many years of consistent nesting, while unfortunate, presented a unique opportunity to measure the response of a well-studied wildlife population to the loss of its primary food supply. Previous studies on the ecology of the cottonmouth population on Seahorse Key found the distribution, demographics, and food habits of these snakes to be heavily influenced by the bird rookery, which historically concentrated on the west end of the island (Wharton 1969; Lillywhite and McCleary 2008).

The spatial and temporal distribution of resources plays a significant role in an-imal ecology, and fluctuations in resource availability can affect consumer pop-ulation dynamics, behavior, and community structure (Anderson et al. 2008; Holt 2008). The loss of a major food source was predicted as far back as the late 18th century to have significant negative effects on the abundance and health of populations (Malthus 1798). This can be particularly true of geographically con-strained populations, such as those on islands. Animals can respond to environ-mental change in a variety of ways. Physiological and morphological changes are thought to be gradual and permanent, while behavioral changes can be immediate and usually reversible.

Studies on the behavioral responses of snakes to shifts in food resources have fo-cused primarily on foraging behaviors of populations within the context of seasonal fluctuations in prey type and abundance (e.g., Ujvari et al. 2011). Studies of pred-atory generalists have found these species to utilize prey in relation to their abun-dance in the field, and they will shift prey selection in response to seasonal changes (Hirai 2004; Luiselli 2006). However, while Brown et al. (2002) found that trop-ical seasonality induced strong fluctuations in many attributes of several Australian snake populations, the patterns of response were variable both among and within species. These variable responses to natural fluctuations in prey make it difficult to predict the response of insular cottonmouths to a complete and prolonged loss of a primary food resource.

Studies of snake responses to sudden or dramatic losses of food resources may be more informative. For example, when an extreme flooding event virtually eliminated the preferred prey item (native Dusky Rats, *Rattus colletti*) for Water Pythons (*Liasis fuscus*) during a three-year period in tropical Australia, the local pythons were unable to respond and did not disperse despite the presence of abundant rats only 8 km away (Ujvari *et al.* 2011). Long seasonal migrations (>10 km) were previously common in these pythons during annual flooding-induced migration of prey, but the absence of rodents over a prolonged period did not induce migration. And, ultimately, many pythons starved to death (Ujvari *et al.* 2011). Similarly, the decimation of frog populations around the world by the amphibian fungal pathogen *Batrachochytrium dendrobatidis* has been linked to recent declines in diversity and body condition of tropical snakes (Zipkin *et al.* 2020). The majority of species that historically relied upon amphibian prey affected by chytrid are experiencing decline. However, a few "winning" species appear to be more abundant with no clear explanation as to why, and at least one winning species is a pitviper (Zipkin *et al.* 2020). An increase in abundance following the loss of frogs suggests the species was able to find alternative prey to frogs, or it could be the result of reduced competition with other frog-feeding species that have experienced a decline in abundance.

Pitvipers such as cottonmouths might be able to develop an adaptive energy-saving phenotype in response to food restriction because they have a generalist diet, feed infrequently, and have low-energy life histories (McCue and Lillywhite 2002). However, the sudden and dramatic loss of food resources over a short time span is predicted to cause declines in measures of energy balance and individual health while also causing altered foraging behavior. Auspiciously, a neighboring island, Snake Key (Figure 10.1), functions as a positive control in this natural ecological experiment, as an estimated 3,000–5,000 waterbirds shifted nesting activities to that island following the abandonment of Seahorse Key in 2015. While Snake Key has previously had some nesting activity as a result of spillover from Seahorse Key when it became too crowded, historic levels were much lower than the concentration of waterbird nesting that has occurred there following the abandonment of Seahorse Key. The aim of this chapter is to summarize published investigations into the ecological and physiological responses of insular cottonmouth snakes inhabiting Seahorse Key and Snake Key to the loss or gain, respectively, of bird-provided food resources and provide a new synthesis and discussion of those findings here.

We can hypothesize that cottonmouths on Snake Key will experience some positive effects from the increase in available food resources. We expect both population-level changes and variation in the individual responses of snakes based on differences in age, sex, and individual behavior (Mathot *et al.* 2012). Our ability to characterize the response of cottonmouths is possible only because of the prior collection of biological data on measures of abundance, body condition, diet, and home range of cottonmouths on Seahorse Key over the past century (Carr 1936; Wharton 1958; Lillywhite and McCleary 2008; Lillywhite and Sheehy 2019).

Before we delve into the science, I encourage the reader to take a moment here to consider the extreme challenge that the population of cottonmouths on Seahorse Key has faced following the unprecedented decline in their available food resources and the toll it must take on an individual to experience the loss of its paradise. The words of John Milton in *Paradise Lost* describing the response of Satan following his fall from heaven might give us some insight into the "mindset" of a cottonmouth on Seahorse Key.

> How overcome this dire Calamity,
> What reinforcement we may gain from Hope,
> If not what resolution from despair.
> —Milton, Paradise Lost (1667)

Body Condition

All animals continuously expend energy to survive, but most, including cottonmouths, do not constantly process food, which leads to a reliance on energy reserves to maintain basic processes of life (McCue 2010). In general, when food is available, animals maintain body mass as a state referred to as *dynamic equilibrium*, whereby energy input matches energy requirements. Therefore, the loss of food resources for cottonmouths on Seahorse Key was predicted to negatively affect the body condition of individuals by limiting the energy available (Sandfoss *et al.* 2018). Body condition is a useful measure of energy availability and is often used to document changes in resource availability and individual health (Bonnet and Naulleau 1995; Beaupre 2008). The residuals from the linear regression of measures of log-snout-vent-length and log-body mass provide an easy calculation of body condition (BCI) in snakes.

Assuming the primary input of energy for cottonmouths on Seahorse Key was provided by nesting waterbirds, Sandfoss *et al.* (2018) made several straightforward predictions including an expectation of declines in body condition following the cessation of bird nesting. In general, ectothermic animals can tolerate starvation for a longer duration than endotherms because of their reduced need to expend energy to maintain an elevated body temperature. However, starvation tolerance varies considerably across taxonomic groups and is determined by the overall energy requirement, which is related to body mass, body temperature, phylogenetic history, and ability to ration fuel resources during starvation (McCue 2010).

The historical locations of waterbird nests, and subsequently food resources, have not been uniformly distributed across Seahorse Key. Sandfoss *et al.* (2018) predicted that the decline in body condition of cottonmouths would be more pronounced for those snakes inhabiting the west portion of Seahorse Key, where nesting historically concentrated. On the other hand, cottonmouths on Snake Key were expected to be in relatively good body condition, if taking advantage of the increased food provided by new numbers of nesting waterbirds.

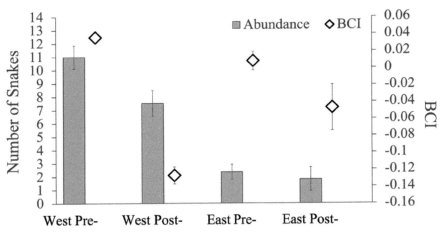

Transect Location and Bird Abandonment Period

Figure 10.2 Plot of body condition index (BCI) values of Florida Cottonmouths (*A. conanti*) sampled in the pre- (*n* = 258) and post-abandonment (*n* = 111) periods on the west or east ends of Seahorse Key, scaled on the right-hand y-axis. Bar plot of the mean number of individual Florida Cottonmouth snakes observed along beach transects on Seahorse Key during the pre- and post- abandonment periods. Error bars represent standard error of the mean. Data include the west transect during the pre-abandonment period ("West Pre," *n* = 86), west transect during the post-abandonment period ("West Post," *n* = 23), east transect during the pre-abandonment period ("East Pre," *n* = 23), and the east transect during the post-abandonment period ("East Post," *n* = 10).

To answer these questions, body condition data were analyzed for 461 cotton-mouth snakes captured throughout Seahorse Key starting in 1999 and continuing through two years after the abandonment event (2015 to 2016) (Sandfoss *et al.* 2018). In addition, BCI was calculated for 29 snakes captured on Snake Key during the post-abandonment period.

Unsurprisingly, the majority of snakes captured on Seahorse Key after the abandonment event (>91%, *n* = 115) were in relatively poor body condition and had negative BCI values (BCI mean ± standard deviation (SD), −0.13 ± 0.10, *n* = 115). In comparison, the BCI of snakes measured in the pre-abandonment period were much higher (BCI mean ± SD, 0.04 ± 0.10, *n* = 346) (Sandfoss *et al.* 2018) (Figure 10.2). Similarly, the BCI of cottonmouths captured on Snake Key during the post-abandonment period were elevated (0.11 ± 0.08, *n* = 29) (Figure 10.3). This is strong evidence that nesting waterbirds provide a significant source of food to cottonmouths and supports the prediction that individuals on Snake Key were able to take advantage of novel resources within a relatively short window of time. Similarly, adders (*Vipera berus*)

Figure 10.3 Two images of adult Florida Cottonmouth snakes, *A. conanti*, representative of the body condition of snakes observed on Seahorse Key (left) and Snake Key (right) following the shift in nesting activities to Snake Key. Both snakes are of similar length, but the cottonmouth from Snake Key is in much better body condition, as revealed by the rounded body, stretched dorsal scales, and lack of a visible backbone. A description of the quantitative difference in body condition between populations can be found in the section on body condition.
Photographs by Mark Sandfoss.

have been shown to grow faster and be in better body condition when prey densities increase (Forsman and Lindell 1997).

The reduced availability of fish carrion on Seahorse Key post-abandonment likely affected annual reproduction (Nowak *et al.* 2008). Body condition of adult snakes of both sexes decreased (Sandfoss *et al.* 2018), and while males have been shown to reduce their reproductive effort when in low body condition (Lind and Beaupre 2015), the reproductive output of snake populations is generally considered to be dictated by the condition of females rather than males (Wharton 1966, 1969; Naulleau and Bonnet 1996; Bonnet *et al.* 1998; Gregory 2006). Females may not be able to reproduce below a threshold level of body condition, and those females that do reproduce can experience increased mortality (Shine 1980; Madsen and Shine 1993, 1999). Previous studies of Seahorse Key cottonmouths indicated that females give birth during early fall following the summer input of fish carrion that provides energy for reproduction (Wharton 1969). Despite observed declines in BCI, it was not possible to quantify changes in reproductive output, such as clutch size or offspring size, as no gravid cottonmouths were found in the post-abandonment period (Sandfoss *et al.* 2018).

There was no significant variation detected in the BCI of cottonmouths based on location: individual BCI from cottonmouths on the east and west ends of Seahorse Key did not differ (Sandfoss *et al.* 2018). This lack of difference is difficult to explain, but it is possible that the loss of energy input from the thousands of nesting water-birds caused a large enough depression in food resources that all snakes on the island were affected regardless of their location.

Abundance

Animal populations are regulated in part by resource availability, and when food resources become limited a decline in population size is expected (Malthus 1798). To quantify changes in the population size of cottonmouths on Seahorse Key following the bird abandonment event, an abundance index was calculated by counting the number of snakes observed along two established beach transects (east and west transect, Figure 10.4) (Lillywhite and Brischoux 2012; Sandfoss *et al.* 2018).

Beach transect surveys were completed at least once in every year between March and November (the snakes' active season), from 2000 through 2016, with a mean of 8 (± 5.16 SD, *n* = 142) surveys completed each year. Each snake observed was recorded and grouped into an age class based on estimated total length (neonates born that year <45 cm; juveniles 45–75 cm; and adults >75 cm) (Sandfoss *et al.* 2018). Abundance data were compared based on bird abandonment period (pre and post), location (east

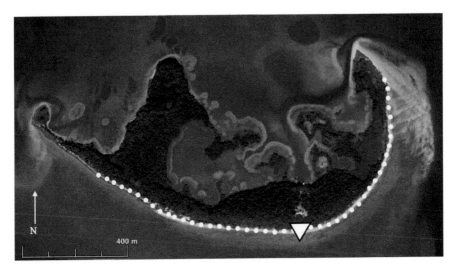

Figure 10.4 Aerial view of Seahorse Key with beach transects used for estimating snake abundance. The white triangle indicates the "midpoint" between east and west transects and where snake counts were initiated. The white sand beach along the island's southern perimeter is obstructed from view by the dotted line representing the transects.

and west), and age class (neonate, juvenile, adult) to characterize changes in abundance related to the loss of bird-provided food resources.

It was hypothesized that abundance would decline across Seahorse Key following the abandonment and that the decline would be more dramatic on the west end of the island where waterbird nesting historically concentrated (Sandfoss *et al.* 2018). Indeed, snake

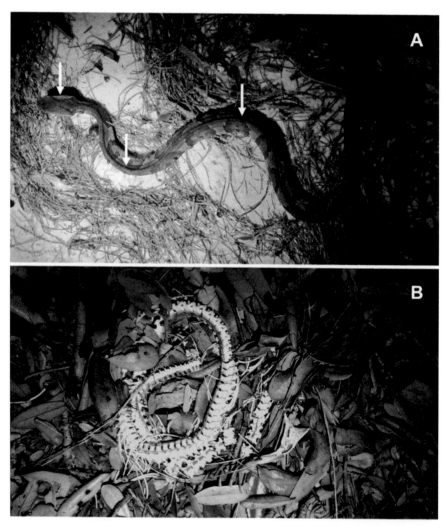

Figure 10.5 (A) Image of a Florida Cottonmouth, *A. conanti*, found in very poor condition foraging along the beach at night. The level of starvation of this snake is indicated by the arrows pointing to the noticeable outline of the backbone and sunken head. (B) A loose coil of bones presumed to be the remains of a recently deceased Florida Cottonmouth. There were no immediate signs of disturbance, suggesting that this snake died of starvation, but this cannot be confirmed.
Photographs by Mark Sandfoss.

abundance on the west transect did significantly decrease from an average of 10.99 ± 8.11 (n = 86) cottonmouths per night during the pre-abandonment period to 7.52 ± 4.64 (n = 23) snakes during the post-abandonment period (Figure 10.2). Sandfoss *et al.* (2018) suggested that the decrease in abundance of cottonmouths on the west transect was attributable to increased mortality from starvation. Numerous very thin (Figure 10.5A), freshly deceased, and skeletonized snakes or carcasses (Figure 10.5B) were found throughout the island in 2015 and 2016 (M. Sandfoss, personal observations).

The abundance of cottonmouths along the east transect of Seahorse Key declined less dramatically between pre- (2.35 ± 2.67; n = 23) and post-abandonment periods (1.80 ± 2.82 SD; n = 10) (Figure 10.2) compared to the west transect (Sandfoss *et al.* 2018). Abundance of cottonmouths on the east transect did not decrease presumably because birds historically did not nest on the east portion of the island in any significant numbers, and, therefore, despite observed declines in BCI, those snakes survived at greater rates by eating food items other than fish carrion. It is also possible that snakes from the west moved to the east in search of food. Whatever the case, the abundance of snakes on the east end has historically been much lower than on the west end (Wharton 1969; Lillywhite and McCleary 2008). And this finding may provide a basis for predicting the future density of the cottonmouth population over the entirety of the island if waterbirds fail to resume nesting in the future.

Declines in age class varied across the east and west ends of the island. Adult snakes did not significantly decline in the east but did in the west. Neonates declined in both the east (from 1.78 ± 1.88, n = 23 to 0.20 ± 0.63 SD, n = 10) and west (from 4.98 ± 6.03, n = 83 to 0.35 ± 0.57, n = 23). Interestingly, the average number of juveniles on the west transect increased from 1.18 ± 1.41 SD (n = 83) to 3.00 ± 2.63 (n = 23) following the abandonment event, while no similar change in juveniles was observed on the east end (Sandfoss *et al.* 2018). The increase in juvenile snakes on the west end is unexpected but could be related to reduced competition with larger cottonmouths for fish carrion and the ability of younger (and smaller) snakes to survive with less energy input than large, adult cottonmouths. Overall, reduced numbers of neonates combined with declines in female BCI and no observed gravid females on Seahorse Key suggest that the loss of food resources has reduced reproductive output (Seigel and Fitch 1985; Ford and Seigel 1989). The rapid decline in abundance of cottonmouths observed on Seahorse Key is alarming.

Trophic Collapse

The trophic niche of a species consists of the sum of all the trophic interactions that link it to other species in an ecosystem, or more simply, the overall trophic role of a species (Elton 1927). The dietary niche of wild animals can be characterized using the ratio of certain stable isotopes in the tissues of consumers and their potential food items (Pilgrim 2005). Specifically, $\delta^{13}C$ and $\delta^{15}N$ values can be used to differentiate the terrestrial versus marine origin of food items and identify the trophic level at which an animal is feeding, respectively. In addition, the combined values for carbon and nitrogen isotopes from a subset of individuals can be used to estimate the trophic niche for a population (Jackson *et al.* 2011).

Cottonmouth snakes are generalist feeders as a species, and their ability to take advantage of the allochthonous resources provided by nesting waterbirds is thought to be a major factor in their successful establishment on islands (Carr 1936; Wharton 1958; Lillywhite and McCleary 2008). Bird-provided fish carrion is the primary food resource for the majority of cottonmouths on Seahorse Key during the waterbird nesting and fledgling season (March to November) (Carr 1936; Wharton 1958). Outside of the active waterbird season, cottonmouths are thought to survive on occasional invasive rats (*R. rattus*), dead fish that are sometimes scavenged within the intertidal zone (Lillywhite and McCleary 2008), or nothing (McCue and Lillywhite 2002). However, the full range of potential food items available to cottonmouths on Seahorse Key includes skinks (*Plestiodon inexpectatus* and *Scincella lateralis*), anoles (*Anolis carolinensis*), and an introduced species of frog (*Eleutherodactylus planirostris*) (Wharton 1969; Lillywhite and Sheehy 2004).

Sandfoss *et al.* (2021) were able to use stable isotope analyses to characterize the trophic niche of cottonmouths on Seahorse Key and document changes in the percent contribution of each prey type to the diet of *A. conanti* between location (east vs west) and bird-abandonment period. It was predicted that the loss of bird-provided food would shift the dietary contributions of available food items and alter the trophic niche of cottonmouths on Seahorse Key. Tissue samples for isotope analyses were collected from all potential prey items (anoles, fish, frogs, and lizards; Figure 10.6) more than 11 months after the shift in waterbird nesting location to account for potential time lag in isotope assimilation.

Unsurprisingly, dietary analyses using stable isotopes revealed that cottonmouths in the pre-abandonment period (east, $n = 15$; west, $n = 173$) on Seahorse Key ate primarily marine fish, and it did not matter which end of the island the snakes inhabited (Figure 10.7) (Parnell *et al.* 2013; Sandfoss *et al.* 2021). In the post-abandonment period, cottonmouths on Seahorse Key (east, $n = 9$; west, $n = 39$) decreased their dietary proportion of marine fish in both the west and east ends of the island, but, unexpectedly, fishes remained their primary food resource.

Although no shift was detected in the primary food item consumed, there was a significant difference in values of $\delta^{13}C$ and $\delta^{15}N$ observed between abandonment periods (Sandfoss *et al.* 2021). This is likely the result of reduced scavenging of marine-derived fish and increased predation on terrestrial prey items with lower values of $\delta^{13}C$ and higher values of $\delta^{15}N$ (Figure 10.6) (Sandfoss *et al.* 2021). The largest increase in mean proportion of prey consumed between pre- and post-abandonment periods was frogs (+27.6%) in the west and rats (+14.4%) in the east (Figure 10.7) (Sandfoss *et al.* 2021). The primary evidence for this shift comes from observed increases in $\delta^{15}N$ values. Increased ratio of $\delta^{15}N$ indicates feeding at a higher trophic level, which could also be the result of cannibalism (Greenwood *et al.* 2010). It is not possible to estimate rates of cannibalism using stable isotope analyses because prey

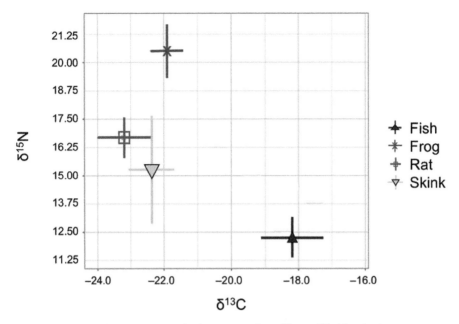

Figure 10.6 Plot of the mean (± SD) of isotopic values (^{13}C and ^{15}N) for the four primary prey items (fish $n = 12$, frog $n = 2$, rat $n = 4$, skink $n = 4$) of insular Florida Cottonmouth snakes sampled during the pre- and post-abandonment periods.

items and consumers would have the same isotopic profile. However, two instances of cannibalism were observed for the first time on Seahorse Key following the abandonment (Sheehy *et al.* 2017) (Figure 10.8).

Cottonmouths might continue to forage on marine fish that wash up on the beach (Lillywhite and McCleary 2008; Lillywhite *et al.* 2008) or scavenge below trees used by resident raptors that eviscerate prey. We might also speculate that the isotopic profile of cottonmouths on post-abandonment Seahorse Key could continue to indicate that fish remain the primary prey item even months after last eating fish due to the low metabolic rate of cottonmouths (McCue and Lillywhite 2002; McCue and Pollock 2008). Cottonmouths on Snake Key appear to be making use of bird-provided marine resources as the stable isotope values of cottonmouths from Snake Key post-abandonment ($n = 34$) were similar to values for cottonmouths sampled on Seahorse Key during the pre-abandonment period (Sandfoss *et al.* 2021).

Ecological characteristics that are predicted to predispose a species to successful invasion of novel habitats often include flexibility of feeding strategies and dietary breadth (Ehrlich 1989; Vasquez 2005; Lillywhite and Martins 2019). Similarly, insular populations surviving on subsidies are thought to require a wide trophic niche (Ruffino *et al.* 2011) to persist when the abundance and availability of food subsidies varies with time (Stapp and Polis 2003). Interestingly, there is emerging evidence that

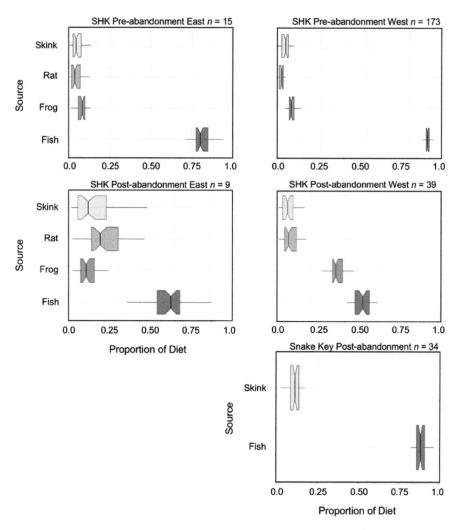

Figure 10.7 Analyses of dietary composition for insular Florida Cottonmouth snakes (*Agkistrodon conanti*). Proportion of *A. conanti* diet made up of four prey types, estimated by Simmr using ^{13}C and ^{15}N stable isotope values from the west and east ends of Seahorse Key (SHK) during the pre- and post-abandonment periods. Cottonmouths sampled on Snake Key during the post-abandonment period are also included.

some populations of generalist feeding species with a wide trophic niche at the population level can be comprised of individuals with relatively narrow diets that feed on a small range of preferred food items (Bolnick *et al.* 2007; Quevedo *et al.* 2009). Therefore, the ability of a species to take advantage of a novel food resource does not necessarily predispose individuals of that species to high levels of dietary plasticity (Ruffino *et al.* 2011). Furthermore, there is evidence from other reptile species

Figure 10.8 Cannibalism involving two adult cottonmouth snakes (*Agkistrodon conanti*) on Seahorse Key, Florida, USA, July 1, 2015. The larger snake is biting and holding the smaller snake behind the head, presumably as venom is injected. Note that both snakes are in relatively poor body condition.
Photograph by Coleman M. Sheehy III.

(*Emydura macquarii*) that generalist feeders that are able to switch prey when preferred food items are not available still endure a direct cost to their energy budgets (Petrov *et al.* 2020).

Insular cottonmouths do appear to maintain a feeding response to active, non-fish prey in a laboratory setting (Lillywhite *et al.* 2015) despite their long history of consuming fish carrion at bird rookeries (Carr 1936; Wharton 1969). This retention of feeding response is probably important for long-term survival in environments where allochthonous subsidy of resources from birds is seasonal as well as temporally and spatially unpredictable (Lillywhite and McCleary 2008). The dietary shifts observed on Seahorse Key to invasive frogs and cannibalism indicates some ability to use alternative food resources, although neither of those food items is likely to be a sustainable resource and will likely result in the continued decrease in abundance of cottonmouths on the island.

Home Range

Home range, as originally defined (for mammals), is the area traversed by an individual in the performance of its normal activities of feeding, mating, and caring for young (Burt 1943). A reduction in the availability of food resources is predicted to increase the size of an individual's home range as individuals will have to travel farther to find food to meet their energy needs (McNab 1963). Wharton (1969) reported that cottonmouths on Seahorse Key had limited activity ranges and often remained under the same rookery tree for extended periods of time. After the loss of bird-provided food, Sandfoss *et al.* (2021) predicted that the home range of cottonmouths on Seahorse Key would increase.

The home range for cottonmouth snakes was calculated using the minimum complex polygon method (Mohr 1947) based on radio-telemetry data. Five adult snakes (female, $n = 1$; male, $n = 3$; unknown, $n = 1$) were tracked during the pre-abandonment period (2002–2004) and six adult cottonmouths (female $n = 1$, male $n = 5$) were tracked during the post-abandonment period (2018–2019). In contrast to predictions, there was no general trend toward an increase in the size of home range for cottonmouths on Seahorse Key between pre- (3.21 ± 4.3 ha) and post-abandonment (2.29 ± 2.7 ha) periods (Figure 10.9) (Sandfoss *et al.* 2021).

Results from a limited number of studies on the relationship between size of home range and energy demands in reptiles, including snakes, have found contradicting results (Turner *et al.* 1969; Christian and Waldschmidt 1984; Taylor *et al.* 2005; Carfagno and Weatherhead 2008; Christy *et al.* 2017). The results from Sandfoss *et al.* (2021), combined with those studies using supplemental feeding to test the effects of energy availability on size of home range (Taylor *et al.* 2005; Wasko and Sasa 2012), do not support the notion that food availability has a direct effect on the size of home range, particularly in pitvipers.

A majority of the cottonmouths that were tracked in the period following the bird-abandonment died of apparent starvation ($n = 3$) or were in poor body condition at the termination of the study. Without fish carrion to scavenge, cottonmouths must find alternative food sources, and the results of Sandfoss *et al.* (2021) indicate insular cottonmouths are unable or unwilling to increase the size of their home range to do so. Cottonmouths are considered to be ambush predators (Gloyd and Conant 1990), and, for decades, insular cottonmouths have been characterized by their sit-and-wait behavior beneath rookery trees containing waterbird nests (Wharton 1969; Lillywhite and McCleary 2008). Perhaps this behavior overrides the tendency of even starved snakes to wander in search of other resources.

Chronic Stress in the Cottonmouths on Seahorse Key

The sudden cessation of bird nesting on Seahorse Key and partial shift of nesting activities to Snake Key provided an unusual opportunity to assess the effects of resource availability on the stress physiology and immune performance of free-ranging cottonmouth snakes. Food provides animals with the energy required for maintenance, reproduction, growth, and storage. A reduction in food availability can lead to tradeoffs between competing functions and cause animals to divert energy away from essential physiological processes, such as immune function (French *et al.* 2007; Dhabhar 2009). This may be particularly true when an animal is responding to a

A

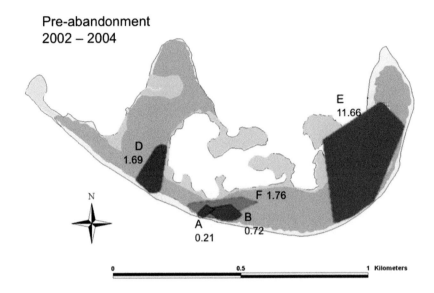

B

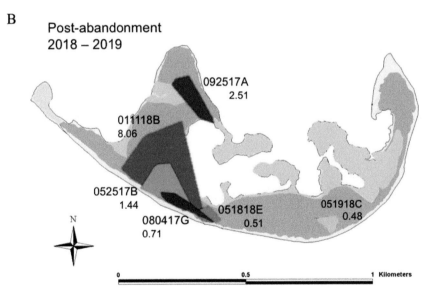

Figure 10.9 Aerial view of estimated home ranges of cottonmouths on Seahorse Key. (A) Minimum complex polygon (MCP) home ranges for five cottonmouths tracked for ~12 months on Seahorse Key during the pre-abandonment period. Each letter represents a unique snake and the numbers correspond to the MCP measured in hectares. (B) MCP home ranges for six cottonmouths tracked for ~12 months on Seahorse Key during the post-abandonment period. Each 7-digit label represents a unique snake and the numbers correspond to the MCP measured in hectares.

stressor. Sandfoss *et al.* (2020) investigated how the loss of food resources on Seahorse Key acts as a stressor to cottonmouths and how this affects aspects of their health and physiology using Snake Key animals for comparison.

Stressors disrupt or disturb an animal's homeostatic balance (Sapolsky *et al.* 2000), and the stress response is thought to allow individuals to ameliorate the effects of a stressor (Romero *et al.* 2009; Wingfield *et al.* 2011). The stress response in reptiles involves physiological and behavioral changes initiated in part by the hypothalamus-pituitary-adrenal axis, which results in several responses including the release of the glucocorticoid hormone, corticosterone (Sapolsky *et al.* 2000). A short-term or acute increase in corticosterone when responding to a stressor is thought to mediate energetic tradeoffs among physiological mechanisms by suppressing processes not immediately critical for survival, thus allowing an individual to endure a stressor (Wingfield 2003; Romero and Wikelski 2010). Traditionally, these acute increases in corticosterone were thought to be adaptive, while prolonged or chronic release of corticosterone was thought to have negative impacts on individual health although what timespan constitutes a chronic response is unclear, particularly in a natural setting (McEwen and Wingfield 2003; Romero *et al.* 2009). Furthermore, complex effects of stress and the release of corticosterone should not be oversimplified, and interpretation of circulating levels of corticosterone appears to be context-dependent (Romero and Beattie 2022). Therefore, it is recommended that corticosterone be used to assess stress when measured with concurrent metrics of physiological health (Breuner *et al.* 2013; Romero and Beattie 2022).

Glucose is commonly measured in studies of starvation to characterize energy mobilization. The blood glucose response to starvation in snakes has been shown to vary among species (McCue *et al.* 2012). To further characterize chronic stress in wild animals, measures of immune capacity, such as natural antibody (NAb) ability, can provide further insight into possible tradeoffs between physiological and immune performance in cottonmouths on Seahorse Key. Sandfoss *et al.* (2020) hypothesized that cottonmouths with limited food resources on Seahorse Key would have reduced physiological and immune performance relative to cottonmouths on Snake Key.

Three years post-abandonment (2018), cottonmouths from Seahorse Key (female $n = 2$, male $n = 4$, mass 323–1,432 g) and Snake Key (female $n = 6$, male $n = 7$, mass 440–2,178 g) were sampled to assess physiological and immune performance (Sandfoss *et al.* 2020). Circulating levels of corticosterone were measured in blood samples collected from individuals immediately upon capture *in situ* ("baseline" blood sample) (Romero and Wikelski 2006) and then following exposure to a 60-min acute stress test consisting of confinement inside a 10-gallon bucket ("acute stress" sample) (Claunch *et al.* 2017a). Corticosterone levels (in ng/mL) of baseline samples were similar for both populations (Seahorse Key $\bar{x} = 15.97 \pm 6.6$, Snake Key $\bar{x} = 14.37 \pm 6.5$), but cottonmouths on Snake Key produced a much larger magnitude stress response (acute stress sample; Seahorse Key $\bar{x} = 76.58 \pm 19.6$, Snake Key $\bar{x} = 138.27 \pm 51.1$) (Sandfoss *et al.* 2020).

There is increasing evidence that baseline plasma corticosterone alone provides little insight into response to chronic stress in pitvipers (Taylor *et al.* 2004; Lind *et al.* 2010; Sandfoss *et al.* 2020). Chronic stress related to food limitation can be a natural experience for pitvipers, and the repetition of such periodic phenomena might have led to development of resistance to increased baseline levels of corticosterone and, ultimately, evolutionary adaptation to limited availability of resources (Boonstra 2013). However, because of the taxonomic consistency and lack of an association between baseline corticosterone and chronic stress in pitvipers thus far examined (Claunch *et al.* 2017b), the observed results may be a taxonomic effect rather than an effect of adaptation to the loss of food. The effects of starvation are complex, and studies have shown that extended periods of fasting may increase or decrease corticosterone depending on the stage of starvation—increased circulating glucocorticoids is thought to be the proximate mechanism of death during the final stage of starvation (McCue 2010; Romero and Wikelski 2010).

Interestingly, cottonmouths from Seahorse Key showed a dampened corticosterone response to acute stress relative to cottonmouths from Snake Key (Sandfoss *et al.* 2021). This finding contrasts with literature on reptiles that reports an inverse relationship between acute corticosterone response and food availability (see Dunlap and Wingfield 1995; Romero 2001; Romero and Wikelski 2010; Neuman-Lee *et al.* 2015). It is also assumed that lower, not higher, plasma corticosterone concentrations during acute stress are associated with greater fitness (Breuner *et al.* 2008). One mechanism for vertebrates to endure chronic stress is to attenuate the stress response by downregulating corticosterone release when encountering acute stressors (Rich and Romero 2005). This strategy may prevent corticosterone itself from disrupting normal functions (Wingfield and Romero 2001) and is partially supported by our findings. However, it cannot be determined if the dampened corticosterone response observed in cottonmouths on Seahorse Key is adaptive or due to impairment from the loss of food resources.

Cottonmouths on Seahorse Key had significantly lower glucose concentrations relative to cottonmouths on Snake Key when responding to acute stress (Sandfoss *et al.* 2020). Previous investigations of glucose levels in reptiles have primarily found glucose values to increase with acute stress (Gangloff *et al.* 2017; Neuman-Lee *et al.* 2019) but see Flower *et al.* (2015). Sparkman *et al.* (2018) suggested that lower glucose concentrations were influenced by limited prey availability in insular relative to mainland populations of *Coluber constrictor* and *Pituophis catenifer*. Interestingly, the viperids tested to date (*Crotalus atrox* and *Bitis gabonica* (McCue *et al.* 2012; Webb *et al.* 2017) experienced a decrease in blood glucose following food restriction, supporting the results of Sandfoss *et al.* (2020).

In agreement with predictions, NAb agglutination ability was better for cottonmouths on Snake Key ($\bar{x} = 6.31 \pm 1.2$) than snakes on Seahorse Key ($\bar{x} = 5.01 \pm 1.3$) (Matson *et al.* 2005; Sandfoss *et al.* 2020). NAb ability is indicative of the ability of animals to effectively respond to a novel antigen (Matson *et al.* 2005) and may be the most ecologically relevant immune measure in ectotherms (Moeller *et al.* 2013). Increased

NAb ability has been linked to increased survival in fish and poultry (Kachamakova *et al.* 2006; Star *et al.* 2007).

There were clear differences in the physiological and immune performance of cottonmouths from the two populations, with snakes from Seahorse Key found to exhibit a dampened stress response, lower glucose values, and reduced immune performance (Sandfoss *et al.* 2020). These responses may demonstrate an energy-saving phenotype for chronic stress following long-term food restriction in cottonmouths on Seahorse Key. It could benefit a food-restricted cottonmouth to dampen its corticosterone responses to acute stress when energy stores are already low; however, chronic stress typically has immuno-suppressive effects (Dhabhar 2009), and the context-dependent nature of stress on the immune system is important to consider when interpreting results.

Insular Cottonmouths in the Cedar Keys: What Does the Future Hold?

The response of insular cottonmouths to the abandonment of Seahorse Key for nesting purposes is complex, and the loss of bird-provided food resources affected multiple aspects of the ecology and physiology of the population. In environments where food resources naturally fluctuate, organisms are commonly food-limited and have to cope with the risk of starvation (*e.g.*, Paulay *et al.* 1985). Food limitation can be one of the strongest environmental pressures that influence behavioral ecology (Holm *et al.* 2019).

Behavioral plasticity of feeding behaviors can allow organisms to survive in variable environments, and foraging strategy can significantly influence starvation tolerance in some species (Holm *et al.* 2019). We did not observe any significant changes in the foraging ecology of cottonmouths on Seahorse Key following the abandonment event. There was no substantial increase in size of home range. And, contrary to predictions, there was no distinguishable difference in diet (isotopic values) as fish remained the primary source of energy and was consistent between the east and west ends of the island. We predicted, and found, that while waterbird nesting has not historically concentrated on Snake Key, previous overflow nesting that occurred on the island appears to have primed cottonmouths on Snake Key to quickly take advantage of newly abundant resources brought by waterbirds.

Foraging strategies are commonly manifested as tradeoffs between maximizing food intake and minimizing metabolic costs (Ydenberg *et al.* 1994). Under limited resources, behaviors that maximize food intake can be energetically too costly and lead to starvation mortality. For example, the Northern Death Adder (*Acanthophis praelongus*) responds to seasonal fluctuations in prey availability in its tropical environment by behaviorally reducing foraging activity during the dry season which, in turn, significantly lowers their field metabolic rate (Christian *et al.* 2007). Physiologically, these snakes do not show a seasonal difference in their resting metabolism, and approximately 94% of the decrease in energy expended in the dry season is simply due to changes in behavior—a decrease in activity and digestion—with lower body temperatures accounting for the remainder (Christian *et al.* 2007). The reliance of cottonmouths on ephemeral allochthonous resources on Seahorse Key means that these snakes were often subjected to periods without food resources when annual bird nesting activities ended; however, this occurred in the winter, when temperatures were lower and when some of the snakes could survive on their stored fat deposits while reducing energy expenditure attributable to activity (Wharton 1969).

One possible physiological response to the loss of food resources observed in the cottonmouth population on Seahorse Key is a reduced stress response. Dampening the acute stress response can be a sign that cottonmouths are conserving energy, but it could also be the result of impaired performance as vital functions shut down without sufficient inputs of energy. Interestingly, while abundance across the island has decreased in the post-abandonment period, smaller snakes (juveniles) were found to be more resilient to decline than larger snakes (adults). Smaller body size in ectothermic organisms can be energetically cheaper to maintain (McNab 1963), and cottonmouths of smaller size may become common on Seahorse Key. Boback and Guyer (2006) analyzed patterns of body size for snake species across islands and found that species of large body size (>1 m) on the mainland will respond to insular habitats, which presumably have reduced availability of food resources, with dwarfism.

Pitvipers seem particularly well-suited for developing an adaptive energy-saving phenotype in response to food restriction because they feed infrequently and have slow life histories (McCue and Lillywhite 2002). Under natural conditions of low food intake, Timber Rattlesnakes (*Crotalus horridus*), also pitvipers, spend more time foraging, experience reduced growth, have lower field metabolic rate, have lower body condition, and do not engage in reproductive activity (Beaupre 2008). Furthermore, Zipkin *et al.*'s (2020) recent study of the response of a tropical snake community to the collapse of amphibian populations found the pitviper *Bothriechis schegelii* to be more abundant following the amphibian collapse although the only other pitviper included in the study, *Bothrops asper*, was not.

Wharton (1958) described how temporal and spatial shifts in waterbird nesting activity dramatically affected the health of individual cottonmouths, and he hypothesized that starvation was the primary cause of mortality for cottonmouths inhabiting Seahorse Key. Based on the results presented here, the majority of snakes on Seahorse Key have been unable to survive multiple active seasons without the energetic input of fish carrion provided by birds.

The colonial bird rookery on Seahorse Key had been one of the largest and more consistent nesting sites for waterbirds in the southeastern United States (Carr 1936; Lillywhite and McCleary 2008; US Fish and Wildlife Service [USFWS], unpublished records). While there is hope that nesting birds will return to Seahorse Key, we do not know whether, or how soon, nesting rookeries might be reestablished. Furthermore, to the extent that cottonmouth snakes might deter potential nest predators and thereby confer a protective advantage to birds nesting at Seahorse Key (Lillywhite and Zaidan 2004; Lillywhite and Sheehy 2019), a significant decrease in the number of snakes might hinder successful nesting and reestablishment of viable rookeries on this island in the future.

While robust nesting colonies of waterbirds occurred historically on Snake Key and Seahorse Key in the 1930s and 1940s (Carr 1936), shifts in concentration of nesting activities have occurred between the Cedar Keys (USFWS unpublished data). Although such changes can last for many years, there is always hope for a return to nesting on Seahorse Key. And, to end this chapter on a positive note with another quote from *Paradise Lost*, "All is not lost" (Milton 1667).

Acknowledgments

I would like to thank Harvey Lillywhite and Marcio Martins for their invitation to contribute this chapter. There are several colleagues with whom I collaborated on this research, and this work could not have been completed without numerous field assistants, visitors, and volunteers that contributed their time to the project. I would also like to acknowledge the staff at the Cedar Keys National Wildlife Refuge and boat captain Kenny McCain. And, last, my work relied heavily on the previous scientific efforts of Charles Wharton and Archie Carr, Jr.

References

Anderson, W. B., D. A. Wait, and P. Stapp. 2008. Resources from another place and time: Responses to pulses in a spatially subsidized system. *Ecology* 89:660–670.

Aubret, F., G. M. Burghardt, S. Maumelat, X. Bonnet, and D. Bradshaw. 2006. Feeding preferences in two disjunct populations of tiger snakes, *Notechis scutatus*. *Austral Ecology* 17:716–725.

Beaupre, S. J. 2008. Annual variation in time-energy allocation by timber rattlesnakes (*Crotalus horridus*) in relation to food acquisition. In W. K. Hayes, K. R. Beaman, M. D. Cardwell, and S. P. Bush (eds.), *The Biology of Rattlesnakes*. Loma Linda University Press, pp. 111–122.

Boback, S. M., and C. Guyer. 2003. Empirical evidence for an optimal body size in snakes. *Evolution* 57:345–351.

Bolnick, D. L., R. Svanbäck, M. S. Araujo, and L. Persson. 2007. Comparative support for the niche variation hypothesis that more generalized populations also are more heterogeneous. *Proceedings of the National Academy of Science* 104:10075–10079.

Bonnet, X., D. Bradshaw, and R. Shine. 1998. Capital versus income breeding: An ectothermic perspective. *Oikos* 83:333–342.

Bonnet, X., and G. Naulleau. 1995. Estimation of body reserves in living snakes using a body condition index. In G. Llorente, X. Santos, and M. A. Carretero (eds.), *Scientia Herpetologica*. Association of Herpetologists Española, pp. 237–240.

Boonstra, R. 2013. Reality as the leading cause of stress: Rethinking the impact of chronic stress in nature. *Functional Ecology* 27:11–23.

Breuner, C. W., B. Delehanty, and R. Boonstra. 2013. Evaluating stress in natural populations of vertebrates: Total CORT is not enough. *Functional Ecology* 27:24–36.

Breuner, C. W., S. H. Patterson, and T. P. Hahn. 2008. In search of relationships between the acute adrenocortical response and fitness. *General Comparative Endocrinology* 157:288–295.

Brown, G., R. Shine, and T. Madsen. 2002. Responses of three sympatric snake species to tropical seasonality in northern Australia. *Journal of Tropical Ecology* 18:549–568.

Burt, W. H. 1943. Territoriality and home range concepts as applied to mammals. *Journal of Mammalogy* 24:346–352.

Carfagno, G. L. F., and P. J. Weatherhead. 2008. Energetics and space use: Intraspecific and interspecific comparisons of movements and home ranges of two colubrid snakes. *Journal of Animal Ecology* 77:416–424.

Carr Jr., A. F. 1936. The Gulf-island cottonmouths. *Proceedings of the Florida Academy of Science* 1:86–90.

Christian, K. A., and S. Waldschmidt. 1984. The relationship between lizard home range and body size: A reanalysis of the data. *Herpetologica* 40:68–75.

Christian, K., J. K. Webb, T. Schultz, and B. Green. 2007. Effects of seasonal variation in prey abundance on field metabolism, water flux, and activity of a tropical ambush foraging snake. *Physiological and Biochemical Zoology* 80:522–533.

Christy, M. T., J. A. Savidge, A. A. Yackel Adams, J. E. Gragg, and G. H. Rodda. 2017. Experimental landscape reduction of wild rodents increases movements in the invasive brown treesnake (*Boiga irregularis*). *Management of Biological Invasions* 8:455–467.

Claunch, N. M., J. A. Frazier, C. Escallon, B. J. Vernasco, I. T. Moore, and E. N. Taylor. 2017a. Physiological and behavioral effects of exogenous corticosterone in a free-ranging ectotherm. *General Comparative Endocrinology* 248:87–96.

Claunch, N. M., M. L. Holding, C. Escallon, B. Vernasco, I. T. Moore, and E. N. Taylor. 2017b. Good vibrations: Assessing the stability of snake venom composition after researcher-induced disturbance in the laboratory. *Toxicon* 133:127–135.

Collins, T. C. 2011. *Cedar Keys Light Station*. Suwannee River Publishing.

Dhabhar, F. S. 2009. Enhancing versus suppressive effects of stress on immune function: Implications for immunoprotection and immunopathology. *Neuroimmunomodulation* 16:300–317.

Dunlap, K. D., and J. C. Wingfield. 1995. External and internal influences on indices of physiological stress. I. Seasonal and population variation in adrenocortical secretion of free-living lizards, *Sceloperus occidentalis*. *Journal of Experimental Zoology* 271:36–46.

Ehrlich, P. 1989. Attributes of invaders and the invading processes: Vertebrates. In J. A. Drake, H. A. Mooney, F. di Castri, R. H. Groves, F. J. Kruger, M. Rejmanek, and M. Williamson (eds.), *Biological Invasions: A Global Perspective*. Wiley, pp. 315–328.

Elton, C. S. 1927. *Animal Ecology*. Sidgwick and Jackson.

Flower, J. E., T. M. Norton, K. M. Andrews, S. E. Nelson, Jr, C. E. Parker, L. M. Romero, and M. A. Mitchell. 2015. Baseline plasma corticosterone, haematological and biochemical results in nesting and rehabilitating loggerhead sea turtles (*Caretta caretta*). *Conservation Physiology* 3:cov003.

Ford, N. B., and R. A. Seigel. 1989. Phenotypic plasticity in reproductive traits: Evidence from a viviparous snake. *Ecology* 70:1768–1774.

Forsman, A., and L. E. Lindell. 1997. Responses of a predator to variation in prey abundance: Survival and emigration of adders in relation to vole density. *Canadian Journal of Zoology* 75:1099–1108.

French, S. S., R. McLemore, B. Vernon, G. I. H. Johnston, and M. C. Moore. 2007. Corticosterone modulation of reproductive and immune systems trade-offs in female tree lizards: Long-term corticosterone manipulations via injectable gelling material. *Journal of Experimental Biology* 210:2859–2865.

Gangloff, E. J., A. M. Sparkman, K. G. Holden, C. J. Corwin, M. Topf, and A. M. Bronikowski. 2017. Geographic variation and within-individual correlations of physiological stress markers in a widespread reptile, the common garter snake (*Thamnophis sirtalis*). *Comparative and Biochemical Physiology A* 205:68–76.

Gloyd, H. K. 1969. Two additional subspecies of North American crotalid snakes, genus Agkistrodon. *Proceedings of the Biological Society of Washington* 82: 219–232.

Gloyd, H. K., and R. Conant. 1990. *Snakes of the* Agkistrodon *Complex: A Monographic Review*. Society for the Study of Amphibians and Reptiles, Contributions to Herpetology 6.

Greenwood, M. J., A. R. McIntosh, and J. S. Hardin. 2010. Disturbance across an ecosystem boundary drives cannibalism propensity in a riparian consumer. *Behavioral Ecology* 21:1227–1235.

Gregory, P. T. 2006. Influence of income and capital on reproduction in a viviparous snake: Direct and indirect effects. *Journal of Zoology* 270:414–419.

Hirai, T. 2004. Dietary shifts of frog eating snakes in response to seasonal changes in prey availability. *Journal of Herpetology* 38:455–460.

Holm, M. W., R. Rodríguez-Torres, B. W. Hansen, and R. Almeda. 2019. Influence of behavioral plasticity and foraging strategy on starvation tolerance of planktonic copepods. *Journal of Experimental Marine Biology and Ecology* 511:19–27.

Holt, R. D. 2008. Theoretical perspectives on resource pulses. *Ecology* 89:671–681.

Jackson, A. L., R. Inger, A. C. Parnell, and S. Bearhop. 2011. Comparing isotopic niche widths among and within communities: SIBER – Stable Isotope Bayesian Ellipses in R. *Journal of Animal Ecology* 80:595–602.

Kachamakova, N. M., I. Irnazarow, H. K. Parmentier, H. F. J. Savelkoul, A. Pilarczyk, and G. F. Wiegertjes. 2006. Genetic differences in natural antibody levels in common carp (*Cyprinus carpio* L.). *Fish and Shellfish Immunology* 21:404–413.

Lillywhite, H. B., and F. Brischoux. 2012. Is it better in the moonlight? Nocturnal activity of insular cottonmouth snakes increases with lunar light levels. *Journal of Zoology* 286:194–199.

Lillywhite, H. B., and M. Martins (eds.). 2019. *Islands and Snakes. Isolation and Adaptive Evolution*. Oxford University Press.

Lillywhite, H. B., and R. J. R. McCleary. 2008. Trophic ecology of insular cottonmouth snakes: A review and perspective. *South American Journal of Herpetology* 3:175–185.

Lillywhite, H. B., J. B. Pfaller, and C. M. Sheehy, III. 2015. Feeding preferences and responses to prey in insular neonatal Florida cottonmouth snakes. *Journal of Zoology* 297:156–163.

Lillywhite H. B., and C. M. Sheehy, III. 2019. The unique insular population of cottonmouth snakes at Seahorse Key. In H. B. Lillywhite and M. Martins (eds.), *Islands and Snakes: Isolation and Adaptive Evolution*. Oxford University Press, pp. 201–240.

Lillywhite H. B., C. M. Sheehy, III, and M. D. McCue. 2002. Scavenging behaviors of cottonmouth snakes at island bird rookeries. *Herpetological Review* 33:259–261.

Lillywhite, H. B., C. M. Sheehy, III, and F. Zaidan, III. 2008. Pitviper scavenging at the intertidal zone: An evolutionary scenario for invasion of the sea. *Bioscience* 58:947–955.

Lillywhite, H. B., and F. Zaidan, III. 2004. Mutualism between birds and snakes at Seahorse Key, Florida. *Abstracts of the Annual Meeting of the Society for Integrative and Comparative Biology*, p. 302.

Lind, C. M., and S. J. Beaupre. 2015. Male snakes allocate time and energy according to individual energetic status: Body condition, steroid hormones, and reproductive behavior in timber rattlesnakes, *Crotalus horridus*. *Physiological and Biochemical Zoology* 88:624–633.

Lind, C. M., J. F. Husak, C. Eikenaar, I. T. Moore, and E. N. Taylor. 2010. The relationship between plasma steroid hormone concentrations and the reproductive cycle in the norther Pacific rattlesnake, *Crotalus oreganus*. *General Comparative Endocrinology* 166:590–599.

Luiselli, L. 2006. Interspecific relationships between two species of sympatric Afrotropical water snake in relation to a seasonally fluctuating food resource. *Journal of Tropical Ecology* 22:91–100.

Madsen, T., and R. Shine. 1993. Costs of reproduction in a population of European adders. *Oecologia* 94:488–495.

Madsen, T., and R. Shine. 1999. The adjustment of reproductive threshold to prey abundance in a capital breeder. *Journal of Animal Ecology* 68:571–580.

Malthus, T. R. 1798. *An Essay on the Principle of Population as It Affects the Future Improvement of Society*. J. Johnson.

Marques, O. A., M. Martins, P. F. Develey, A. Macarrao, and I. Sazima. 2012. The golden lance-head *Bothrops insularis* (Serpentes: Viperidae) relies on two seasonally plentiful bird species visiting its island habitat. *Journal of Natural History* 46:885–895.

Mathot, K. J., J. Wright, B. Kempenaers, and N. J. Dingemanse. 2012. Adaptive strategies for managing uncertainty may explain personality-related differences in behavioral plasticity. *Oikos* 121:1009–1020.

Matson, K. D., R. E. Ricklefs, and K. C. Klasing. 2005. A hemolysis-hemagglutination assay for characterizing constitutive innate humoral immunity in wild and domestic birds. *Developmental and Comparative Immunology* 29:275–286.

McCarthy, K. 2007. *Cedar Key, Florida. A History*. History Press Library Editions.

McCue, M. D. 2010. Starvation physiology: Reviewing the different strategies animals use to survive a common challenge. *Comparative and Biochemical Physiology* 156A:1–18.

McCue, M. D., and H. B. Lillywhite. 2002. Oxygen consumption and the energetics of island-dwelling Florida cottonmouth snakes. *Physiological and Biochemical Zoology* 75:165–178.

McCue, M. D., H. B. Lillywhite, and S. J. Beaupre. 2012. Physiological responses to starvation in snakes: Low energy specialists. In M. D. McCue (ed.), *Comparative Physiology of Fasting, Starvation, and Food Limitation*. Springer-Verlag, pp. 103–132.

McCue, M. D., and E. D. Pollock. 2008. Stable isotopes may provide evidence for starvation in reptiles. *Rapid Communications in Mass Spectrometry* 22:2307–2314.

McEwen, B. S., and J. C. Wingfield. 2003. The concept of allostasis in biology and biomedicine. *Hormones and Behavior* 43:2–15.

McNab, B. K. 1963. Bioenergetics and the determination of home range size. *American Naturalist* 97:133–140.

Milton, J. 1968. *Paradise Lost, 1667*. Scolar Press.

Moeller, K. T., M. W. Butler, and D. F. DeNardo. 2013. The effect of hydration state and energy balance on innate immunity of a desert reptile. *Frontiers in Zoology* 10:23.

Mohr, C. O. 1947. Table of equivalent populations of North American small mammals. *American Midland Naturalist* 37:223–449.

Naulleau, G., and X. Bonnet. 1996. Body condition threshold for breeding in a viviparous snake. *Oecologia* 107:301–306.

Neuman-Lee, L. A., H. B. Fokidis, A. R. Spence, M. Van der Walt, G. D. Smith, S. Durham, and S. S. French. 2015. Food restriction and chronic stress alter energy use and affect immunity in an infrequent feeder. *Functional Ecology* 29:1453–1462.

Neuman-Lee, L. A., S. B. Hudson, A. C. Webb, and S. S. French. 2019. Investigating the relationship between corticosterone and glucose in a reptile. *Journal of Experimental Biology* 2019:jeb.203885.

Nowak, E. M., T. C. Theimer, and G. W. Schuett. 2008. Functional and numerical responses of predators: Where do vipers fit in the traditional paradigms? *Biological Reviews* 83:601–620.

Parnell, A. C., D. L. Phillips, S. Bearhop, B. X. Semmens, E. J. Ward, J. W. Moore, A. L. Jackson, J. Grey, *et al.* 2013. Bayesian stable isotope mixing models. *Environmetrics* 24:387–399.

Paulay, G., L. Boring, and R. R. Strathmann. 1985. Food limited growth and development of larvae: Experiments with natural sea water. *Journal of Experimental Marine Biology and Ecology* 93:1–10.

Petrov, K., R. J. Spencer, N. Malkiewicz, J. Lewis, C. Keitel, and J. U. Van Dyke. 2020. Prey-switching does not protect a generalist turtle from bioenergetic consequences when its preferred food is scarce. *BMC Ecology* 20:11.

Pilgrim, M. A. 2005. Linking Microgeographic Variation in Pigmy Rattlesnake (*Sistrurus miliarius*) Life History and Demography with Diet Composition: A Stable Isotope Approach. PhD thesis, University of Arkansas.

Quevedo, M., R. Svanbäck, and P. Eklo. 2009. Intrapopulation niche partitioning in a generalist predator limits food web connectivity. *Ecology* 90:2263–2274.

Rich, E. L., and L. M. Romero. 2005. Exposure to chronic stress downregulates corticosterone responses to acute stressors. *America Journal of Physiology* 288:R1628–R1636.

Romero, L. M. 2001. Mechanisms underlying seasonal differences in the avian stress response. In A. Dawson and C. M. Chaturvedi (eds.), *Avian Endocrinology*. Narosa Publishing House, pp. 373–384.

Romero, L. M., and U. K. Beattie. 2022. Common myths of glucocorticoid function in ecology and conservation. *Journal of Experimental Zoology* A 337:7–14.

Romero, L. M., M. J. Dickens, and N. E. Cyr. 2009. The reactive scope model–a new model integrating homeostasis, allostasis, and stress. *Hormones and Behavior* 55:375–389.

Romero, L. M., and M. Wikelski. 2006. Diurnal and nocturnal differences in hypothalamic-pituitary-adrenal axis function in Galápagos marine iguanas. *General Comparative Endocrinology* 145:177–181.

Romero, L. M., and M. Wikelski. 2010. Stress physiology as a predictor of survival in Galapagos marine iguanas. *Proceedings of the Royal Society B* 277:3157–3162.

Ruffino, L., J. C. Russell, B. Pisanu, S. Caut, and E. Vidal. 2011. Low individual-level dietary plasticity in an island-invasive generalist forager. *Population Ecology* 53:535–548.

Sandfoss, M. R., N. M. Claunch, N. I. Stacy, C. M. Romagosa, and H. B. Lillywhite. 2020. A tale of two islands: Evidence for reduced stress response and impaired immune functions in an insular pit-viper following ecological disturbance. *Conservation Physiology* 8:coaa031.

Sandfoss, M. R., and H. B. Lillywhite. 2019. Water relations of an insular pit viper. *Journal of Experimental Biology* 222:jeb204065.

Sandfoss, M. R., M. D. McCue, and H. B. Lillywhite. 2021. Trophic niche and home range of an insular pit viper following loss of food resources. *Journal of Zoology* 314:296–310.

Sandfoss, M. R., C. M. Sheehy III, and H. B. Lillywhite. 2018. Collapse of a unique insular bird-snake relationship. *Journal of Zoology* 304:276–283.

Sapolsky, R. M., L. M. Romero, and A. U. Munck. 2000. How do glucocorticoids influence stress responses? Integrating permissive, suppressive, stimulatory, and preparative actions. *Endocrinology Review* 21:55–89.

Seigel, R. A., and H. S. Fitch. 1985. Annual variation in reproduction in snakes in a fluctuating environment. *Journal of Animal Ecology* 54:497–505.

Sheehy III, C. M., M. R. Sandfoss, and H. B. Lillywhite. 2017. Cannibalism and changing food resources in insular cottonmouth snakes. *Herpetological Review* 48:310–312.

Shine, R. 1980. "Cost" of reproduction in reptiles. *Oecologia* 46:92–100.

Shine, R., L. X. Sun, E. Zhao, and X. Bonnet. 2002. A review of 30 years of ecological research on the Shedao piviper, *Gloydius shedaoensis*. *Herpetological Natural History* 9:1–14.

Sparkman, A. M., A. D. Clark, L. J. Brummett, K. R. Chism, L. L. Combrink, N. M. Kabey, and T. S. Schwartz. 2018. Convergence in reduced body size, head size, and blood glucose in three island reptiles. *Ecology and Evolution* 8:6169–6182.

Stapp, P., and G. A. Polis. 2003. Marine resources subsidize insular rodent populations in the Gulf of California, Mexico. *Oecologia* 134:496–504.

Star, L., K. Frankena, B. Kemp, M. G. B. Nieuwland, and H. K. Parmentier. 2007. Innate immune competence and survival in pure bred layer lines. *Poultry Science* 86:1090–1099.

Taylor, E. N., D. F. DeNardo, and D. H. Jennings. 2004. Seasonal steroid hormone levels and their relation to reproduction in the western diamond-backed rattlesnake, *Crotalus atrox* (Serpentes: Viperidae). *General Comparative Endocrinology* 136:328–337.

Taylor, E. N., M. A. Malawy, D. M. Browning, S. V. Lemar, and D. F. DeNardo. 2005. Effects of food supplementation on the physiological ecology of female western diamond-backed rattlesnakes (*Crotalus atrox*). *Oecologia* 144:206–213.

Turner, F. B., R. I. Jennrich, and J. D. Wein-Traub. 1969. Home ranges and body size of lizards. *Ecology* 50:1076–1081.

Ujvari, B., R. Shine, and T. Madsen. 2011. How well do predators adjust to climate-mediated shifts in prey distribution? A study on Australian water pythons. *Ecology* 92:777–783.

Vasquez, D. P. 2005. Exploring the relationship between invasion success and niche breadth. In M. W. Cadotte, S. M. McMahon, and T. Fukami (eds.), *Conceptual Ecology and Invasions Biology*. Springer, pp. 317–332.

Wasko, D. K., and M. Sasa. 2012. Food resources influence spatial ecology, habitat selection, and foraging behavior in an ambush-hunting snake (Viperida: *Bothrops asper*): An experimental study. *Zoology* 115:179–187.

Webb, A. C., L. D. Chick, V. A. Cobb, and M. Klukowski. 2017. Effects of moderate food deprivation on plasma corticosterone and blood metabolites in common watersnakes (*Nerodia sipedon*). *Journal of Herpetology* 51:134–141.

Wharton, C. H. 1958. The Ecology of the Cottonmouths *Agkistrodon piscivorus piscivorus* Lacepede of Sea Horse Key, Florida. PhD dissertation, University of Florida.

Wharton, C. H. 1966. Reproduction and growth in the cottonmouths, *Agkistrodon piscivorus* Lacépède, of Cedar Keys, Florida. *Copeia* 1966:149–161.

Wharton, C. H. 1969. The cottonmouth moccasin on Sea Horse Key, Florida. *Bulletin of the Florida State Museum* 14:227–272.

Wingfield, J. C. 2003. Control of behavioral strategies for capricious environments. *Animal Behavior* 66:807–816.

Wingfield, J. C., J. P. Kelley, F. Angelier, O. Chastel, F. Lei, S. E. Lynn, B. Miner, *et al.* 2011. Organism–environment interactions in a changing world: A mechanistic approach. *Journal of Ornithology* 152:279–288.

Wingfield, J. C., and L. M. Romero. 2001. Adrenocortical responses to stress and their modulation in free-living vertebrates. In B. S. McEwen and H. M. Goodman (eds.), *Handbook of Physiology: Section 7: The Endocrine System; Volume IV: Coping with the Environment: Neural and Endocrine Mechanisms*. Oxford University Press, pp. 211–234.

Ydenberg, R. C., C. V. J. Welham, R. Schmid-Hempel, P. Schmid-Hempel, and G. Beauchamp. 1994. Time and energy constraints and the relationships between currencies in foraging theory. *Behavioral Ecology* 5:28–34.

Zipkin, E. F., G. V. Di Renzo, J. M. Ray, S. Rossman, and K. R. Lips. 2020. Tropical snake diversity collapses after widespread amphibian loss. *Science* 367:814–816.

11

Marooned Snakes

Ecology of Freshwater *Natrix maura* Isolated on the Atlantic Islands of Galicia (Spain)

Pedro Galán and François Brischoux

Ons Island from the Atlantic Islands of Galicia

The Atlantic Islands of Galicia are formed by several archipelagos—the main ones being Sálvora, Ons, and Cíes—situated along the northwestern coast of Spain (Figure 11.1). Rising sea levels occurring after the Last Glacial Period isolated these islands from mainland Spain approximately 8,000 years ago (Dias *et al.* 2000).

Among these islands, the Ons archipelago is formed by a main island, Ons, which is the largest of the Atlantic Islands of Galicia (4.14 km²), and a small islet, Onza (0.32 km²), separated from Ons by 650 m (Figure 11.1). The isolation of the Ons archipelago from mainland Spain has led to the geographic separation (distance to the continent ~3.5 km) of populations of numerous species from their continental counterparts (Vilas *et al.* 2005; Bernárdez-Villegas 2006; Mouriño *et al.* 2007), including a relatively rich herpetofauna (Galán 2003). For instance, the amphibian community of Ons Island includes three species: the Fire Salamander (*Salamandra salamandra*), the Iberian Newt (*Lissotriton boscai*), and the Iberian Painted Frog (*Discoglossus galganoi*). Although the presence of the two latter species seems to be restricted to the proximity of freshwater springs (Figure 11.1, Cordero-Rivera *et al.* 2007; Galán 2003, 2012), the Fire Salamander can be found across the entire island (Galán 2003, Velo-Antón *et al.* 2007). The reptilian community of Ons Island is composed of the Lusitanian Wall Lizard (*Podarcis lusitanicus*), the Ocellated Lizard (*Timon lepidus*), the Western Three-toed Skink (*Chalcides striatus*), the Glass Lizard (*Anguis fragilis*), the Ladder Snake (*Zamenis scalaris*), the Southern Smooth Snake (*Coronella girondica*), and the Viperine Snake (*Natrix maura*) (Galán 2003).

The Viperine Snake, *Natrix maura*

The viperine Snake (*N. maura*) is a semi-aquatic freshwater natricine widely distributed across Western Europe and Northern Africa, from France to Morocco (Miras *et al.* 2015). This relatively small-sized species (up to ~80 cm total length) typically

Pedro Galán and François Brischoux, *Marooned Snakes* In: *Islands and Snakes*. Edited by: Harvey B. Lillywhite and Marcio Martins, Oxford University Press. © Oxford University Press 2023. DOI: 10.1093/oso/9780197641521.003.0011

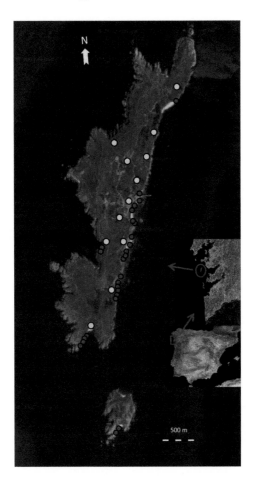

Figure 11.1 Maps of Ons Archipelago and its general location on the northwestern coast of Spain. Insets show the location of Ons Archipelago relative to Spain. The main map shows the relatively large Ons Island and the smaller Onza islet separated by ~650 m. Red dots indicate locations of Viperine Snake observations and/or captures during several field trips (see also Galán 2012). Blue dots indicate the locations of freshwater springs where two amphibian species occur (Iberian Newt, *Lissotriton boscai,* and Iberian Painted Frog, *Discoglossus galganoi*).
Pictures modified from GoogleEarth.

forages for fish and amphibians in aquatic environments such as streams, rivers, marshes, and lakes (Miras *et al.* 2015; Santos and Llorente 2009).

Among the organisms that were marooned on the Ons archipelago, Viperine Snakes are remarkable in their foraging ecology. On Ons Island, aquatic ecosystems are scarce and restricted to few freshwater springs (Figure 11.1), most of which have been transformed because of past human activity related to clothes washing. These environmental conditions contrast with those available on the continent, where Viperine Snakes rely on freshwater environments to forage for aquatic prey. In this chapter, we provide complementary ecological data on the Viperine Snake population from Ons Island to illustrate their shifts in foraging ecology.

Distribution of Viperine Snakes on Ons Island

In continental populations, amphibians are a dominant part of the diet of Viperine Snakes (Miras *et al.* 2015; Santos and Llorente 2009; Lemaire *et al.* 2018). On Ons Island, most of the freshwater sources shelter some amphibians (*i.e.*, Iberian Newts and Iberian Painted Frogs). In addition, Fire Salamanders can be found across the island because of their independence from water bodies (Ons Fire Salamanders are remarkably pueriparous; Velo-Antón *et al.* 2007; Mulder *et al.* 2022). In contrast, the distribution of the Viperine Snake is largely restricted to the seashore, especially along the eastern coast of the island (Figure 11.1). It is important to stress that our searching effort was not focused on coastal areas solely and that the entire island was searched for Viperine Snakes (including freshwater springs) during field trips that regularly took place in spring and summer since the early 1980s (Galán 2003, 2004, 2012; Lemaire *et al.* 2018). The distribution of insular Viperine Snakes suggests an independence from freshwater bodies and a spatial disconnection from their potential amphibian prey.

This peculiar spatial distribution of Viperine Snakes, along with field observations made by the authors, further suggest that this insular population does not forage on amphibian prey and that isolation on Ons Island has been accompanied by a shift in their foraging ecology. Indeed, Viperine Snakes are often observed swimming in tidal pools (Figure 11.2) and presumably foraging for marine fish that are ensconced in such pools. Accordingly, during the course of our surveys, an individual regurgitated a recently caught blennid (Shanny, *Lipophrys pholis*, Figure 11.3; Galán 2004, 2012). Although such observation remains anecdotal, the spatial distribution of Viperine Snakes, the numerous observations of individuals apparently foraging (moving slowly, swimming, with frequent tongue-flicks and crevice probing) in tidal pools, and the abundance of Blenniidae in such environments (Shanny, *Lipophrys pholis* and Tompot Blenny, *Parablennius gattorugine*, Figure 11.3) clearly suggest that insular Viperine Snakes forage for blennies in the intertidal area.

This conclusion is further supported by the peculiar distribution of Viperine Snakes along the coasts of Ons Island. Indeed, most snakes were observed on the eastern shore while very few individuals have been captured on the western coast of

Figure 11.2 Photograph of a Viperine Snake (*Natrix maura*) swimming—presumably foraging—in a tidal pool on the eastern coast of Ons Island. In contrast to the continental populations of this species, insular snakes of the Atlantic Islands of Galicia forage for Blenniidae in such pools.
Photo by Pedro Galán.

the island (Figure 11.1). The topography of the island, with steep cliffs on the western coast versus gradual slopes and beaches on the eastern shore, is likely to influence fishing success of Viperine Snakes. Fish are arguably relatively fast-moving and difficult to catch in open waters, and Viperine Snakes seem to specifically target species that are ensconced in tidal pools, which should increase their foraging success.

Taken together, these observations suggest that insular populations of Viperine Snakes have shifted from foraging on freshwater amphibian prey to foraging on marine fish. This shift in prey corresponds to their distribution along the eastern shore, where tidal pools might facilitate the capture of these fast-moving fish.

Complementary Evidences for Foraging at Sea

In addition to these observations, we have collected two independent sets of evidences that support foraging for marine prey in this insular population of Viperine Snakes. First, because seawater is hyperosmotic relative to the internal fluids of most vertebrates, contact with seawater can induce water loss and salt gain across permeable surfaces (Schmidt-Nielsen 1983). Additionally, incidental drinking of seawater (*e.g.*, during prey capture) will impose a supplementary salt load (Costa 2002; Houser *et al.* 2005). As a consequence, foraging in seawater entails a significant risk of salt load and/or dehydration (Brischoux and Kornilev 2014; Brischoux *et al.* 2013, 2017; Lillywhite *et al.* 2012, 2014a, 2014b; Rash and Lillywhite 2019).

Figure 11.3 Photographs of identified prey of insular Viperine Snakes (*Natrix maura*) from Ons Island. Upper photo: Shanny, *Lipophrys pholis*. Lower photo: Tompot Blenny, *Parablennius gattorugine*. Both species are Blenniidae and are relatively abundant in the tidal pools where snakes forage.

Photos by Pedro Galán.

To test whether the peculiar foraging ecology of insular Viperine Snakes affects their osmotic balance, we collected blood samples from 10 individuals caught on the eastern shore and assessed their plasmatic osmolality (blood was centrifuged and osmolality [mOsmol.kg^{-1}] was measured from 10 µL aliquots on a Vapro2 osmometer [Elitech group, France]). We found that all individuals had osmolality above normosmolality in snakes (Figure 11.4). Indeed, mean osmolality of snakes from Ons was 342.2 ± 11.7 mOsm.kg^{-1} (max 362 mOsm.kg^{-1}, Figure 11.4), a value that exceeds normosmolality in freshwater snakes (~310 mOsm.kg^{-1}; Brischoux *et al.* 2017). These data support the hypothesis that insular Viperine Snakes forage at sea for fish and that contact with seawater and/or incidental drinking during prey capture induce a significant salt load. In addition, our result accords well with studies performed in a similar context on a closely related species, the Dice Snake (*Natrix tessellata*, Brischoux and Kornilev 2014; Brischoux *et al.* 2017). Dice Snakes foraging in the Black Sea similarly displayed significant salt loads (Brischoux and Kornilev 2014; Brischoux *et al.* 2017). It remains to be assessed whether such salt gain and thus elevated osmolality of body fluids can have detrimental physiological consequences for insular Viperine Snakes. Climatic conditions on Ons Island, specifically frequent rainfall (Rodríguez-Guitián and Ramil-Rego 2007), may allow these

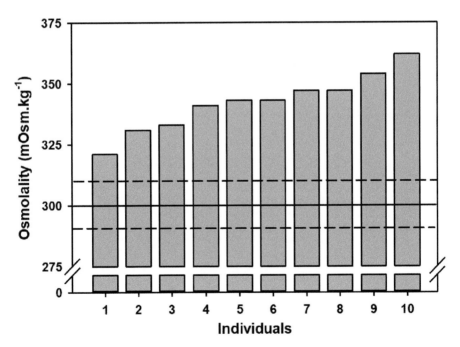

Figure 11.4 Plasmatic osmolality (mOsm.kg^{-1}) of 10 free-ranging individual *Natrix maura* captured on the eastern coast of Ons Island. The dashed lines indicate the range of normosmolality and the horizontal black line indicates mean normosmolality in freshwater snakes (see Brischoux *et al.* 2017). All individuals have osmolality values that are above normosmolality, indicating that foraging at sea induce salt gain and/or dehydration. For clarity, individuals are ranked by ascending order of osmolality.

snakes to regularly drink freshwater in order to equilibrate their osmotic balance (Brischoux *et al.* 2017; Dezetter *et al.* 2022).

Second, marine fish are characterized by relatively elevated mercury (Hg) concentrations as compared to freshwater species (Hansson *et al.* 2017). Because Hg bioaccumulates within organisms and biomagnifies through the trophic webs, it is expected that snakes foraging on marine prey should display higher Hg concentrations than those foraging in freshwater environments. Because Hg binds to keratins (Hopkins *et al.* 2013), we collected scale clips from 13 individuals caught on the eastern shore and assessed their Hg concentrations (see Lemaire *et al.* 2018 for details). We found that insular Viperine Snakes had significantly higher Hg concentrations as compared to individuals ($N = 26$) captured in a coastal marsh along the Atlantic coast of France (t-test, $t = -3.96$, $p < 0.001$, Figure 11.5). Indeed, the mean Hg concentration of snakes from Ons was 0.43 ± 0.24 $\mu g.g^{-1}$ (max 0.78 $\mu g.g^{-1}$, Figure 11.5), a value greater than Hg concentration found in continental individuals (0.18 ± 0.16 $\mu g.g^{-1}$, maximum 0.66 $\mu g.g^{-1}$, Figure 11.5; see also Lemaire *et al.* 2018). In addition to these higher Hg concentrations in insular Viperine Snakes, we found that continental individuals were characterized

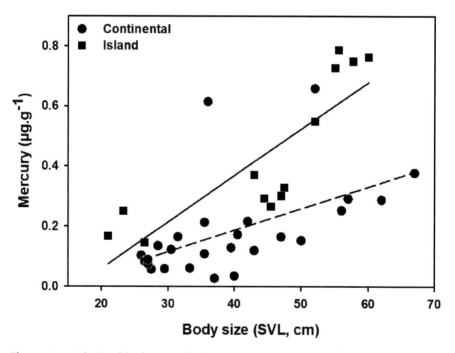

Figure 11.5 Relationships between body size (snout to vent length [SVL] cm) and Hg concentrations ($\mu g.g^{-1}$) measured in scale-clips of individual *Natrix maura* from coastal marsh from the Atlantic coast of France (circles) and from insular snakes from Ons Island (squares). Insular snakes display both higher Hg concentrations and a steeper slope of the body size–Hg relationship (slope of 0.007 for continental individuals vs. 0.015 for insular snakes; see Lemaire *et al.* 2018 for statistical details).
Modified from Lemaire *et al.* 2018.

not only by lower mean values of Hg concentrations, but also by a weaker slope of the body size–Hg relationship as compared to insular Viperine Snakes (Figure 11.5, see Lemaire *et al.* 2018 for details), suggesting that differences in Hg accumulation rates are attributable to food resources (Lemaire *et al.* 2018). Possible toxic effects of Hg (Tan *et al.* 2009; Scheuhammer *et al.* 2015) in insular Viperine Snakes deserve further investigation.

Phenotype of Insular Viperine Snakes

Finally, we explored whether the phenotype of the insular population of Viperine Snakes from Ons Island differed from their continental counterparts. During the course of several field trips on Ons Island and continental Galicia, we measured the body size (snout to vent length [SVL]) of $N = 25$ individuals from Ons Island, a single individual ($N = 1$) from Onza islet, and $N = 57$ from 18 continental sites spread across Galicia. We found that individuals from Ons Island were significantly larger than continental snakes (t-test, t = 3.12, p = 0.002, Figure 11.6),

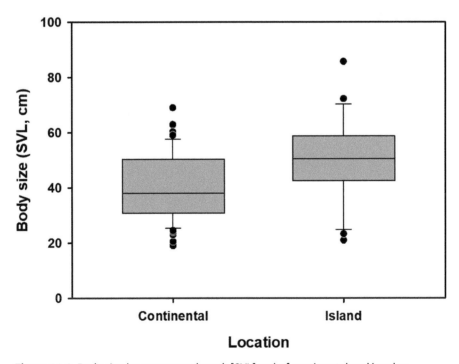

Figure 11.6 Body size (snout to vent length [SVL] cm) of continental and insular Viperine Snakes. Body size of the smallest individuals is similar, but insular individuals reach larger body size than their continental counterparts. The top and bottom of the boxes represent the first and last quartiles, the line across the box represents the median, the whiskers represent the fifth and ninety-fifth percentiles, and the circles represent outliers (see text for statistical differences).

Figure 11.7 Photograph of a large female Viperine Snake from Onza islet. The coloration of insular individuals is darker than that of continental populations, and such melanism seems to increase with body size (see Figure 11.2 for a comparison with a smaller individual).
Photo by Pedro Galán.

an effect that was found in both males and females (see below). Interestingly, this effect was not likely to be linked to differences in size at birth as the body size of the smallest individuals from both populations was similar (minimum SVL = 19 cm for a continental individual versus 21 cm for an insular snake, Figure 11.6). Conversely, this difference was obvious in larger individuals of both sexes (females: max SVL = 69 cm for a continental individual versus 85 cm for an insular snake, males: max SVL = 52 cm for a continental individual versus 61.5 cm for an insular snake; Figure 11.6). Such increased body size in insular Viperine Snakes might be related to the diet alteration hypothesis (Boback 2003) and seems to resemble the situation highlighted for Insular Tiger Snake (*Notechis scutatus*) in Australia (reviewed in Aubret 2019).

Insular Viperine Snakes seem to be characterized by a different color pattern than their continental counterparts, although this impression is anecdotal and without quantitative verification. Indeed, insular individuals are arguably darker (Figure 11.7) as compared to the zigzag dorsal pattern or the striped dorsal patterns found in continental populations (pictures available in Santos *et al.* 2018). This dichotomy seems especially the case in larger—hence older—individuals (Figure 11.7), while the zigzag dorsal pattern can still be detectable in smaller snakes (Figure 11.2). Such ontogenic melanism has been documented in several snake species (Lorioux *et al.* 2008; Lillywhite and Sheehy 2019) and the underlying significance—whether linked to thermoregulation, antipredation, protection against sun damage, industrial melanism, or indeed a combination thereof (Lorioux *et al.* 2008; Goiran *et al.* 2017)—for insular Viperine Snakes remains to be investigated (Santos *et al.* 2018).

Conclusion

Overall, our complementary sets of data show that the relatively recent isolation of Viperine Snakes (*N. maura*) on the Atlantic Islands of Galicia has profoundly modified their ecology. Most notably, our data show that Viperine Snakes occur almost exclusively close to the sea (rather than around the few freshwater sources spread on the islands) and forage on marine fish (rather than on amphibians occurring in freshwater sources). Physiological markers of foraging in seawater (osmolality and Hg concentration) confirm these field observations. Our results also suggest that the peculiar insular ecology of this species may have influenced body size and coloration. Taken together, these different elements suggest that insular Viperine Snakes are a promising model organism by which to investigate evolutionary issues such as those linked to the transition to marine life in vertebrates and/or those linked to insular isolation and ecology.

Acknowledgments

We thank all the people who were involved in snake surveys, the staff of the Parque Nacional das Illas Atlanticas de Galicia on Ons Island and especially Luis Martínez and Gustavo Cochon. We also thank José Antonio Fernández Bouzas and Vicente Piorno (Parque Nacional das Illas Atlanticas de Galicia) for providing research permits and helping with logistics during field work. Funding was provided by the CNRS.

References

Aubret, F. 2019. Pleasure and pain: Island tiger snakes and sea-birds in Australia. In H. B. Lillywhite and M. Martins (eds.), *Islands and Snakes: Isolation and Adaptive Evolution.* Oxford University Press, 138–155.

Bernárdez-Villegas, J. G. 2006. *Estudio florístico de la Isla de Ons. Parque Nacional Marítimo-Terrestre de las Islas Atlánticas de Galicia. Naturaleza y Parques Nacionales.* Serie Técnica.

Boback, S. M. 2003. Body size evolution in snakes: Evidence from island populations. *Copeia* 2003:81–94.

Brischoux, F., M. J. Briand, G. Billy, and X. Bonnet. 2013. Variations of natremia in sea kraits (*Laticauda* spp.) kept in seawater and fresh water. *Comparative Biochemistry and Physiology Part A* 166:333–337.

Brischoux, F., and Y. Kornilev. 2014. Hypernatremia in dice snakes (*Natrix tessellata*) from a coastal population: Implications for osmoregulation in marine snake prototypes. *PLoS ONE* 9:e9261.

Brischoux, F., Y. Kornilev, and H. B. Lillywhite. 2017. Physiological and behavioral responses to salinity in coastal dice snakes. *Comparative Biochemistry and Physiology Part A* 214:13–18

Cordero-Rivera, A., G. Velo-Antón, and P. Galán. 2007. Ecology of amphibians in small Holocene islands: Local adaptations and the effect of exotic tree plantations. *Munibe* (Suplemento) 25:94–103.

Costa, D. P. 2002. Osmoregulation. In W. F. Perrin *et al.* (eds.), *Encyclopedia of Marine Mammals*. Academic Press, pp. 337–342.

Dezetter, M., J.-F. Le Galliard, M. Leroux-Coyau, F. Brischoux, F. Angelier, and O. Lourdais. 2022. Two stressors are worse than one: Combined heatwave and drought affect hydration state and glucocorticoid levels in a temperate ectotherm. *Journal of Experimental Biology* 225:jeb243777.

Dias, J. M. A., T. Boski, A. Rodrigues, and F. Magalhaes. 2000. Coastal line evolution in Portugal since the Last Glacial Maximum until present: A synthesis. *Marine Geology* 170:177–186.

Galán, P. 2003. *Anfibios y reptiles del Parque Nacional de las Islas Atlánticas de Galicia. Faunística, biología y conservación. Naturaleza y Parques Nacionales.* Serie Técnica. Ministerio de Medio Ambiente. Organismo Autónomo de Parques Nacionales.

Galán, P. 2004. *Natrix maura* (viperine snake): Marine inhabitation. *Herpetological Review* 35(1):71.

Galán P. 2012. *Natrix maura* en el medio marino de las Islas Atlánticas de Galicia. *Boletín de la Asociación Herpetológica Española* 23:38–43.

Goiran, C., P. Bustamante, and R. Shine. 2017. Industrial melanism in the seasnake *Emydocephalus annulatus. Current Biology* 27:2510–2513.

Hansson, S. V., J. Sonke, D. Galop, G. Bareille, S. Jean, and G. Roux. 2017. Transfer of marine mercury to mountain lakes. *Scientific Report* 7(1):12719.

Hopkins, B. C., M. J. Hepner, and W. A. Hopkins. 2013. Non-destructive techniques for bio-monitoring of spatial, temporal, and demographic patterns of mercury bioaccumulation and maternal transfer in turtles. *Environmental Pollution* 177:164–170.

Houser, D. S., D. E. Crocker, and D. P. Costa. 2005. Ecology of water relations and thermoregulation. *eLS published online.* http://dx.doi.org/10.1038/npg.els.0003216.

Lemaire, J., P. Bustamante, A. Olivier, O. Lourdais, B. Michaud, A. Boissinot, P. Galán, and F. Brischoux. 2018. Determinants of mercury contamination in viperine snakes, *Natrix maura*, in Western Europe. *Science of the Total Environment* 635:20–25.

Lillywhite, H. B., F. Brischoux, C. M. Sheehy III, and J. B. Pfaller. 2012. Dehydration and drinking responses in a pelagic sea snake. *Integrative and Comparative Biology* 52:227–234.

Lillywhite, H. B., H. Heatwole, and C. M. Sheehy III. 2014a. Dehydration and drinking behaviour of the marine file snake *Acrochordus granulatus. Physiological and Biochemical Zoology* 87:46–55.

Lillywhite, H. B., and C. M. Sheehy III. 2019. The unique insular population of cottonmouth snakes at Seahorse Key. In H. B. Lillywhite and M. Martins (eds.), *Islands and Snakes. Isolation and Adaptive Evolution*. Oxford University Press, pp. 201–240.

Lillywhite, H. B., C. M. Sheehy III, F. Brischoux, and A. Grech. 2014b. Pelagic sea snakes dehydrate at sea. *Proceedings of the Royal Society B* 281:2014119.

Lorioux, S., X. Bonnet, F. Brischoux, and M. De Crignis. 2008. Is melanism adaptive in sea kraits? *Amphibia-Reptilia* 29:1–5.

Miras, J. A. M., M. Cheylan, M. S. Nouira, U. Joger, P. Sá-Sousa, V. Pérez-Mellado, B. Schmidt, *et al.* 2015. *Natrix maura.* IUCN Red List of Threatened Species 2009. https://doi.org/10.2305/IUCN.UK.2009.RLTS.T61538A12510365.en

Mouriño, J., F. Arcos, J. C. Aldariz, and R. Salvadores. 2007. *Guía da fauna terrestre do Parque Nacional Illas Atlánticas de Galiza. Vertebrados. Parque Nacional Illas Atlánticas de Galicia*. Vigo.

Mulder, K. P., L. Alarcón-Ríos, A. G. Nicieza, R. C. Fleischer, R. C. Bell, and G. Velo-Antón. 2022. Independent evolutionary transitions to pueriparity across multiple timescales in the viviparous genus *Salamandra, Molecular Phylogenetics and Evolution* 167:107347.

Rash, R., and H. B. Lillywhite. 2019. Drinking behaviors and water balance in marine vertebrates. *Marine Biology* 166:122.

Rodríguez-Guitián, M. A., and P. Ramil-Rego. 2007. Clasificaciones climáticas aplicadas a Galicia: Revisión desde una perspectiva biogeográfica. *Recursos Rurais* 3:31–53.

Santos, X., J. S. Azor, S. Cortes, E. Rodriguez, J. Larios, and J. M. Pleguezuelos. 2018. Ecological significance of dorsal polymorphism in a Batesian mimic snake. *Current Zoology* 64:745–753

Santos, X., and G. A. Llorente. 2009. Decline of a common reptile: Case study of the viperine snake *Natrix maura* in a Mediterranean wetland. *Acta Herpetologica* 4(2):161–169.

Scheuhammer, A. M., B. Braune, H. M. Chan, H. Frouin, A. Krey, R. Letcher, L. Loseto, *et al.* 2015. Recent progress on our understanding of the biological effects of mercury in fish and wildlife in the Canadian Arctic. *Science of the Total Environment* 509–510:91–103.

Schmidt-Nielsen, K. 1983. *Animal Physiology: Adaptations and Environments*. Cambridge University Press.

Tan, S. W., J. C. Meiller, and K. R. Mahaffey. 2009. The endocrine effects of mercury in humans and wildlife. *Critical Review in Toxicology* 39:228–269.

Velo-Antón, G., M. García-París, P. Galán, and A. Cordero-Rivera. 2007. The evolution of viviparity in Holocene islands: Ecological adaptation versus phylogenetic descent along the transition from aquatic to terrestrial environments. *Journal of Zoological Systematics and Evolutionary Research* 45(4):345–352.

Vilas, A., B. Gamallo, J. Framil, J. Bonache, K. Sanz, M. Lois, and M. Toubes. 2005. *Guía de visita del Parque Nacional Marítimo-Terrestre de las Islas Atlánticas de Galicia*. Organismo Autónomo de Parques Nacionales.

12

Golem Grad

From a Ghost Island to a Snake Sanctuary

Xavier Bonnet, Dragan Arsovski, Ana Golubović, and Ljiljana Tomović

Introduction

Despite what its name might suggest, Golem Grad (*Golem* means "large" and *Grad* "city") is a small uninhabited island (~18 ha, ~850 asl), the single one of North Macedonia (Figure 12.1). Broadly situated 2 km off the nearest western shore in Great Prespa Lake, this lake island is a strictly protected zone within Galičica National Park. Home to the largest colony of Dalmatian Pelicans (*Pelecanus crispus*, the largest pelican species), the fish-bearing waters of the Prespa Lakes support an abundant fauna (Catsadorakis *et al.* 2013, 2021). Great Prespa Lake is shared by three countries, North Macedonia, Albania, and Greece (the triple frontier lies ~1.3 km south of Golem Grad). With an estimated age of 3.5–4.0 My, it is one of the oldest lakes in the world (Popovska and Bonacci 2007). Human population densities are low in this mountainous region, there is no large industry or mass tourism, agricultural practices are relatively traditional, and the lake shores have been spared from concrete cover. Consequently, Prespa Lake ecosystems are relatively well preserved, with Golem Grad standing out from the water as a jewel of biodiversity in a limestone case. However, continuous drops in water level driven by climate change and irrigation (water surface covered 27,300 ha 20 years ago, but currently less than 25,000 ha; Soria and Apostolova 2022), overfishing, and poaching (Ajtić *et al.* 2013), as well as pollution (*e.g.*, pesticides from local agriculture) represent serious threats for the functioning of Golem Grad ecosystems.

The physiognomy of Golem Grad is spectacular. Steep cliffs separate a narrow rocky shore from a plateau, and a few small craggy paths connect these two habitats (Figure 12.1). These features resemble those of a Medieval castle. Golem Grad is a natural fortress that has been used for thousands of years (earliest human presence dated at 4th century BCE) by human populations seeking a safe or sacred place (Bitrakova-Grozdanova 1985; Bouzek *et al.* 2017). Indeed, Macedonia is a region of crossroads and has been a theater of migrations and confrontations since immemorial times. Alexander the Great, Cyril, and Methodious or Suleiman the Magnificent provide examples of famous characters involved in the complex history of the southern Balkans. Depending upon the outcomes of conflicts and political and religious

Xavier Bonnet, Dragan Arsovski, Ana Golubović, and Ljiljana Tomović, *Golem Grad* In: *Islands and Snakes*.
Edited by: Harvey B. Lillywhite and Marcio Martins, Oxford University Press. © Oxford University Press 2023.
DOI: 10.1093/oso/9780197641521.003.0012

Figure 12.1 Left: Golem Grad is a small (18 ha) island in the south of Macedonia (image Google Earth 2021). Right: Steep cliffs separate the plateau from narrow rocky shores. Photos by Google Earth (left) and X. Bonnet (right).

dominant currents, minorities or persecuted populations settled on Golem Grad (*e.g.*, early Christians [the Bogomils] built houses and churches there). Populated in the past, the plateau of Golem Grad was partly covered by stone constructions. Over time, people—particularly monks—took up residence on the island as late as the mid-20th century, after which Golem Grad has been completely abandoned. As a non-intentional heritage, human inhabitants left many comfortable houses to reptiles: piles of stones and roof tiles are distributed across the island and form a dense network of artificial refuges. A forest overstory dominated by tall *Juniperus* trees (>10 m) now replaces red-tiled roofs. Dominant *Juniperus excelsa* trees are associated with *Celtis glabrata*, *Cornus mas*, *Prunus mahaleb*, and *Prunus cerasifera* (both plum trees were cultivated in the past). Various scrubs (*Rosa dumalis*, *Rubus ulmifolius*, *Ephedra campylopoda*, and *Asparagus acutifolius*) and herbaceous species provide further microhabitats for reptiles and their prey. On the shores, the dominant trees are *Prunus mahaleb*, *Prunus cerasifera*, *Ficus carica* (previously cultivated), *Ostrya carpinifolia*, and *Fraxinus ornus*. Shrubby vegetation (*Euphorbia characias*, *Campanula versicolor*, *Corydalis ochroleuca*, *Rubus ulmifolius*, *Phragmites australis*, *Carex hordeistichos*, *Vitis vinifera*) and other herbaceous species are flourishing. Overall, humans profoundly influenced the amount and nature of shelters that are used by reptiles.

Currently, Golem Grad offers both archaeological wealth and unique biological richness. Ancient ruins colonized by various plants and chaotic karstic limestone formations provide suitable habitats to a wide range of animals. A relatively mild Mediterranean climate contributes to the zoological and botanical prosperity of Golem Grad. Four snake species occur on Golem Grad (Figure 12.2). One large male Four-lined Snake was spotted once (*Elaphe quatorlineata*) and thus should be considered as an irregular visitor. The beautiful Persian Grass Snake (*Natrix natrix persa*) is a common guest although our data suggest that no population is established.

Figure 12.2 Four snake species occur on Golem Grad. Top left: The Four-lined Snake (*Elaphe quatorlineata*) was sighted only once. Top right: Persian Grass Snake (*Natrix natrix persa*) is regularly observed. Bottom left: The Nose-horned Viper (*Vipera ammodytes*) is abundant. Bottom right: The Dice Snake (*Natrix tessellata*) occurs in huge numbers.
Photos by X. Bonnet.

The two other species, the Dice Snake (*Natrix tessellata*) and the Nose-horned Viper (*Vipera ammodytes*), are abundant and their populations are resident. The extremely dense population of piscivorous Dice Snakes (sustained by huge schools of endemic Prespa Bleaks, *Alburnus belvica*) explains why Golem Grad is locally known as Snake Island. Although data are lacking, it is unlikely that humans have ever cohabited with abundant snakes. Vipers might have been systematically killed to limit envenomation risks, and dice snakes were perhaps also killed, but almost certainly frequently disturbed by humans and domestic animals. We speculate that the abandonment of Golem Grad facilitated the development of robust snake populations. Considering the global decline of snakes (Reading *et al.* 2010), Golem Grad now represents a sanctuary.

Phenotypic and demographic peculiarities of the reptiles that live on Golem Grad (snakes, wall lizards, and tortoises) suggest that local conditions markedly shaped these organisms. Here we focus on vipers and Dice Snakes and on several important environmental factors expected to influence their day-to-day life. Annual and seasonal climatic fluctuations may affect vipers and Dice Snakes in different ways. Vipers are terrestrial and isolated on the island; most individuals that live on the plateau have no access to fresh water. No springs or even ephemeral water bodies exist, and the soil does not retain water during rainfalls, or only briefly. We never observed any viper in the water or drinking at the shore (in contrast to tortoises, for example). A swimming distance of 2 km represents a barrier. The island vipers are thus separated from continental populations; they live in a dry place that is often parched during summer. In contrast, the amphibious dice snakes benefit from access to fresh water year-round. Their swimming ability allows them to move from or to the shore. Overall, the two snake species we studied experience contrasted ecological conditions. We used this contrast to explore several life history traits that are classically considered in island versus mainland comparisons (*e.g.*, body size). Our goal is to highlight selected morphological aspects that seem to be unique to Golem Grad snakes. A tightly linked objective is to promote the protection of an exceptional site that hosts exceptional snake populations.

Dwarf Vipers

A mark and recapture study of vipers began in 2007. Fifteen years later, in 2022, nearly 600 individuals have been marked and more than 300 recaptures collected (Figure 12.3). An immediate impression one can get with animals in the hands is that Golem Grad vipers are particularly small and lightweight. This impression is supported by the data. Compared to mainland conspecifics, Golem Grad vipers are dwarf (Figure 12.4; Tomović *et al.* 2022). Their mean body size is 20% lower than what is generally recorded in continental vipers, and they never reach a large body size (maximal total body length <60 cm compared to ~80 cm in large mainland specimens). Data from another site (Cyclades Islands in the Aegean Sea, L. Tomović; unpublished data) suggest that insularity is associated with small body size in this snake species (Figure 12.4). Golem Grad vipers also have a relatively short jaw and a short tail (Tomović *et al.* 2022). Morphometric traits tend to be correlated (*e.g.*, head size vs. snout-to-vent length [SVL]). It would be interesting to examine if the peculiarities observed in Golem Grad vipers extend to other characteristics such as head width, horn length and width, snout length and width, scalation, etc.

As expected from their small body size, dwarf island vipers are less fecund than larger mainland conspecifics. On average, Golem Grad females produce a litter size of four offspring, nearly half the mean litter size (approximately seven) observed in mainland populations (Tomović *et al.* 2022). Pooling all populations of Nose-horned Vipers, there is a positive correlation between maternal size and fecundity. Yet the fecundity of Golem Grad vipers is significantly lower than estimated when using this relationship (Tomović *et al.* 2022). Very small litters are unusual in vipers; however, in Golem Grad, one-third of the litters range from one to three offspring. Golem Grad dwarf vipers exhibit a particularly low fecundity, both in absolute and SVL-corrected terms. However, relative offspring size (*i.e.*, offspring SVL scaled by maternal SVL) is not affected by maternal size, and offspring are proportionally normal-sized on the island (Tomović *et al.* 2022). Compared with mainland vipers, which reach maturity

Figure 12.3 Top: Adult male Nose-horned Viper (*Vipera ammodytes*). Bottom: Adult male in ambush for Wall Lizards.
Photos by X. Bonnet.

at size of about 44 cm (Anđelković *et al.* 2021), island conspecifics become mature at smaller body size (35–37 cm SVL). Overall, in island vipers, a set of reproductive traits are markedly different from what is observed in mainland populations.

Food intake is a major driver of growth, adult body size, and reproductive output in snakes (Madsen and Shine 2000; Bonnet *et al.* 2001; Seigel and Ford 2001; Taylor *et al.* 2005; Filippakopoulou *et al.* 2014). Therefore, we examined the diet of the island

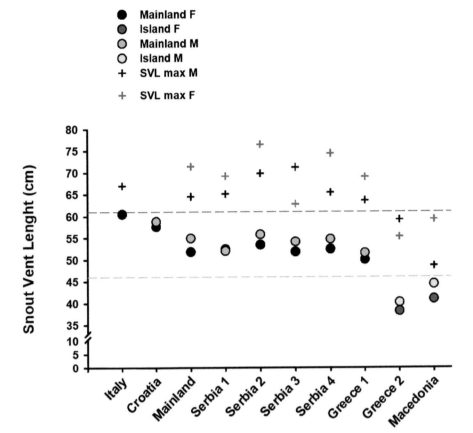

Figure 12.4 Comparison of the mean body size (SVL) among eight mainland (black and gray circles) and two island populations (color circles) of Nose-horned Vipers (*Vipera ammodytes*). "Mainland" stands for a pool of mainland populations. "Greece 2" represents a population sampled in the Cyclades. "Macedonia" represents Golem Grad. Mainland females are represented with black circles, mainland males with gray circles; island females with green circles and island males with yellow circles. Crosses indicate maximal values. The dashed lines facilitate comparisons, light gray for females and dark gray for males

Source of the data is Tomović *et al.* 2022.

vipers and found strong differences compared to mainland snakes. Terrestrial micro-mammals, including small rodents (field or domestic mice, voles, rats) and shrews are usual prey for nose-horned vipers, but they are absent from Golem Grad. The only abundant potential prey for island vipers (*i.e.*, prey also found in the diet of Nose-horned Vipers in other populations) are small reptiles such as Common Wall Lizards (*Podarcis muralis*) and juvenile Dice Snakes, as well as centipedes (*Scolopendra cingulata*) (Figure 12.5). Vipers are observed in ambush in most places, even on

Figure 12.5 Centipedes (*Scolopendra cingulata*, top left and top right) and Common Wall Lizards (*Podarcis muralis*, bottom left) are the main prey of Golem Grad vipers. Centipedes can be dangerous prey, however. A small female viper was killed by the prey she ingested: the centipede forced its way through the body wall of the snake but it died (centipede was partly pulled out). Birds are major snake predators (two middle left photos): Fresh carcasses of female Dice Snakes abandoned by raptors disturbed by researchers (the birds targeted the eggs and the liver); below, a heron caught a viper. Bottom right: A female Dice Snake bringing an endemic bleak (*Alburnus belvica*), the main prey of this snake species, to the shore.
Photographs by X. Bonnet, except the heron with viper photograph by J. M. Ballouard.

low branches, both on the plateau and among boulders on the shore (except in the northern very rocky parts beneath the cliffs). We collected information from 133 ingested meals. Island vipers mainly feed on Common Wall Lizards ($N = 94$; 71%) and centipedes ($N = 27$; 20%). For unknown reasons, Dice Snakes are avoided by vipers despite their abundance ($N = 5$ vipers with a Dice Snake in the stomach). Other types of prey are also rarely consumed by the vipers, and only by the adults: newborn rabbits ($N = 3$), birds ($N = 2$), and frogs ($N = 1$). Mainland vipers feed largely on lizards ($N = 67$; 62%) and small mammals ($N = 38$; 35%) (Tomović et al. 2022).

In all populations (mainland and island), there is a marked ontogenetic shift in the diet. Preliminary proteome analysis of venom reveals differences between mainland and island vipers and shows that venom composition changes ontogenetically (M. Lakušić unpublished data). These results offer a clear parallel to diet analyses. Small (young) snakes essentially feed on lizards (>68% of the prey). Centipedes are the secondary prey in small island vipers (32% of the prey) versus micromammals in small mainland vipers (19%). In Golem Grad, the proportion of centipedes in the diet (more than one-third) is remarkably elevated, suggesting that myriapods represent an essential resource for neonatal vipers. Feeding on prey that are themselves predators can be dangerous, even for venomous snakes (Bonnet et al. 2010). Feeding on large prey fitted with formidable weapons (i.e., strong toxicognaths; Figure 12.5), such as centipedes, can be a risky choice. Some centipedes equal the size and mass of the smallest vipers. A centipede swallowed by a small female viper forced its way through the stomach and body wall of the reptile; both animals eventually died, likely from envenomation and wounds, respectively (Arsovski et al. 2014; Figure 12.5). Meals made of lizards and centipedes bring many keratinous scales and chitinous plates, but little flesh and entrails compared to mammals or birds. In addition, feeding frequency is significantly lower in island than in mainland vipers: 77% of island snakes were found with empty stomach versus 61% of mainland snakes (Tomović et al. 2022). Low prey quality combined with relatively lower feeding frequency may explain why island vipers are dwarf.

Centipedes and wall lizards pullulate, however; prey numbers may compensate prey quality, but the proportion of vipers with a full stomach is relatively low. A number of factors not yet investigated (e.g., hydro-mineral balance, local adaption) probably play important roles in the specificities of the ecology and energy budget of Golem Grad vipers (Lillywhite et al. 2008; Ma et al. 2019). Thus, how dwarfism syndrome was established in Golem Grad vipers remains unknown. Genetic and experimental investigations are needed to evaluate the respective contribution of phenotypic plasticity and local adaptation. Differences in relative tail length—a trait presumably not subjected to selection associated with energy intake—suggest that adaptation is involved (Tomović et al. 2022). Yet body size and relative jaw length are particularly sensitive to plasticity associated with food intake (Queral-Regil and King 1998; Bonnet et al. 2021).

Despite their small size and low fecundity, vipers flourish on Golem Grad. Several hundred of them live on a small 18 ha surface. A number of factors may explain this prosperity. Among them, the lack of mammalian predators (*e.g.*, foxes, dogs, cats) is likely determinant. Annual adult survival rate is estimated at 0.7 (unpublished data), a value within the range documented in snakes from temperate climates (Lind *et al.* 2005; Lelièvre *et al.* 2013; King *et al.* 2018; Bauwens and Claus 2019; Jolly *et al.* 2021). Whatever the case, our data show that life history traits of Golem Grad vipers are set on a low energy income, a strategy that works well apparently. However, low fecundity and delayed maturity (slow-pace life history traits) combined with the fact that the vipers are isolated on a small island suggest the population is extremely sensitive to any threat. One strong risk is represented by snake collectors; Golem Grad vipers have been seen for sale on the Internet (X. Bonnet unpublished). The introduction (voluntary or accidental) of snake predators such as feral cats represents another major danger; an attempt to introduce mongooses in the late 18th century fortunately failed (see the threat of introducing mongooses in islands in Henderson 1992). Golem Grad is easy to survey, and the favorable periods to find vipers are known. Providing that local authorities honor their mission to protect Golem Grad, vipers should thrive there in the future.

Giant Dice Snakes

Golem Grad is the home of a very large population of Dice Snakes, a phenomenon observed in other insular natricine snakes (King *et al.* 2018). Dice Snakes have been observed almost everywhere on Golem Grad; densities are higher near the shore, however. Since 2008, we captured and measured more than 7,000 individuals. This number reflects our capacity to process animals as well as the actual number of snakes that we may have captured during each field session. Although not precisely evaluated, population density was incredibly high at the dawn of the 21st century. Frightened by human observers visiting the island, numerous snakes were cascading from trees and rocks, heading toward the water, bushes, or boulders (this amphibious snake uses small trees and bushes to rest; Figure 12.6). Presumably, tens of thousands of snakes were living on Golem Grad at that time and likely earlier (L. Tomović pers. obs.). Later, at the onset of the study in 2008, despite a perceptible decrease of numbers, thick carpets of snakes were still covering large parts of the shores; many trees were full of snakes. Numbers were so high that the friction of crawling snakes produced sounds loud enough to be audible from distance (~5 m away). Likely, tens of thousands of snakes were still present in 2008. The population dropped drastically during the past decade (Sterijovski *et al.* 2014), and the vision of living carpets of snakes belongs to the past. Nonetheless, Dice Snakes are still very abundant, and hundreds can be captured during a single day under favorable conditions, notably in April or May. Our efforts to mark a substantial proportion of the population in order to estimate its size and other demographic traits were thwarted by numbers ($N > 6,000$ marked individuals, $N > 680$ recaptures, $N > 1,000$ snakes captured, measured, but not marked). Therefore, although several hundreds of snakes were captured more than once, the low recapture rate (<10%) prevents estimating accurately demographic

Figure 12.6 Top: An adult Dice Snake (*Natrix tessellata*) uniformly gray. Bottom: An adult female dice snake displaying the typical blotched coloration.
Photos by X. Bonnet.

parameters. Yet large datasets enable comprehensive pictures to be made of variations of the phenotypic traits under focus. As with vipers, we examined body size, fecundity, and the diet of Golem Grad Dice Snakes, and we compared the results with those obtained in other populations.

The distribution range of *N. tessellata* is wide, covering a large proportion of the central-south Palearctic, from Italy to China. Across such a broad area, strong phenotypic variations have been reported, notably regarding color and scalation patterns (Mebert 2011a, 2011b). Mean and maximal body size have been estimated in more than 10 populations spread across a substantial part of the distribution range of the species, from Italy to Iran. Compared to other populations, Golem Grad Dice Snakes are large (Figure 12.7). Although the term "gigantism" might be excessive, mean and maximal body sizes of females and males exceed those from all the other populations sampled. A lack of published data prevents comparisons of other major morphological traits (*e.g.*, jaw length, tail length). Thus, we cannot assess if Golem Grad Dice Snakes exhibit other morphological peculiarities in addition to size.

We were able to determine the reproductive status of 1,484 females and count the number of eggs in 1,026 gravid females. All Golem Grad gravid females were larger than 63 cm; this value might be considered as the onset for maturity in this

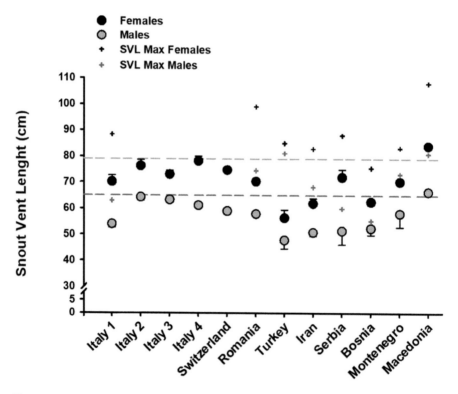

Figure 12.7 Comparison of the mean body size (SVL ± SE) among 12 populations of Dice Snakes (*Natrix tessellata*). Golem Grad is represented by Macedonia. Females are represented with black circles, males with gray circles. Crosses indicate maximal values. The dashed lines facilitate comparisons, light gray for females and dark gray for males. Sources of data are Mebert 2011b; Ajtić *et al.* 2013; Šukalo *et al.* 2019.

population. However, not all females reaching 63 cm were reproductive. In fact, only 23% of the females with SVL ranging from 65 to 70 cm SVL were gravid. This proportion increased rapidly with SVL to reach a plateau: 70% of the females larger than 75 cm SVL were gravid. This suggests that maturity takes place progressively with increasing SVL from 63 cm to 75 cm. We found a positive effect of SVL on clutch size ($r = 0.39$, $F_{1, 1024} = 187.67$, $P < 0.001$). In addition to the classical positive effect of body size on fecundity (Shine 2003), plotting the data for insular Dice Snakes revealed a scattered pattern (Figure 12.8). Body size is a poor determinant of the number of eggs that a female produces, explaining only 15% of the regression between fecundity and body size ($r^2 = 0.15$). The large sample size was also useful to visualize the area that contains a range of variation of clutch size as a function of body size; it is delimited by lower and upper thresholds (Figure 12.8). The upper limit might represent the maximal distention capacity of the abdomen to accommodate the clutch or might be associated with the maximal filling tolerance of the abdomen compatible with swimming and terrestrial locomotion. To our knowledge, these hypotheses have not been

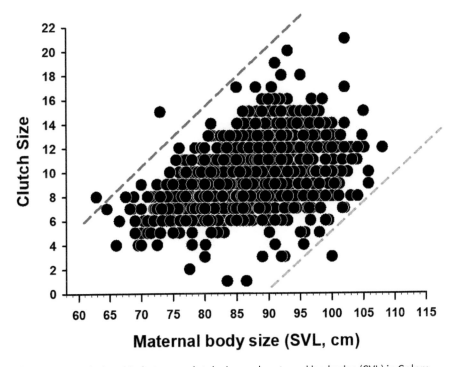

Figure 12.8 Relationship between clutch size and maternal body size (SVL) in Golem Grad Dice Snakes (*Natrix tessellata*). The almost total lack of data above the upper and below the lower dashed lines suggests morphological or physiological limits. For example, a female smaller than 80 cm SVL might not be able to accommodate more than 16 eggs in her abdomen, whereas a female larger than 100 cm SVL seems able to systematically produce more than 6 eggs.

explored. The lower limit suggests that large females manage to not produce a very small clutch (clutch size of fewer than four eggs, $N = 1$). More generally, clutch sizes of less than six eggs are rare ($N = 28$, 2.7%). On the other hand, females also exceptionally produce more than 14 eggs ($N = 32$, 3.1% of the gravid females). The benefits for the mother to lay few eggs might not offset the costs of engaging reproduction, while producing large clutches might be mechanistically constrained or too costly. The tradeoff between offspring size and offspring number could not be implemented in the analyses due to a lack of data on egg mass.

Irrespective of body size, the vast majority of clutches range from 6 to 14 with a mean of 9.6 ± 2.5 (standard deviation [SD]). This mean value contrasts with published data for Dice Snakes in Italy (14.5 ± 7.4 eggs, $N = 22$) and in Germany (12.1 ± 3.9 eggs, $N = 8$) that were recorded in relatively smaller females (mean SVL of gravid females was 70.2 ± 9.1 cm in Italy, 84.3 ± 6.9 cm in Germany) compared to Golem Grad (87.2 ± 7.4 cm SVL). The body size–clutch size relationship was tight in Italian and German snakes ($r^2 = 0.89$ and $r^2 = 0.57$, respectively, in each correlation of clutch size against SVL). Thus, in the two other populations studied, body size is a strong predictor of fecundity.

Overall, Golem Grad female Dice Snakes are large, they produce a relatively small number of eggs, and their body size weakly predicts their fecundity. Food intake positively influences growth and fecundity in snakes; the low fecundity associated with large maternal size suggests that these traits are largely dissociated in Golem Grad Dice Snakes. The loose correlation between clutch size and maternal SVL means that, in a great proportion of females, the space available in the abdomen is not filled up by the clutch, possibly due to resource limitation during vitellogenesis. Strong fluctuations in prey availability may explain such dissociations between body size and reproductive output. During feast, females may grow rapidly to reach a large size, but later they may not be able to produce the maximal number of eggs during each reproductive bout when prey are scarcer. One tenet of this assumption is that prey availability fluctuates strongly among seasons and years (see below).

Golem Grad offers excellent refuges to snakes and also supports the energy demand of thousands of snakes. Dice Snakes feed on fish they capture near the shore. The main prey is represented by an endemic fish, *Alburnus belvica* (Figure 12.5), a local ablet (Ajtić *et al.* 2013). Occasional prey include other local fish (mostly *Alburnus* spp.) and introduced species such a Rainbow Perch (*Lepomis gibbosus*). Schools of small fish that seek safety among the rocks around Golem Grad are easy prey for snakes in ambush. Dice Snakes sometimes fight for the same prey, occasionally triggering accidental cannibalism (X. Bonnet personal observation). Most meals contain one to three prey, but as many as 11 fish can be present in the stomach. In contrast to myriapods and lizards, fish are highly profitable prey, especially when they are easy to find. The proportion of snakes with a full stomach varies over time and across years. In some years, more than 60% of the snakes have a prey in the stomach during summer following vitellogenesis, but this proportion can drop to less than 4% during vitellogenesis in the spring (Ajtić *et al.* 2013).

The species name of "Dice Snakes" comes from the word "tessellatum," a small piece of ceramic or stone used to make mosaics. Typical individuals exhibit a blotched color pattern that evokes the appearance of ancient mosaics. Three color morphs of Dice Snakes coexist on Golem Grad—diced (~square blotches), black, and uniformly gray. Each color morph is represented by substantial numbers of individuals; 58% of the individuals are blotched, 13% are black, and 29% are gray (Ajtić *et al.* 2013). Black and uniformly gray snakes are rare in other populations (Mebert 2011a, 2011b). Protection against predators is a principal function of coloration in snakes (Farallo and Forstner 2012). On Golem Grad, strong predation pressure is exerted on Dice Snakes. We witnessed cases of predation and collected abundant evidence (feces examination) that a variety of predators (birds, otters) feed on Dice Snakes (Ajtić *et al.* 2013, Figure 12.5). Presumably, during sun basking, black snakes can reach high body temperature more rapidly than paler snakes but are conspicuous, especially in white limestone and bushy habitats. The opposite situation is expected for blotched snakes. The maintenance of the color polymorphism might be explained by the balance between the thermoregulatory versus antipredator (crypsis) advantages of each of these morphs (Tanaka 2007). Yet explanation for the third color—uniformly gray pattern— is lacking. Possible roles of coloration when the snakes are fishing (camouflage during ambush) or moving between land and water might be involved.

Dice Snakes employ a rich repertoire of antipredator behaviors following their capture, including struggling, hissing, head flattening, striking, spreading cloacal secretion, or death feigning. Death feigning is quite common in Golem Grad Dice Snakes (31.5% of the snakes tested), but it is more rarely displayed (18.8–6.4%) in the four other populations sampled (Golubović *et al.* 2021). Individuals that use death feigning also exhibit an attenuated stress response (acute increase of corticosterone and glucose blood levels), suggesting a physiological link with the expression of this behavior (Lakušić *et al.* 2020). Thus, Golem Grad Dice Snakes developed a specific antipredator response with frequent death feigning displays, possibly promoted by strong predation pressures by raptors, herons, and other diurnal avian predators.

Overall, Dice Snakes from Golem Grad have accumulated evolutionary peculiarities. Prespa Lake Dice Snakes are abundant on the mainland shores and on a small Albanian Island 9 km southwest of Golem Grad. Several populations have been sampled. Preliminary analyses suggest that snakes on Golem Grad are significantly larger (unpublished data) compared to other Prespa Lake populations. Moreover, the frequency of black (<2%) and gray (<15%) morphs is considerably lower in other populations. Thus, Dice Snakes on Golem Grad are different from those of neighboring populations. Phenotypic and genetic investigations are needed to unravel the factors that promote these differences.

A Fragile Sanctuary

Unsustainable fishing, bush fires, and tourism are immediate threats to snakes on Golem Grad (Figure 12.9). Water level changes (especially dropping), temperature increase, and

Figure 12.9 Two fishermen illegally setting a fishing net too close to the shore of Golem Grad. Many snakes drown while attempting to feed on fish entangled in the mesh. Photos by X. Bonnet.

pollution are medium-term threats. Limiting the first-named factors should promote the capacities of Golem Grad ecosystems and snake species to cope with the longer-term problems. Rapid and effective responses of organisms to global climatic changes are more likely to occur in large and healthy populations. Regulating fishing is likely the most important action to undertake in the near term. The estimated amounts of financial resources to achieve this objective are small, considering an initial input of less than 50,000 euros to recruit rangers plus a small boat devoted to Golem Grad; several tens of thousands of euros per year to run the system would be sufficient. Unfortunately, after 15 years of recommendations, no actions have been implemented.

Conclusion

Golem Grad is a small lake island situated at short distance from the surrounding mainland (~2 km). Different populations of Nose-horned Vipers live on the mainland shores of the lake. Dice Snakes are found in the different sites we surveyed, and they may occur almost everywhere in Prespa Lake. Consequently, Golem Grad snakes might not be genetically isolated from mainland populations; we speculate that dispersing individuals maintain a substantial gene flow between the island and neighboring populations, at least for Dice Snakes.

In addition, proximity to a mainland confers similarity of climatic conditions prevailing on Golem Grad and other nearby areas. Nonetheless, reptiles that live on Golem Grad exhibit features that are specific to the island. Moreover, our observations show that both dwarfism and gigantism can coexist in snakes residing on a single small island. Gigantism does not translate into greater relative fecundity in Dice Snakes, whereas dwarfism is associated with low fecundity in vipers. Any attempt to apply the island rule to snakes (but see Meiri *et al.* 2008) is thwarted by their variability of body size at the population level. Tight morpho-functional linkages that constrain body size and metabolism in endothermic vertebrates (birds, mammals) appear to be relaxed in snakes. For example, a large snake may well survive a long period of total food restriction (>6 months) via down-regulation (Secor 2008) without deleterious effect, something not easily conceivable for a bird or mammal. Although a very small island cannot sustain a large biomass of endothermic vertebrates, it may well host large populations (or large individuals) of reptiles. For example, the estimated total mass of the tortoises living on Golem Grad (N ~1,500 individuals) is more than 500 kg, corresponding to a herd of approximately 20 small goats. The carrying capacity of the ecosystem would be overflowed by the ungulates, but it is not by the chelonians. This illustrates that small islands may not exert identical pressures on ectothermic and endothermic vertebrates, possibly explaining why many islands shelter large populations of snakes.

Vipers and Dice Snakes do not exploit the same ecosystem. Terrestrial vipers are trapped on the island by the lake, while the Dice Snakes are semi-aquatic and can disperse over water to other habitats. Thus, only the vipers should be considered as pure island dwellers. The comparison between vipers and Dice Snakes poses the question of what is really an insular snake.

Acknowledgments

We warmly thank numerous students and collaborators; without them, achieving the task of collecting so many data would have been impossible. Special thanks to B. Sterijovski, R. Ajtić, J. Crnobrnja-Isailović, M. Djuraki, Nikolić S., and J. M. Ballouard, who were in the May 2008 funding trip to set up the population monitoring of dice snakes and tortoises. M. Tasevski provided essential hospitality and logistics. The authors were supported by the Ministry of Education, Science, and Technological Development of the Republic of Serbia; the Ecological Society of Macedonia; and the Centre National de la Recherche Scientifique.

References

Ajtić, R., L. Tomović, B. Sterijovski, J. Crnobrnja-Isailović, S. Jordjević, M. Djurakić, J. M. Ballouard, and X. Bonnet. 2013. Unexpected life history traits in a very dense population of dice snakes. *Zoologischer Anzeiger* 252:350–358.

Anđelković, M., S. Nikolić, and L. Tomović. 2021. Reproductive characteristics, diet composition and fat reserves of nose-horned vipers (*Vipera ammodytes*). *Herpetological Journal* 31:151–161.

Arsovski, D., R. Ajtić, A. Golubović, I. Trajčeska, S. Đorđević, M. Anđelković, X. Bonnet, and L. Tomović. 2014. Two fangs good, a hundred legs better: Juvenile viper devoured by an adult centipede it had ingested. *Ecologica Montenegrina* 1:6–8.

Bauwens, D., and K. Claus. 2019. Intermittent reproduction, mortality patterns and lifetime breeding frequency of females in a population of the adder (*Vipera berus*). *PeerJ* 7: e6912.

Bitrakova-Grozdanova, G. G. 1985. L'agglomération antique et médiévale de Golem Grad sur le lac de Prespa. *Macedoniae Acta Archaeologica* 10:101–133.

Bonnet, X., F. Brischoux, M. Briand, and R. Shine. 2021. Plasticity matches phenotype to local conditions despite genetic homogeneity across 13 snake populations. *Proceedings of the Royal Society B* 288:20202916.

Bonnet, X., F. Brischoux, and R. Lang. 2010. Highly venomous sea kraits must fight to get their prey. *Coral Reefs* 29:379–379.

Bonnet, X., G. Naulleau, R. Shine, and O. Lourdais. 2001. Short-term versus long-term effects of food intake on reproductive output in a viviparous snake, *Vipera aspis*. *Oikos* 92:297–308.

Bouzek, J., V. Čisťakova, P. Tušlová, and B. Weissová 2017. New Studies in Black Sea and Balkan Archaeology (2014–2016). *Studia Hercynia* 21:149–163.

Catsadorakis, G., P. Aleks, O. Avramoski, T. Bino, A. Bojadzi, Z. Brajanoski, W. Fremuth, *et al.* 2013. Waterbirds wintering at the Prespa lakes as revealed by simultaneous counts in the three adjoining littoral states. *Macedonian Journal of Ecology and Environment* 15:23–31.

Catsadorakis, G., V. Roumeliotou, I. Koutseri, and M. Malakou. 2021. Multifaceted local action for the conservation of the transboundary Prespa lakes Ramsar sites in the Balkans. *Marine and Freshwater Research*. CSIRO. https://doi.org/10.1071/MF21123

Farallo, V. R., and M. R. Forstner. 2012. Predation and the maintenance of color polymorphism in a habitat specialist squamate. *PloS One* 7: e30316.

Filippakopoulou, A., X. Santos, M. Feriche, J. M. Pleguezuelos, and G. A. Llorente. 2014. Effect of prey availability on growth-rate trajectories of an aquatic predator, the viperine snake *Natrix maura*. *Basic and Applied Herpetology* 28:35–50.

Golubović A., M. Anđelković, L. Tomović, D. Arsovski, S. Gvozdenović, G. Šukalo, R. Ajtić, and X. Bonnet. 2021. Death-feigning propensity varies within dice snake populations but not with sex or colour morph. *Journal of Zoology* 314:203–210.

Henderson, R. W. 1992. Consequences of predator introductions and habitat destruction on amphibians and reptiles in the post-Columbus West Indies. *Caribbean Journal of Science* 28:1–10.

Jolly, C. J., B. Von Takach, and J. K. Webb. 2021. Slow life history leaves endangered snake vulnerable to illegal collecting. *Scientific Reports* 11:1–11.

King, R. B., K. M. Stanford, and P. C. Jones. 2018. Sunning themselves in heaps, knots, and snarls: The extraordinary abundance and demography of island watersnakes. *Ecology and Evolution* 8:7500–7521.

Lakušić, M., G. Billy, V. Bjelica, A. Golubović, M. Anđelković, and X. Bonnet. 2020. Effect of capture, phenotype, and physiological status on blood glucose and plasma corticosterone levels in free-ranging dice snakes. *Physiological and Biochemical Zoology* 93:477–487.

Lelièvre, H., P. Rivalan, V. Delmas, J. M. Ballouard, X. Bonnet, G. Blouin-Demers, and O. Lourdais. 2013. The thermoregulatory strategy of two sympatric colubrid snakes affects their demography. *Population Ecology* 55:585–593.

Lillywhite, H. B., L. S. Babonis, C. M. Sheehy III, and M. C. Tu. 2008. Sea snakes (*Laticauda* spp.) require fresh drinking water: Implication for the distribution and persistence of populations. *Physiological and Biochemical Zoology* 81:785–796.

Lind, A. J., H. H. Welsh Jr, and D. A. Tallmon. 2005. Garter snake population dynamics from a 16-year study: Considerations for ecological monitoring. *Ecological Applications* 15:294–303.

Ma, L. I., P. Liu, S. Su, L. G. Luo, W. G. Zhao, and X. Ji. 2019. Life-history consequences of local adaptation in lizards: *Takydromus wolteri* (Lacertidae) as a model organism. *Biological Journal of the Linnean Society* 127:88–99.

Madsen, T., and R. Shine. 2000. Silver spoons and snake body sizes: Prey availability early in life influences long-term growth rates of free-ranging pythons. *Journal of Animal Ecology* 69:952–958.

Mebert, K. 2011a. Geographic variation of morphological characters in the dice snake (*Natrix tessellata*). *Mertensiella* 18:11–20.

Mebert, K. 2011b. *The Dice Snake, Natrix Tessellata: Biology, Distribution and Conservation of a Palaearctic Species*. Mertensiella 18, Deutsche Gesellschaft für Herpetologie und Terrarienkunde.

Meiri, S., N. Cooper, and A. Purvis. 2008. The island rule: Made to be broken? *Proceedings of the Royal Society B* 275:141–148.

Popovska, C., and O. Bonacci. 2007. Basic data on the hydrology of Lakes Ohrid and Prespa. *Hydrological Processes: An International Journal* 21:658–664.

Queral-Regil, A., and R. B. King. 1998. Evidence for phenotypic plasticity in snake body size and relative head dimensions in response to amount and size of prey. *Copeia* 1998:423–429.

Reading, C. J., L. M. Luiselli, G. C. Akani, X. Bonnet, G. Amori, J. M. Ballouard, E. Filippi, *et al.* 2010. Are snake populations in widespread decline? *Biology Letters* 6:777–780.

Secor, S. M. 2008. Digestive physiology of the Burmese python: Broad regulation of integrated performance. *Journal of Experimental Biology* 211:3767–3774.

Seigel, R. A., and N. B. Ford. 2001. Phenotypic plasticity in reproductive traits: Geographical variation in plasticity in a viviparous snake. *Functional Ecology* 15:36–42.

Shine, R. 2003. Reproductive strategies in snakes. *Proceedings of the Royal Society of London. Series B* 270:995–1004.

Soria, J., and N. Apostolova. 2022. Decrease in the water level of Lake Prespa (North Macedonia) studied by remote sensing methodology: relation with hydrology and agriculture. *Hydrology* 9:99.

Sterijovski, B., R. Ajtić, L. Tomović, and X. Bonnet. 2014. Conservation threats to dice snakes (*Natrix tessellata*) in Golem Grad Island (FYR of Macedonia). *Herpetological Conservation and Biology* 9:468–474.

Šukalo, G., S. Nikolić, D. Dmitrović, and L. Tomović. 2019. Population and ecological characteristics of the dice snake, *Natrix tessellata* (Laurenti, 1768), in lower portions of the Vrbanja River (Republic of Srpska, Bosnia and Herzegovina). *Turkish Journal of Zoology* 43:657–664.

Tanaka, K. 2007. Thermal biology of a colour-dimorphic snake, *Elaphe quadrivirgata*, in a montane forest: Do melanistic snakes enjoy thermal advantages?. *Biological Journal of the Linnean Society* 92:309–322.

Taylor, E. N., M. A. Malawy, D. M. Browning, S. V. Lemar, and D. F. DeNardo. 2005. Effects of food supplementation on the physiological ecology of female western diamond-backed rattlesnakes (*Crotalus atrox*). *Oecologia* 144:206–213.

Tomović, L., M. Anđelković, A. Golubović, D. Arsovski, R. Ajtić, B. Sterijovski, S. Nikolić, *et al.* 2022. Dwarf vipers on a small island: Body size, diet and fecundity correlates. *Biological Journal of the Linnean Society* 20:1–13.

13

Welcome to Paradise

Snake Invasions on Islands

Natalie M. Claunch, Keara L. Clancy, Madison E. A. Harman, Kodiak C. Hengstebeck, Diego Juárez-Sánchez, Daniel Haro, Arik Hartmann, Mariaguadalupe Vilchez, Rebecca K. McKee, Amber Sutton, and Christina M. Romagosa

Introduction

Species invasions can have multifaceted impacts, including disrupting the ecology of areas where introductions occur (Pimentel *et al.* 2005; Reaser *et al.* 2007). In addition to studying the impact of invaders, examinations of invasions can elucidate the effects of long-distance dispersal and novel environments on rapid evolution and adaptations of both invasive and native species (Herrel *et al.* 2008; Stuart *et al.* 2014). In this way, invasion ecology integrates with the broader field of island biogeography, where dispersal, isolation, and diversity shape the resulting ecological communities and associations of both islands and invaded areas. Indeed, these fields often overlap completely as some of the more dramatic documented impacts of invasive species are of those introduced to islands (Fritts and Rodda 1998; Towns *et al.* 2006).

Perhaps the best studied of island invasions is that of the Brown Treesnake (*Boiga irregularis*) on Guam, a formerly snake-free island, where it has demonstrated economic, human, and ecological impacts (reviewed in Rodda *et al.* 1999; Kahl *et al.* 2012; Figure 13.1). Though infamous, *B. irregularis* is neither the first nor last snake introduced to islands, and much can be gleaned from a review of other snake invasions on islands. Our goal herein is to synthesize the current knowledge of snake invasions on islands worldwide to identify patterns and gaps that can inform future studies and management efforts, building on and updating the work of Kraus (2009), which examined herpetofauna invasions more broadly. While we acknowledge invasive species are by definition dispersed by humans to areas where they are not native and have some demonstrated impact in that new area (Beck *et al.* 2008), we also acknowledge that, for many invasions, impacts may be unknown, unmeasured, or imperceptible, lagging behind current observation (Parker *et al.* 1999; Jeschke *et al.* 2014). In the case of *B. irregularis*, for example, its impact on native bird populations was not formally acknowledged until 35 years after its introduction (Rodda *et al.*

Natalie M. Claunch, Keara L. Clancy, Madison E. A. Harman, Kodiak C. Hengstebeck, Diego Juárez-Sánchez, Daniel Haro, Arik Hartmann, Mariaguadalupe Vilchez, Rebecca K. McKee, Amber Sutton, and Christina M. Romagosa, *Welcome to Paradise* In: *Islands and Snakes*. Edited by: Harvey B. Lillywhite and Marcio Martins, Oxford University Press.

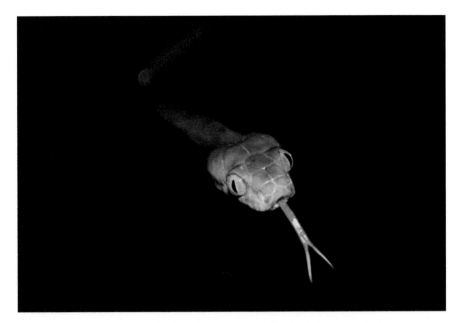

Figure 13.1 The longest-studied snake invasive to islands, *Boiga irregularis,* is responsible for extinction and extirpation of endemic birds, bats, and lizards from Guam.
Photograph by Natalie Claunch.

1999). Thus, we have opted to summarize all known introductions of snakes to islands, even where impacts are not currently assessed. A compendium of available information used to prepare this chapter, with more in-depth information on each species and introduction, is available on Zenodo (Claunch *et al.* 2022).

Overview of Snakes Invading Islands

Snake introductions to islands have been recorded as early as the 4th century BCE (*Zamenis scalaris*; Vigne and Alcover 1985) to the present (*Zamenis longissimus*; Di Nicola and Vaccaro 2020), on islands ranging in size from 0.4 to 785,783 km². Snakes from seven families (Colubridae, Pythonidae, Boidae, Lamprophiidae, Viperidae, Typhlopidae, Elapidae) have been introduced to islands in the Atlantic, Pacific, and Indian Oceans, the Gulf of Mexico, and 11 of the world's seas: the Arabian, Aegean, Balearic, Caribbean, East China, South China, Ionian, Mediterranean, North, Tyrrhenian, and Philippine Seas (Figure 13.2). The most widespread snake introduced to islands is the diminutive Brahminy Blind Snake, or Flower-pot Snake (*Indotyphlops braminus*), with introductions in all of the aforementioned seas and oceans except the North Sea (see Wallach 2020; Claunch *et al.* 2022; Figure 13.2; Figure 13.3). Most other snake species established on islands are restricted to a single

Figure 13.2 The Brahminy Blind Snake (*Indotyphlops braminus*), one of the most widely distributed snakes introduced to islands worldwide, is also known as a "Flowerpot Snake" because many of its introductions are attributed to the nursery trade.
Photographs by Arik Hartmann (left) and Natalie Claunch (right).

ocean or sea, excepting *Z. longissimus* (Langton *et al.* 2011; Di Nicola and Vaccaro 2020), *Lycodon aulicus* (O'Shea *et al.* 2018), and *Xenocrophis vittatus* (Herrera-Montes 2015), which have been introduced to two distinct oceans or seas, and *Telescopus fallax*, introduced to several seas feeding into the Mediterranean (Warnecke 1988; Schembri and Lanfranco 1996).

Observations of introductions are, of course, biased toward those that have resulted in successful establishment as they are more noticeable and more often documented (Zenni and Nuñez 2013). However, there are several records of introductions that have apparently failed to establish. These include species where several individuals were initially reported, but recent records are lacking (*e.g.*, *Naja kaouthia* to Okinawa-jima, Shiroma *et al.* 1994; *Hemorrhois hippocrepis* to Formentera, Silva-Rocha *et al.* 2018; *Malpolon monspessulanus* to Ibiza and Formentera, Febrer-Serra *et al.* 2021; *Leioheterodon madagascariensis* in the Comoros, Meirte 1993; and *Boa constrictor* to the Florida Keys, Hanslowe *et al.* 2018). *Boa constrictor* is also mentioned as an introduced species to Curaçao in a field guide (Van Buurt 2005), but there are no voucher records or recent sightings to confirm this. Due to the cryptic nature of snakes and their associated low detectability in field surveys, it is a challenge to confirm their absence from an area.

Some more recently reported potential introductions where sightings are limited and establishment is yet uncertain include a number of species to the Canary Islands

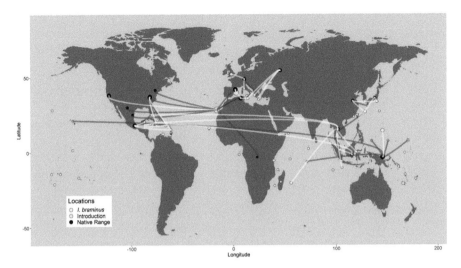

Figure 13.3 Independent snake introductions to new island ranges (yellow) from native ranges (purple). Size of points reflects the number of introductions. Lines reflect the connections between introductions from respective native ranges; white lines indicate introductions that have been documented as established from those native ranges, while gray lines indicate recorded introductions without evidence for establishment. Gray points without connections represent additional locations of recorded introductions of *Indotyphlops braminus*.

(*Lampropeltis triangulum sinaloa, L. triangulum hondurensis, L. mexicana thayeri, Pantherophis obsoletus, Crotaphopeltis hotamboeia, Natrix maura, H. hippocrepis, Z. scalaris*; Ministerio para la Transición Ecológica 2018), *B. irregularis* to Rota and several other Pacific islands (Perry and Vice 2009), *Python regius* in the Florida Keys (Hanslowe *et al.* 2018), and *Malayopython reticulatus* to Puerto Rico (Endangered and Threatened Wildlife and Plants 2022), for which monitoring will be crucial. It is near-certain that non-native snakes on other islands will be recorded.

Geographic Distribution and Pathways

The principles of dispersal in island biogeography state that organisms dispersing naturally are likely to originate from the nearest large landmass. In the case of snakes introduced to islands, only 54% of established populations are native to the nearest landmass, attributable to the various human-mediated modes of dispersal that facilitate far-ranging introductions. All of the reported ancient introductions (4th century, BCE to 13th century AD) are of snakes introduced from mainland Europe and north Africa into nearby seas within the Mediterranean Sea (*Z. scalaris, N. maura,* and *Macroprotodon mauritanicus,* Pleguezuelos 2002; *T. fallax,* Warnecke 1988), highlighting the relatively limited capacity of human travel in ancient compared to modern times. Some recently introduced island snakes are native to nearby islands in the same body of water, such as *Elaphe climacophora* on Hachijo-jima of the Izu Islands (Ohno 1994), *Protobothrops elegans* on Okinawa-jima of the Nansei-shotō (Ryukyu) Islands (Ota 1999), *Mastigodryas bruesi* in Barbados (Underwood *et al.* 1999), and *B. constrictor* in several locations (Reynolds and Henderson 2018).

There are some species that are established relatively far from their native range(s). The farthest documented introductions include *X. vittatus*, introduced to Puerto Rico roughly 17,000 km from its native range in Indonesia (Herrera-Montes 2015); *Python bivittatus*, introduced to Key Largo, Florida, roughly 15,000 km from its native range in Southeast Asia (Reed and Rodda 2011); *Lampropeltis californiae*, native to western North America and established roughly 9,300 km away in Isla Gran Canaria, Spain (Cabrera-Pérez *et al.* 2012); and *L. aulicus*, introduced to New Guinea roughly 8,600 km from its native range in Southeast Asia (O'Shea *et al.* 2018). *Indotyphlops braminus* is a likely contender for some of the farthest introductions, but it is so widely established that its native origin is difficult to determine; it may be native to India and Sri-Lanka (Wallach 2009).

The pathways of snake introductions to islands vary widely. The majority of documented introductions are attributed to nursery trade and cargo (41% and 31% of all introduction events, respectively; Figure 13.4). Almost all introductions of *I. braminus* are associated with cargo or nursery trade, excepting one potential natural dispersal event to Krakatoa (Rawlinson *et al.* 1992). When excluding *I. braminus*, 28% of documented snake introductions to islands are attributed to cargo. Considering only established populations, the top pathway is pet trade, accounting for 23% of established introductions, and cargo falls to 21% (Figure 13.4). Some snake introductions to islands are also attributed to intentional releases, medicine and religious reasons, and entertainment, medical, and liquor industry escapes. There are no documented instances of snakes introduced from islands due to research or food pathways (excepting snake liquor, which is considered medicinal by some importers).

A lack of detailed historical records makes the pathway(s) of many of the ancient introductions uncertain. The oldest introductions, those of *Z. scalaris* to Menorca and *N. maura* to Mallorca, may be a result of a psychological warfare technique, wherein Phoenicians and Carthaginians are reported having dropped vases full of serpents on enemy ships to cause panic among the sailors (Pleguezuelos 2002). Çabej (2012) claims *N. maura* may have been introduced to Mallorca due to the use of snakes in Roman fertility rituals (though the author does not give a source). One of the other oldest documented introductions was of *T. fallax* to several Greek islands in the 13th century as a religious ritual release (Warnecke 1988). More recent introductions are better documented.

Intentional release (*i.e.*, release with the explicit purpose of establishing non-native snake populations on islands) has been implicated in the introductions of *Z. longissimus* to Great Britain (Langton *et al.* 2011) and *Hebius vibakari* to Miyake-jima (Sengoku 1979). Boas (*B. imperator*) were introduced to Cozumel after their use as

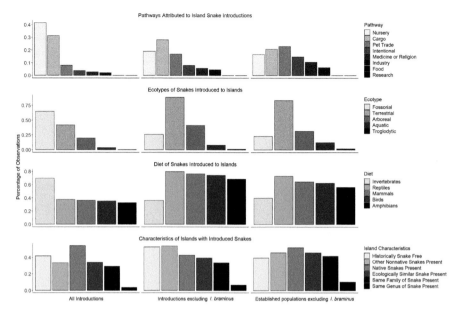

Figure 13.4 A summary of the introduction pathways of snakes to islands, the ecotypes and diets of snakes introduced to islands, and the ecological characteristics of islands where snakes have been introduced. Data are shown for all documented introductions of snakes to islands, introductions excluding *Indotyphlops braminus* (which accounts for the majority of snake introductions to islands), and established populations of snakes on islands excluding *I. braminus*.

props in a movie filmed there (Martínez-Morales and Cuarón 1999). Some other industrial escapes and releases are also implicated, such as *Protobothrops elegans* and *P. mucrosquamatus*, which were imported for use in snake liquor and subsequently introduced to Okinawa-jima, Japan (Terada 2011; Figure 13.5); *Elaphe taeniura*, which was imported for use in medicinal snake powder and display in zoos and was subsequently introduced to Okinawa-jima (Terada 2011); and *N. kaouthia*, which was imported in the thousands for entertainment purposes to Okinawa-jima but appears not to have established (Shiroma *et al.* 1994; H. Ota, *pers. comm.*).

Many snakes are introduced to islands unintentionally when hiding in various types of shipments and as stowaways in cargo. Nursery and plant trade are likely responsible for recent introductions of *Z. scalaris*, *Hemmorhois hippocrepis*, and *M. monspessulanus* to various Balearic Islands via imported olive trees (Silva-Rocha *et al.* 2015). Plant shipments from Florida, in the United States, led to the introduction of *Pantherophis guttatus* to the Bahamas (Johnson and Woods 2016). Consolidation of military assets and cargo during and following World War II is implicated in the introductions of *B. irregularis* to Guam (Kahl *et al.* 2012) and *T. fallax* and *Hemorrhois algirus* to Malta, although the non-native status of the latter two is contentious (Schembri and

Figure 13.5 Various *Protobothrops* species in snake liquor for sale in Okinawa-jima, Japan.
Photograph by Emily Taylor.

Lanfranco 1996). Other cargo-associated introductions include *Z. longissimus* to Elba Island in Italy (Di Nicola and Vaccaro 2020) and *L. aulicus* to Mauritius, Christmas Island, Reunion Island, and New Guinea through various trade routes.

In recent years, the release or escape of captive pet snakes has been implicated in many introductions. In line with Kraus (2009), a greater proportion of recent introductions of snakes to islands are attributed to pet trade (37% of established non-blindsnake introductions after 1970 compared to 6% of established non-blindsnake introductions before 1970). A popular pet trade snake, *P. bivittatus*, established in Key Largo, Florida, thus expanded its invasive range from mainland Florida (Greene *et al.* 2007; Hanslowe *et al.* 2018; Figure 13.6); *P. bivittatus* may expand farther into the southern Keys and Florida Bay, as this snake has the capacity to disperse across saltwater (Bartoszek *et al.* 2018). Other snakes introduced through pet trade include *Storeria dekayi* to the Bahamas (Lee 2005) and *B. constrictor* to Aruba, Puerto Rico, and St. Croix (Martínez-Morales and Cuarón 1999; Reynolds *et al.* 2013; Bushar *et al.* 2015). On Gran Canaria, multiple color patterns of introduced *L. californiae* that are common in the pet trade are commonly observed (Cabrera-Pérez *et al.* 2012). Continued introductions of snakes to islands are likely due to the growing popularity and availability of snakes in the pet trade (Valdez 2021).

Figure 13.6 Introduced to Florida, USA, from pet trade, *Python bivittatus* has established on Key Largo, Florida by expanding its introduction from the mainland. Photograph by Kodiak Hengstebeck.

The pathway of arrival may influence the establishment likelihood of snakes on islands and the potential to prevent or manage these introductions. Many snakes arrive to islands in cargo and via the pet trade, but snakes may be more likely to establish if purposefully imported—where snakes may arrive in greater numbers with appropriate sex ratios (*i.e.*, importers may have goals of establishing captive breeding populations). While escapees from pet trade are not intentional introductions, all species of snakes that were intentionally introduced to islands successfully established populations on those islands. Intentional and pet trade introductions of snakes to islands may be preventable with clear regulations and diligent screening of animal imports. In contrast, introductions from unintentional pathways such as nursery trade or cargo are difficult to prevent as intense screening of nearly all shipments becomes necessary. Predicting impacts of snake introductions to islands based on arrival pathways is challenging because of the sparse data available, and the impacts appear more closely tied to island characteristics than arrival pathways.

Ecology and Impacts

The impacts of non-native species can be broadly classified into ecological, human health, or economic impacts (Jeschke *et al.* 2014). It can often take decades after establishment for impacts of introductions to become apparent (known as a *lag effect*; Parker *et al.* 1999; Jeschke *et al.* 2014). Most snake introductions to islands have not been formally evaluated for associated impacts. This is partly due to the time and cost associated with researching impacts and a general lack or availability of data pre-introduction to conduct appropriate comparisons—a consistent challenge for most biological invasions (Parker *et al.* 1999). The lack of or lag in data collection makes it particularly difficult to compare severity of impacts attributable to different snake species' introductions to islands. Perhaps unsurprisingly, these data challenges can stagnate management action when funding for snake eradications depends on demonstrated impacts. Here, we summarize the demonstrated and potential impacts of documented snake introductions to islands.

Evaluating actual or potential ecological impacts of introduced snakes depends on knowledge of both the ecology of the introduced snake and of the island before its introduction. To date, most snakes introduced to islands are non-venomous, terrestrial, generalist predators averaging 1,659 mm maximum SVL (Figure 13.4). However, the most widely introduced snake, *I. braminus*, is a termite and ant specialist (Wallach 2009). Three other snakes with specialist diets have established on islands: *N. maura*, an amphibian specialist (Pinya *et al.* 2016); *E. climacophora*, a mammal specialist (Hasegawa and Moriguchi 1989); and *M. mauritanicus*, a reptile specialist (Pinya and Carretero 2011). Six venomous species have been introduced to islands. Of the documented established introductions (excluding *I. braminus*), 39.5% comprise introductions to historically snake-free islands and 54% to islands that lack snakes of the same ecotype (Figure 13.4); these circumstances may increase the likelihood of ecological impacts following snake introduction.

The most common ecological impact documented among snakes introduced to islands is their role in the decline or extirpation of endemic species. *Boiga irregularis* is a focal example as it is responsible for the extirpation of at least 10 species of endemic birds, reptiles, and mammals on Guam, a historically snake-free island (Fritts and Rodda 1998). The introduction resulted in ecological collapse with the loss of major pollinators and seed-dispersers (Mortensen *et al.* 2008; Caves *et al.* 2013). The saga of this snake is well-recorded elsewhere (Fritts and Rodda 1998). Although Aruba hosts native snakes, the introduced *B. constrictor* is implicated in native bird declines on this island (Derix *et al.* 2013; Wells and Wells 2017). Non-avian species on islands also suffer the effects of snake introductions. *Macroprotodon mauritanicus* and *H. hippocrepis* likely contributed to the extirpation of endemic *Podarcis* lizard species (*P. lilfordi*) from Mallorca and Menorca; *P. pituyensis* from Ibiza (Pinya and Carretero 2011; Montes *et al.* 2022); and *N. maura* is implicated in the decline and extinction of an endemic toad species (*Alytes muletensis* on Mallorca and Menorca; Pantoja Cuadros 2016). In Gran Canaria, *L. californiae* has reduced the abundance of some endemic lizard species by 90% in invaded areas (Piquet and López-Darias 2021). On Christmas Island, *L. aulicus* contributes to ongoing declines of four endemic reptiles, compounding the impacts of invasive cats on these species (Smith *et al.* 2012). *Python bivittatus* consumes federally endangered rodents in Key Largo, Florida (Greene *et al.* 2007), and it will be unsurprising when endemic key deer are documented in their diet.

Introduced snakes also have ecological impacts that are not related to predation. For example, there may be genetic pollution of the endemic Habu Snake (*Protobothrops flavoviridis*), which has hybridized with introduced *P. elegans* (Aird *et al.* 2015) and *P. mucrosquamatus* (Ota 2015) in Okinawa-jima, Japan. Impacts caused by introduction of invasive parasites and pathogens are also possible but may be overlooked due to a poor record of endemic parasites prior to invasion. *B. irregularis* is implicated in the introduction of a parasitic flatworm (*Spirometra erinaceieuropaei*) to Guam, where it is able to complete its complex lifecycle by using other non-native hosts (Holldorf *et al.* 2015). Likewise, *P. bivittatus* is responsible for spillover of pentastome parasites in Florida that capitalize on both native and non-native reptilian hosts (Miller *et al.* 2018; Fieldsend *et al.* 2021), many of which occur in the Florida Keys.

Speculated ecological impacts exist for several species, often where data are lacking or in more recent introductions. For example, *M. monspessulanus* in the Balearic Islands and *Elaphe taeniura* in Okinawa-jima may present threats through depredation of endemic prey species (Ota *et al.* 2004; Pinya and Carretero 2011). *Pantherophis guttatus*, which appears to be spreading throughout islands in the Bahamas, may present a threat to endemic prey species while also competing with native snakes (*e.g.*, *Alsophis vudii picticeps*, *Epicrates* spp.; Worthington-Hill *et al.* 2014; Johnson and Woods 2016). *B. imperator* might be threatening the critically endangered endemic dwarf carnivores of Cozumel and other endemic fauna (González Baca 2006; Romero-Nájera, *et al.* 2007). *Indotyphlops braminus* may compete with native, endemic blindsnakes throughout the Caribbean and Micronesia (Ineich *et al.* 2017; DeVos and Giery 2021). It remains important to collect long-term data to formally evaluate the outcome of speculated impacts.

Several introduced snake species have demonstrated economic impacts on islands. The majority of documented costs arising from reptiles are associated with their invasion on islands, in part due to the economic impacts of *B. irregularis* (Soto *et al.* 2022). For example, in a one-year period *B. irregularis* bridged power lines and caused electrical black-outs on Guam an average of once every 1.8 days, costing millions of dollars (Shwiff *et al.* 2010). Many islands serve as popular tourist destinations, and the introduction of snakes has potential to disrupt the tourist economy. One study estimated the most conservative potential losses in tourism revenue to be $138 million in Hawaii if *B. irregularis* were to establish there (Shwiff *et al.* 2010). The management of invasive snakes on islands also incurs costs. For example, $3.4 million was allocated in fiscal year 2020 for research and management of *B. irregularis* (US Department of the Interior 2020), and over €1 million was spent from 2011 to 2015 on *L. californiae* research and management in the Canary Islands (Ministerio para la Transición Ecológica 2018; Figure 13.7). The impact of invasive snakes on ecosystem services may also incur costs, but the losses of ecosystem services attributable to invasive snakes have not been quantified (Pejchar and Mooney 2009).

Figure 13.7 The California Kingsnake, *Lampropeltis californiae*, established on Gran Canaria, is implicated in the drastic decline in endemic herpetofauna there.
Photograph by Natalie Claunch.

In general, there are few direct impacts of introduced insular snakes to human well-being. Venomous snakes can have some impact if people are bitten and envenomated. Medically significant bites have been documented from introduced *Protobothrops elegans* and *P. mucrosquamatus* on Okinawa-jima (Izumi *et al.* 2014; Anonymous 2020). On Guam, bites from *B. irregularis* lead to around 170 hospital visits/year (Shwiff *et al.* 2010). Introduced snakes may have indirect influence on human health by influencing local disease ecology (*e.g.*, *P. bivittatus*; Hoyer *et al.* 2017), though these have yet to be documented on islands. Snake introductions can also influence human culture—for instance in Chuuk, where no native snakes existed prior to the introduction of *I. braminus*. There were no indigenous words to describe snakes, so existing words for "eel" and "worm" are used as well as adoption of the word "serpent" into Chuukese (*serepenit*; Davis 1999). On Guam, a migratory bird festival commemorates the cultural importance of species extirpated by *B. irregularis* in hopes of raising awareness and support for eventual reintroduction of the birds (Demeulenaere 2022).

From a perspective of rapid evolution, it is interesting to consider that introductions to islands can impact the phenotypes of introduced snakes (see also Chapters 4 and 14 of this volume). This is apparent with *B. irregularis*, which attains larger body sizes compared to native populations (Whittier *et al.* 2000) and has experienced physiological and reproductive stress in conjunction with the extirpation of native prey, fluctuating population sizes, and management pressure (Claunch *et al.* 2021). In the Balearic Islands, introduced *H. hippocrepis* have an increased reproductive season and greater body condition, and females mature at smaller sizes compared with native mainland populations (Montes *et al.* 2020). Similar effects are observed in *L. californiae* in Gran Canaria, where snakes produce larger clutches and have greater body condition than in their native range (Fisher *et al.* 2021). Continued study of insular snake invasions may reveal the relative importance of factors such as prey availability, release from pressures of competition or predation, and niche shifts influencing the resultant phenotypes in invasive populations.

Response and Prevention Efforts

Due to their relative isolation, islands present a realistic opportunity to monitor borders and trade routes to practice pro-active management by preventing invasive snakes from establishing outright. Some islands have strict prevention and interdiction protocols in place that are apparently preventing snake introductions, such as the swift interception of *I. braminus* in New Zealand (Gill *et al.* 2001). It is notable that, because of these efforts, no snakes have been successfully introduced to New Zealand (Gill *et al.* 2001). Where snakes have established, there are interdiction efforts to prevent potential spread to other islands, such as the use of heat fumigation and detector dogs to intercept potential *B. irregularis* transport from Guam to Rota, Hawaii, and other Pacific islands (Kahl *et al.* 2012). Theoretically, eradication of established snakes should be facilitated by the isolation and generally smaller size of islands. However, snakes are notoriously difficult to detect, which is a barrier to eradication wherever they establish. Somewhat counterintuitively, the most useful prevention mechanism—regulation in trade or importation of snake species—often comes well after establishment as a reactive measure. Examples are *B. irregularis* (Nonindigenous Aquatic Nuisance Prevention and Control Act of 1990), *P. bivittatus* (Injurious Wildlife Species 2012), and *L. californiae* (Real Decreto 630/2013) which were restricted in trade approximately 40, 5, and 6 years after likely establishment on islands, respectively (Rodda *et al.* 1999; Greene *et al.* 2007; Pether and Mateo 2007). To our knowledge, there are no examples of successful eradications of snakes introduced to islands.

When prevention efforts (or lack thereof) result in establishment of invasive snakes on islands, management is often required. In the case of invasive *P. mucrosquamatus* on Okinawa-jima, government control measures for removing native venomous snakes, mostly *Protobothrops flavoviridis*, from urban areas were already in place (Terada 2011), which provided existing infrastructure for invasive snake management. Removal numbers of *P. mucrosquamatus* stopped increasing, which may indicate that the population stabilized or that removal programs were somewhat effective (Terada 2011), though the range of occurrence of this snake on Okinawa-jima is still extending gradually (Anonymous 2020). This might be an exception, however, as most islands have had to create and fund snake-specific control programs or integrate snake control into existing wildlife conservation organizations with limited success.

Various control and removal efforts have been employed in managing snake invasions of islands. The first step is often a community-based effort to report and remove snakes as they are encountered using hotlines and rapid-response teams, where members of dedicated agencies, trained community-member volunteers, or paid contractors handle removals (*e.g.*, Cabrera-Pérez *et al.* 2012). In the case of *Z. longissimus* in England, abandonment of such removal efforts may have resulted in its establishment 10 years later (Langton *et al.* 2011). More formal visual searching methods are also used, sometimes with detector dogs trained on invasive snake scent to help narrow down locations (*e.g.*, Romagosa *et al.* 2011). Harris Hawks have also been used to capture *L. californiae* with limited success (Gallo-Barneto 2013). A common method for removal is the use of traps, which are sometimes baited with prey items such as rodents (*e.g.*, Terada 2011; Silva-Rocha *et al.* 2018; Figure 13.8). In the case of *L. californiae*, which is often ophiophagous, shed rattlesnake skins have also been used as a lure (Gallo-Barneto *et al.* 2016). Containment or exclusion fences are also used, though success with this method relies on certainty that snakes are not already present on the other side of the fence and that the fence itself is an efficient barrier to snakes; the latter is an issue where land features such as rivers prevent fencing (Terada 2011). Some creative methods have also been applied, such as the tracking of radio-telemetered *P. bivittatus* as scout snakes to lead observers to other individuals (Smith *et al.* 2016; US Geological Survey Wetland and Aquatic Research Center 2022) and the aerial deployment of acetaminophen-laced mouse baits as toxins on Guam (Siers

Figure 13.8 A trap using live rodent bait to remove *Zamenis scalaris* on the Balearic Islands, where it is responsible for the decline of endemic lizards.
Photograph by Roberto García-Roa.

et al. 2019). Evaluating the effectiveness of each method is beyond the scope of this chapter but warrants investigation.

Future Directions and Gaps

As snakes continue to be introduced to islands and as established species continue to progress in their invasions, there are several gaps of knowledge to address. Science-based evaluations of pro-invasion traits in snakes on islands are necessary to improve screening tools for prioritizing management and regulating trade. This may require a meta-analytic approach that focuses especially on the context of introductions to ev-olutionarily snake-free islands, which may be more ecologically sensitive to invasion. A major caveat is that studies of invasions are biased toward successfully established species as failed invasions are often imperceptible or unreported (Zenni and Nuñez 2013). A lack of data on snake traits from failed invasions makes an unbiased formal evaluation of snake pro-invasion traits particularly difficult. However, as introduc-tions of snakes to islands are likely to continue, such efforts may still yield important predictions.

One potential strategy to address the bias in documentation of established versus failed introductions is the promotion of community science efforts. Data from ecological reporting platforms such as iNaturalist are improving ecological monitoring globally (Brown and Williams 2019), including detection of invasive species (Pawson *et al.* 2020). These and other locality-specific platforms have the potential to increase detections of newly introduced snake species on islands as snakes are often secretive, with low detection rates that prohibit exhaustive scientific surveys of islands. From a management perspective, people may be more likely to use reporting platforms than traditional hotlines (Pawson *et al.* 2020), and platforms where data are publicly available may provide more incentive to participate through tangible contributions to understanding and protecting the local ecology. There are challenges to promoting or adopting this method, especially in areas with limited internet access or where access is restricted by the government, but it may be a useful strategy to pursue.

Another important gap needed to understand island snake invasions and to justify screening and prevention efforts is an improved understanding of impacts. To be classified as invasive species, and in some cases to justify funding for management, impacts of the introduced snakes must be demonstrated and quantified. As impact data are slow to accumulate and some impacts of snakes are irreversible, it may also be more cost-effective to adopt a proactive management approach by assuming negative impacts and employing rapid eradication at the initial discovery of an introduced snake to an island (*e.g.*, Ahmed *et al.* 2022). The majority of snakes introduced to islands have not been evaluated for impacts. The lack of data on impacts might be misinterpreted by agencies to indicate species are not a threat (Beck *et al.* 2008), which may result in the classic reactive management approach—delaying eradication efforts until impacts are obvious, well after species are firmly established and after long-term management becomes cost-prohibitive. This circular reasoning is not present in all countries, however, and some governments may benefit from international collaboration in developing policies for screening and prevention of introduced snakes. For advancing both management efforts and an understanding of how island snake invasions drive ecological and evolutionary change, better data evaluating the lag effect of impacts are necessary. It is important to prioritize long-term ecological monitoring, especially on islands where recent snake introductions have been documented, to formally evaluate management actions and create a record of impacts that can act as justification for proactive management measures in the future.

References

Ahmed D. A., E. J. Hudgins, R. N. Cuthbert, M. Kourantidou, C. Diagne, P. J. Haubrock, B. Leung, *et al.* 2022. Managing biological invasions: The cost of inaction. *Biological Invasions* online preview. https://doi.org/10.1007/s10530-022-02755-0
Aird, S. D., S. Aggarwal, A. Villar-Briones, M. M. N. Tin, K. Terada, and A. S. Mikheyev. 2015. Snake venoms are integrated systems, but abundant venom proteins evolve more rapidly. *BMC Genomics* 16:1–20.

Anonymous. 2020. Distribution data of *Protobothrops mucrosquamatus* from a survey in 2019. *Eiseiken News* 40:5. (in Japanese) https://www.pref.okinawa.jp/site/hoken/eiken/news/documents/40taiwanhabu.pdf

Bartoszek, I. A., M. B. Hendricks, I. C. Easterling, and P. T. Andreadis. 2018. *Python bivittatus* (Burmese python): Dispersal/marine incursion. *Herpetological Review*, 49:554–555.

Beck, K. G., K. Zimmerman, J. D. Schardt, J. Stone, R. R. Lukens, S. Reichard, J. Randall, *et al.* 2008. Invasive species defined in a policy context: Recommendations from the Federal Invasive Species Advisory Committee. *Invasive Plant Science and Management* 1:414–421.

Brown, E. D., and B. K. Williams. 2019. The potential for citizen science to produce reliable and useful information in ecology. *Conservation Biology* 33:561–569.

Bushar, L. M., R. G. Reynolds, S. Tucker, L. C. Pace, W. I. Lutterschmidt, R. A. Odum, and H. K. Reinert. 2015. Genetic characterization of an invasive *Boa constrictor* population on the Caribbean island of Aruba. *Journal of Herpetology* 49:602–610.

Çabej, N. 2012. *Epigenetic Principles of Evolution* Elsevier.

Cabrera-Pérez, M. Á., R. Gallo-Barneto, I. Esteve, C. Patiño-Martínez, and L. F. López-Jurado. 2012. The management and control of the California kingsnake in Gran Canaria (Canary Islands): Project LIFE+ Lampropeltis. *Aliens: The Invasive Species Bulletin* 32:20–28.

Caves, E. M., S. B. Jennings, J. HilleRisLambers, J. J Tewksbury, and H. S. Rogers. 2013. Natural experiment demonstrates that bird loss leads to the cessation of dispersal of native seeds from intact to degraded forests. *PLoS One* 8:e65618.

Claunch, N., K. L. Clancy, M. E. A. Harman, K. C. Hengestebeck, D. Juárez-Sánchez, D. Haro, A. Hartmann, *et al.* 2022. Invasive snakes on islands: Dataset and species vignettes (pre) [Data set]. *Zenodo*:10.5281/zenodo.6629178

Claunch, N., I. Moore, H. Waye, S. J. Oakey, R. N. Reed, and C. M. Romagosa. 2021. Understanding metrics of stress in the context of invasion history: The case of the brown treesnake (*Boiga irregularis*). *Conservation Physiology* 9:coab008.

Davis, A. E. 1999. A preliminary list of animal names in the Chuuk district, Micronesia. *Micronesica* 31:1–245.

Demeulenaere, E. 2022, February 11. Migratory bird festival infuses storytelling, art, and conservation. *Pacific Daily News*. Accessed June 3, 2022. https://www.guampdn.com/lifestyle.

Derix, R., G. Peterson, and D. Marquez. 2013. *The National Bird Count in 2011 in Aruba*. Central Bureau of Statistics (CBS) Aruba, Department of Environmental Statistics.

DeVos, T., and S. Giery. 2021. Establishment of the introduced Brahminy blindsnake (*Indotyphlops braminus*) on Abaco Island, the Bahamas, with notes on potential niche overlap with the native Cuban Brown Blindsnake (*Typhlops lumbricalis*). *Reptiles & Amphibians* 28:555–557.

Di Nicola, M. R., and A. Vaccaro. 2020. New data on the presence of the Aesculapian snake *Zamenis longissimus* on Elba Island (Tuscany, Italy). *Biodiversity Journal* 11:611–614.

Endangered and Threatened Wildlife and Plants; Removal of Puerto Rican Boa from the Federal List of Endangered and Threatened Wildlife. 87 F. R. 133 (proposed July 13, 2022). https://www.govinfo.gov/content/pkg/FR-2022-07-13/pdf/2022-14961.pdf

Febrer-Serra, M., N. Lassnig, E. Perelló, V. Colomar, G. Picó, A. Aguiló-Zuzama, A. Sureda, and S. Pinya. 2021. Invasion of Montpellier snake *Malpolon monspessulanus* (Hermann, 1809) on Mallorca: New threat to insular ecosystems in an internationally protected area. *BioInvasions Records* 10:210–219.

Fieldsend, T., M. Harman, and M. Miller. 2021. First record of an Asian tongueworm, *Raillietiella orientalis* (Pentastomida: Raillietiellidae), parasitizing a Tokay gecko (*Gekko gecko*, Squamata: Gekkonidae): A novel interaction between two non-native species in Florida. *Reptiles and Amphibians* 28:255–256.

Fisher, S., R. N. Fisher, S. E. Alcaraz, R. Gallo-Barneto, C. Patino-Martinez, L. F. López-Jurado, M. Á. Cabrera-Pérez, and J. L. Grismer. 2021. Reproductive plasticity as an advantage of snakes during island invasion. *Conservation Science and Practice* 3:e554.

Fritts, T., and G. Rodda. 1998. The role of introduced species in the degradation of island ecosystems: A case history of Guam. *Annual Review of Ecology and Systematics* 29:113–140.

Gallo-Barneto, R. 2013, June 16–19. Control of invasive alien species *Lampropeltis getula californiae* on the island of Gran Canaria. Conference Proceedings: *Invasive Alien Predators: Policy, research and management in Europe*, 13.

Gallo-Barneto, R., M. Á. Cabrera-Pérez, M. P. Estévez, C. P. Martínez, and C. Monzón-Argüello. 2016. Culebra Real de California: Una intrusa en el Jardín de las Hespérides. *El Indiferente: Centro de Educación Ambiental Municipal* 22:126–141.

Gill, B. J., D. Bejakovich, and A. H. Whitaker. 2001. Records of foreign reptiles and amphibians accidentally imported to New Zealand. *New Zealand Journal of Zoology* 28:351–359.

González-Baca, C. A. 2006. *Ecología de forrajeo de Boa constrictor. Un depredador introducido a la Isla Cozumel.* Wildlife Conservation. Universidad Nacional Autónoma de México.

Greene, D. U., J. M. Potts, J. G. Duquesnel, and R. W. Snow. 2007. *Python molurus bivittatus* (Burmese python). *Herpetological Review* 38:355.

Hanslowe, E. B., J. G. Duquesnel, R. W. Snow, B. G. Falk, A. A. Y. Adams, E. F. Metzger III, M. A. Collier, and R. N. Reed. 2018. Exotic predators may threaten another island ecosystem: A comprehensive assessment of python and boa reports from the Florida Keys. *Management of Biological Invasions* 9:369–377.

Hasegawa, M., and H. Moriguchi. 1989. Geographic variation in food habits, body size and life history traits of the snakes on the Izu Islands. In M. Matsui, T. Hikida, and R. C. Goris (eds.), *Current Herpetology in East Asia*. Herpetological Society of Japan, pp. 414–432.

Herrel, A., K. Huyghe, B. Vanhooydonck, T. Backeljau, K. Breugelmans, I. Grbac, R. Van Damme, and D. J. Irschick. 2008. Rapid large-scale evolutionary divergence in morphology and performance associated with exploitation of a different dietary resource. *Proceedings of the National Academy of Sciences* 105:4792–4795.

Herrera-Montes, A. 2015. Notes on the striped keelback (*Xenochrophis vittatus*) in Puerto Rico: A recently reported non-native snake in the western hemisphere. *Reptiles and Amphibians* 22:178–181.

Holldorf, E. T., S. R. Siers, J. Q. Richmond, P. E. Klug, and R. N. Reed. 2015. Invaded invaders: Infection of the invasive brown treesnakes on Guam by an exotic larval cestode with a life cycle comprised of non-native hosts. *PLoS One* 10:e0143718.

Hoyer, I. J., E. M. Blosser, C. Acevedo, A. C. Thompson, L. E. Reeves, and N. D. Burkett-Cadana. 2017. Mammal decline, linked to invasive Burmese python, shifts host use of vector mosquito towards reservoir hosts of a zoonotic disease. *Biology Letters* 13:20170353.

Ineich I. V., A. D. Wynn, C. H. Giraud, and V. Wallach. 2017. *Indotyphlops braminus* (Daudin, 1803): Distribution and oldest record of collection dates in Oceania, with report of a newly established population in French Polynesia (Tahiti Island, Society Archipelago). *Micronesica* 1:1–3.

Injurious Wildlife Species; Listing Three Python Species and One Anaconda Species as Injurious Reptiles, 77 Fed. Reg. 3329 (January 23, 2012). https://www.govinfo.gov/app/details/FR-2012-01-23/2012-1155

Izumi, Y., K. Terada, N. Morine, and J. Kudaka. 2014. Epidemiology of venomous snakebite in Okinawa Prefecture in 2013. *Okinawa Prefectural Institute of Health and Environment Bulletin* 48:75–77.

Jeschke, J. M., S. Bacher, T. M. Blackburn, J. T. A. Dick, F. Essl, T. Evans, M. Gaertner, *et al.* 2014. Defining the impact of non-native species. *Conservation Biology* 28:1188–1194.

Johnson, S., and D. Woods. 2016. First record of red cornsnakes (*Pantherophis guttatus*) on Andros Island, the Bahamas. *Reptiles and Amphibians* 23:187.

Kahl, S. S., S. E. Henke, M. A. Hall, and D. K. Britton. 2012. Brown treesnakes: A potential invasive species for the United States. *Human-Wildlife Interactions* 6:181–203.

Kraus, F. 2009. *Alien Reptiles and Amphibians: A Scientific Compendium and Analysis.* Springer.

Langton, T. E. S., W. Atkins, and C. Herbert. 2011. On the distribution, ecology and management of non-native reptiles and amphibians in the London Area. *London Naturalist* 90:93–94.

Lee, D. S. 2005. Reptiles and amphibians introduced to the Bahamas: A potential conservation crisis. *Bahamas Journal of Science* 12:2–6.

Martínez-Morales, M. A., and A. D. Cuarón. 1999. *Boa constrictor*, an introduced predator threatening the endemic fauna on Cozumel Island, Mexico. *Biodiversity and Conservation* 8:957–963.

Meirte, D. 1993. New records of *Leioheterodon madagascariensis* (Reptilia: Colubridae) from the Comoros. *Journal of the Herpetological Association of Africa* 42:21–23.

Miller, M. A., J. M. Kinsella, R. W. Snow, M. M. Hayes, B. G. Falk, R. N. Reed, F. J. Mazzotti, *et al.* 2018. Parasite spillover: Indirect effects of invasive Burmese pythons. *Ecology and Evolution* 8:830–840.

Ministerio para la Transición Ecológica. 2018. *Estrategia de gestión, control, y posibleerradicación de ofidios invasores en islas.* Gobierno de España.

Montes, E., M. Feriche, L. Ruiz-Sueiro, E. Alaminos, and J. M. Pleguezuelos. 2020. Reproduction ecology of the recently invasive snake *Hemorrhois hippocrepis* on the island of Ibiza. *Current Zoology* 66:363–371.

Montes, E., F. Kraus, B. Chergui, and J. M. Pleguezuelos. 2022. Collapse of the endemic lizard *Podarcis pityusensis* on the island of Ibiza mediated by an invasive snake. *Current Zoology* 68:295–303.

Mortensen, H. S., Y. L. Dupont, and J. M. Olesen. 2008. A snake in paradise: Disturbance of plant reproduction following extirpation of bird flower-visitors on Guam. *Biological Conservation* 141:2146–2154.

Nonindigenous Aquatic Nuisance Prevention and Control Act of 1990. Pub. L. No. 101-646, title I, 104 Stat. 4761. 1990. https://www.congress.gov/bill/101st-congress/house-bill/5390/actions?r=4&s=2

Ohno, M. 1994. Reptiles and amphibians. In Anonymous (ed.), *Report from a Basic Survey for Environment of the Natural Park around Geothermal Sites of Hachijo-jima Island, the Izu Islands, Central Japan.* Tokyo Electric Power Company and Tokyo Electric Power Service, pp. 256–262. (in Japanese)

O'Shea, M., K. I. Kusuma, and H. Kaiser. 2018. First record of the island wolfsnake, *Lycodon capucinus*, from New Guinea, with comments on its widespread distribution and confused taxonomy, and a new record for the common sun skink, *Eutropis multifasciata. Reptiles and Amphibians: Conservation and Natural History* 25:70–84.

Ota, H. 1999. Introduced amphibians and reptiles of the Ryukyu archipelago, Japan. In G. H. Rodda, Y. Sawai, D. Chizar, and H. Tanaka (eds.), *Problem Snake Management: The Habu and Brown Tree Snake.* Cornell University Press, pp. 439–452.

Ota, H. 2015. Genetic pollution in Japanese reptiles, caused by introduction of non-native species and other artificial changes in habitat environment. *Iden (Heredity, Japan)* 69(2):85–94. (in Japanese)

Ota, H., M. Toda, G. Masunaga, A. Kikukawa, and M. Toda. 2004. Feral populations of amphibians and reptiles in the Ryukyu Archipelago, Japan. *Global Environmental Research* 8:133–143.

Pantoja Cuadros, E. 2016. *Ecology of the viperine snake* (Natrix maura) *as invasive snake in Mallorca: A first approach.* Universitat de les Illes Balears. Final Degree Project Report.

Parker, I. M., D. Simberloff, W. M. Lonsdale, K. Goodell, M. Wonham, P. M. Kareiva, M. H. Williamson, *et al.* 1999. Impact: Toward a framework for understanding the ecological effects of invaders. *Biological Invasions* 1:3–19.

Pawson, S. M., J. J. Sullivan, and A. Grant. 2020. Expanding general surveillance of invasive species by integrating citizens as both observers and identifiers. *Journal of Pest Science* 93:1155–1166.

Pejchar L., and H. A. Mooney. 2009. Invasive species, ecosystem services, and human well-being. *Trends in Ecology and Evolution* 24:497–504.

Perry, G., and D. Vice. 2009. Forecasting the risk of brown tree snake dispersal from Guam: A mixedtransport establishment model. *Conservation Biology* 23:992–1000.

Pether, J., and J. A. Mateo. 2007. La Culebra Real (*Lampropeltis getulus*) en Gran Canaria, otro caso preocupante de reptil introducido en el Archipiélago Canario. *Boletín de La Asociación Herpetológica Española* 18:20–23.

Pimentel, D., R. Zuniga, and D. Morrison. 2005. Update on the environmental and economic costs associated with alien-invasive species in the United States. *Ecological Economics* 52:273–288.

Pinya, S., and M. A. Carretero. 2011. The Balearic herpetofauna: A species update and a review on the evidence. *Acta Herpetologica* 6:59–80.

Pinya, S., S. Tejada, X. Capó, and A. Sureda 2016. Invasive predator snake induces oxidative stress responses in insular amphibian species. *Science of the Total Environment* 566:57–62.

Piquet, J. C., and M. López-Darias. 2021. Invasive snake causes massive reduction of all endemic herpetofauna on Gran Canaria. *Proceedings of the Royal Society B* 288:20211939.

Pleguezuelos J. M. 2002. Las especies introducidas de Anfibios y Reptiles. In Pleguezuelos J. M., R. Marquez and M. Lizana (eds.), *Atlas y Libro Rojo de los Anfibios y Reptiles de España*. Dirección General de Conservación de la Naturaleza-Asociación Herpetológica Española, pp. 503–524.

Rawlinson, P. A., R. A. Zann, S. Van Balen, and I. W. B. Thornton. 1992. Colonization of the Krakatau islands by vertebrates. *GeoJournal* 28:225–231.

Reaser, J. K., L. A. Meyerson, Q. Cronk, M. A. J. De Poorter, L. G. Eldrege, E. Green, M. Kairo, P. Latasi, R. N. Mack, J. Mauremootoo, and D. O'Dowd. 2007. Ecological and socioeconomic impacts of invasive alien species in island ecosystems. *Environmental Conservation* 34:98–111.

Real Decreto 630/2013, de 2 de Agosto, por el que se regula el Catálogo español de especies exóticas invasoras. 2013. https://www.cms.int/sites/default/files/document/cms_nlp_esp_real_decreto_630_2013.pdf

Reed, R. N., and G. H. Rodda. 2011. Burmese pythons and other giant constrictors. In D. Simberloff and M. Rejmanek (eds.), *Encyclopedia of Invasive Introduced Species*. University of California Press, pp. 85–91.

Reynolds, R. G., and R. W. Henderson. 2018. Boas of the world (Superfamily Booidae): A checklist with systematic, taxonomic, and conservation assessments. *Bulletin of the Museum of Comparative Zoology* 163:1–58.

Reynolds, R. G., A. R. Puente-Rolón, R. N. Reed., and L. J. Revell. 2013. Genetic analysis of a novel invasion of Puerto Rico by an exotic constricting snake. *Biological Invasions* 15:953–959.

Rodda, G. H., T. H. Fritts, M. J. McCoid, and E. W. Campbell III. 1999. An overview of the biology of the brown tree snake (*B. irregularis*), a costly introduced pest on the Pacific Islands. In G. H. Rodda, Y. Sawai, D. Chizar, and H. Tanaka (eds.), *Problem Snake Management: The Habu and Brown Tree Snake*. Cornell University Press, pp. 44–80.

Romagosa, C. M., T. D. Steury, M. A. Miller, C. Guyer, B. Rogers, T. C. Angle, and R. L. Gillette. 2011. Assessment of detection dogs as a potential tool for python detection efforts. Unpublished report, Auburn University.

Romero-Nájera, I., A. D. Cuarón, and C. González-Baca. 2007. Distribution, abundance, and habitat use of introduced Boa constrictor threatening the native biota of Cozumel Island, Mexico. *Biodiversity and Conservation* 16(4):1183–1195. https://doi.org/10.1007/s10531-006-9101-2

Schembri, P. J., and E. Lanfranco. 1996. Introduced species in the Maltese Islands. In A. E. Baldacchino and A. Pizzuto (eds.), *Introduction of Alien Species of Flora and Fauna.* [Proceedings of a seminar held at Qawra, Malta, 5 March 1996]. Environment Protection Department, pp. 29–54.

Sengoku, S. 1979. Notes on *Amphiesma v. vibakari* on Miyake-jima Island. *Japanese Journal of Herpetology* 8:64–65. (in Japanese)

Shiroma, H., S. Katsuren, and M. Nozaki. 1994. Cobra species in Nago, Okinawa. *Annual Report Okinawa Prefectural Institute of Health and Environment* 28:89–93.

Shwiff, S. A., K. Gebhardt, K. N. Kirkpatrick, and S. S. Shwiff. 2010. Potential economic damage from introduction of brown tree snakes, *Boiga irregularis* (Reptilia: Colubridae), to the islands of Hawai'i. *Pacific Science* 64:1–10.

Siers, S. R., W. C. Pitt, J. D. Eisemann, L. Clark, A. B. Shiels, C. S. Clark, R. J. Gosnell, and M. C. Messaros. 2019. *In situ* evaluation of an automated aerial bait delivery system for landscape-scale control of invasive brown treesnakes on Guam. In C. R. Veitch, M. N. Clout, A. R. Martin, J. C. Russell and C. J. West (eds.), *Island Invasives: Scaling Up to Meet the Challenge.* Occasional Paper SSC no. 62. IUCN, pp. 348–355.

Silva-Rocha, I., D. Salvi, N. Sillero, J. A. Mateo, and M. A. Carretero. 2015. Snakes on the Balearic Islands: An invasion tale with implications for native biodiversity conservation. *PLoS One* 10:e0121026.

Silva-Rocha, I., E. Montes, D. Salvi, N. Sillero, J. A. Mateo, E. Ayllón, J. M. Pleguezuelos, and M. A. Carretero. 2018. Herpetological history of the Balearic Islands: When aliens conquered these islands and what to do next. In A. I. Queiroz and S. Pooley (eds.), *Histories of Bioinvasions in the Mediterranean.* Springer, pp. 105–131.

Smith, B. J., M. S. Cherkiss, K. M. Hart, M. R. Rochford, T. H. Selby, R. W. Snow, and F. J. Mazzotti. 2016. Betrayal: Radio-tagged Burmese pythons reveal locations of conspecifics in Everglades National Park. *Biological Invasions* 18:3239–3250.

Smith, M., H. Cogger, B. Tiernan, D. Maple, C. Boland, F. Napier, T. Detto, and P. Smith. 2012. An oceanic reptile community under threat: The decline of reptiles on Christmas Island, Indian Ocean. *Herpetological Conservation and Biology* 7:206–218.

Soto, I., R. N. Cuthbert, A. Kouba, C. Capinha, A. Turbelin, E. J. Hudgins, C. Diagne, F. Courchamp, and P. J. Haubrock. 2022. Global economic costs of herpetofauna invasions. *Scientific Reports* 12:10829.

Stuart, Y. E., T. S. Campbell, P. A. Hohenlohe, R. G. Reynolds, L. J. Revell, and J. B. Losos. 2014. Rapid evolution of a native species following invasion by a congener. *Science* 346:463–466.

Terada, K. 2011. The distribution, population density, and controls of *Protobothrops mucrosquamatus, Protobothrops elegans,* and *Elaphe taeniura friesei,* 3 snake species established in Okinawa Island. *Bulletin of the Herpetological Society of Japan* 2011(2):161–168.

Towns, D. R., I. A. E. Atkinson, and C. H. Daugherty. 2006. Have the harmful effects of introduced rats on islands been exaggerated? *Biological Invasions* 8:863–891.

Underwood, G., J. A. Horrocks, and J. C. Daltry. 1999. A new snake from Barbados. *Journal of the Barbados Museum and Historical Society* 45:67–75.

US Department of the Interior. 2020 June 03. Interior announces $3.4 million for brown tree snake control on Guam. https://www.doi.gov/oia/press/interior-announces-34-million-brown-tree-snake-control-guam

US Geological Survey Wetland and Aquatic Research Center. 2022, February 8. Using scout Burmese pythons and detector dogs to protect endangered species in the Florida Keys.

https://www.usgs.gov/centers/wetland-and-aquatic-research-center/science/using-scout-burmese-pythons-and-detector-dogs

Valdez, J. W. 2021. Using Google trends to determine current, past, and future trends in the reptile pet trade. *Animals* 11:676.

Van Buurt, G. 2005. *Field guide to the amphibians and reptiles of Aruba, Curaçao and Bonaire.* Edition Chimaira.

Vigne, J. D., and J. A. Alcover. 1985. Incidence des rélations historiques entre l'homme et l'animal dans la composition actuelle du peuplement amphibien, reptilien et mammalien des îles de Mediterranée occidentale. *Actes du 110ème Congrés National des Sociétés Savantes* 2:79–91.

Wallach, V. 2009. *Ramphotyphlops braminus* (Daudin): A synopsis of morphology, taxonomy, nomenclature and distribution (Serpentes: Typhlopidae). *Hamadryad* 34:34–61.

Wallach, V. 2020. New country and state records for Indotyphlops braminus (Serpentes: Typhlopidae). Part II. *Bulletin of the Chicago Herpetological Society* 55:77–81.

Warnecke, H. 1988. *Telescopus fallax* (Fleischmann, 1831) auf den ozeanischen Strophaden-Inseln? *Salamandra* 24:16–19.

Wells, J. V., and A. C. Wells. 2017. *Birds of Aruba, Bonaire, and Curaçao.* Cornell University Press.

Whittier, J., C. Macrokanis, and R. T. Mason. 2000. Morphology of the brown tree snake, *Boiga irregularis,* with a comparison of native and extralimital populations. *Australian Journal of Zoology* 48:357–367.

Worthington-Hill, J. O., R. W. Yarnell, and L. K. Gentle. 2014. Eliciting a predatory response in the eastern corn snake (*Pantherophis guttatus*) using live and inanimate sensory stimuli: Implications for managing invasive populations. *International Journal of Pest Management* 60:180–186.

Zenni, R. D., and M. A. Nuñez. 2013. The elephant in the room: The role of failed invasions in understanding invasion biology. *Oikos* 122:801–815.

14

Gartersnakes of the Beaver Archipelago

A Story of Plasticity and Adaptation

Gordon M. Burghardt, Mark A. Krause, John S. Placyk Jr., and James C. Gillingham

Introduction

Islands are often home to many animals characterized with labels such as unusual, relic, gigantic, dwarf, tame, invasive, and endangered. Non-avian reptiles, especially snakes, are among these species, and a rather remarkable assemblage of these animals occurs on islands in Lake Michigan, a large freshwater temperate lake in North America. These islands, unlike the Galapagos and many others in saltwater and more southern latitudes where snakes are more often studied (Lillywhite and Martins 2019), are of recent origin and thus a model system for evaluating geographic differences across small distances, gene flow, plasticity, and the role of diet and predators in snake populations.

Situated in northeastern Lake Michigan midway between Michigan's upper (UP) and lower (LP) peninsulas lies the Beaver Archipelago (Figure 14.1). It consists of 10 islands ranging in size from the 14,500-ha Beaver Island to the 1.0-ha Pismire Island. This group of islands is roughly equidistant from the mainland to the east and north (15–20 km) and more than twice the distance to the mainland going west (45–50 km). With a human population of 650 permanent residents, Beaver Island is the largest island in Lake Michigan and the only inhabited island in the archipelago. One municipality, St. James, is located at the extreme northeast corner of the island and is situated on Paradise Bay, a functional natural harbor. The island is accessed from Charlevoix on the mainland by ferry and air taxi services.

Beaver Island is surrounded by three moderate-sized islands, Garden, High, and Hog Islands, which are 2,000, 1,410, and 840 ha in area, respectively, and are managed by the Michigan Department of Natural Resources. The remaining, much smaller, islands are either township property, privately owned, or are a part of the Michigan Islands National Wildlife Refuge.

Beaver Island is ecologically diverse and is largely covered by forests ranging from American Beech–Sugar Maple (*Fagus grandifolia–Acer saccharum*) associations to northern coniferous forest (Eastern White and Red Pine; *Pinus strobus* and *P. resinosa*). The perimeter of the island consists mostly of postglacial dune systems that range up to 2–3 km from the shoreline inland and which support early successional

Gordon M. Burghardt, Mark A. Krause, John S. Placyk Jr., and James C. Gillingham, *Gartersnakes of the Beaver Archipelago*
In: *Islands and Snakes*. Edited by: Harvey B. Lillywhite and Marcio Martins, Oxford University Press.
© Oxford University Press 2023. DOI: 10.1093/oso/9780197641521.003.0014

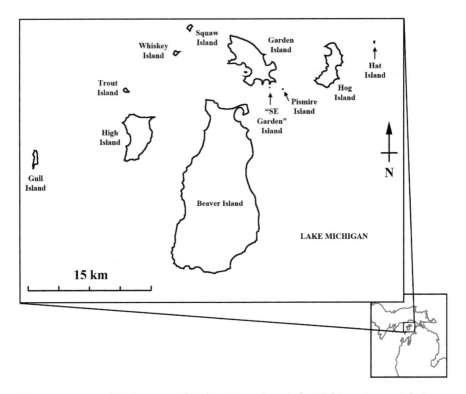

Figure 14.1 Map of the Beaver Archipelago in northern Lake Michigan. Lower right inset depicts the archipelago's relationship to the upper and lower peninsulas of Michigan. Map courtesy of Nancy Seefelt.

stands of aspen (*Populus* sp.), Eastern Hemlock (*Tsuga canadensis*), White Cedar (*Thuja canadensis*), and Balsam Fir (*Abies balsamea*). Open old field tracts resulting from 19th-century abandoned attempts at farming are common, particularly in the northern third of the island, and are dominated by extensive stands of Common Juniper (*Juniperus communis*) shrubs. The northeast corner of Beaver Island and the north side of Garden Island to the north are characterized by exposed limestone bedrock similar to an Alvar ecosystem. Beaver Island contains six inland lakes ranging in size from 14 to 222 ha as well as numerous bog, fen, marsh, and swamp wetland habitats. Two flowing streams, Iron Ore Creek and the Jordan "River," flow from the island's interior into Lake Michigan.

Overview of the Herpetofauna

Beaver Island supports 11 species of amphibians (frogs, toads, and salamanders) and 10 species of reptiles (snakes and turtles). The upper and lower

peninsulas of mainland Michigan contain 26 species of amphibians and 29 species of reptiles (snakes, turtles, and lizards) (Gillingham 1988; Phillips 2016).
The 21 reptile and amphibian species currently listed for Beaver Island are not
all found on each of the outer, smaller islands within the archipelago (Placyk
and Gillingham 2002).

Compared to mainland Michigan's 17 snake species, Beaver Island has eight
species, all harmless. The largest snake found on Beaver Island is the Eastern
Foxsnake (*Pantherophis vulpinus*), which grows to 179 cm total length and is
a constrictor feeding primarily on mice and birds (Harding and Mifsud 2017).
The smallest snake on Beaver Island is the Red-bellied Snake (*Storeria occipitomaculata*), which has a maximum total length of 40 cm (Harding and Mifsud
2017), is color polymorphic, and feeds on slugs (do Amaral 1999). The other
six species include the Common Watersnake (*Nerodia sipedon*), Eastern
Milksnake (*Lampropeltis triangulum*), Eastern Ribbonsnake (*Thamnophis saurita*), Northern Ring-necked Snake (*Diadophis punctatus edwardsii*), Smooth
Greensnake (*Opheodrys vernalis*), and the Common Gartersnake (*Thamnophis
sirtalis*). The most abundant snake on Beaver Island, the entire archipelago, and
the state of Michigan (Holman and Harding 2006) as well as the widest ranging
snake in North America (Rossman *et al.* 1996) is the Common Gartersnake
(*T. sirtalis*; Figure 14.2); its ecology in Michigan is the basis of a seminal monograph on the species (Carpenter 1952).

On Beaver Island, and within the Beaver Archipelago, the Common
Gartersnake occupies virtually every habitat type available. Population densities
for this species on the mainland of Michigan and elsewhere in the Midwest range
from 18 to 89 per ha (Carpenter 1952) but are much higher (230–279 per ha) than
this on Beaver Island, especially in areas with artificial cover (Dickinson 1979).
Similar to the Common Watersnake, the Common Gartersnake is sexually di-
morphic, with the mean snout-vent length (SVL) of females measuring 40 cm and
males 35 cm (Dickinson 1979). As with its habitat preferences, this species shows
a great deal of diet diversity. On Beaver Island, in addition to feeding on all the
11 anuran and caudate amphibian species, 80% of the Common Gartersnakes'
diet consists of earthworms. Larger gartersnakes take nestling birds from the nest
(Carpenter 1951; J. C. Gillingham pers. obs.; Figure 14.3). Also on Beaver Island,
a very large (99 cm total length) gartersnake was documented to feed on a dead
young Snowshoe Hare (*Lepus americanus*), thereby exhibiting scavenging behavior (Casper *et al.* 2015) and indicating they will feed on mammals. It is this
species that is the focus of our chapter, detailing its genetic, behavioral, morphological, and life history variation within and between island and mainland
populations.

Figure 14.2 Common Gartersnake (*Thamnophis sirtalis*) moving onto a snow bank in April on the Johnny Martin Trail, Beaver Island, Michigan.
Photo courtesy of Nancy Seefelt.

Figure 14.3 Adult Common Gartersnake (*Thamnophis sirtalis*) capturing and retreating with a Red-wing Blackbird (*Agelaius phoeniceus*) nestling at Miller's Marsh. The snake successfully swallowed the prey.
Photo courtesy of Raymond Hampton.

Origin of the Gartersnakes on the Beaver Archipelago

The current geological and geographic topography of the Beaver Archipelago was determined by the retreat of the Wisconsinan Laurentide glacier about 10–11 Kya (Dietrich 1988). Following this glacial retreat, gartersnakes from outside the maximum extent of the ice sheet repopulated Michigan's LP as well as Wisconsin and into Michigan's UP (Holman 2000). As the gartersnakes on Beaver Archipelago are both ubiquitous and also common on both the upper and lower peninsulas of Michigan, it is relevant to know which population(s) colonized the Beaver Archipelago after it was formed. To answer this question, as well as broader ones concerning the range expansion and differentiation of the *T. sirtalis* in the Midwestern United States after the glaciers retreated, numerous populations both within and outside of the maximum extent of the Wisconsinan ice sheets were sampled using ND2 mitochondrial DNA (mtDNA) sequences (Placyk Jr. *et al.* 2007). Between 4,000 and 8,000 years ago, the Beaver Archipelago may have been connected to the Michigan LP mainland, and thus some gartersnakes likely arrived via that route. Reconstruction of gene trees support the view that the Wisconsin and UP of Michigan populations were colonized by animals west of Lake Michigan and the LP populations colonized by animals east of Lake Michigan; the results did not support the view that LP animals were the sole source of the Beaver Archipelago gartersnakes. In fact, most of the haplotypes found in the Beaver Archipelago actually were from the UP populations, and thus the Archipelago is a zone of secondary contact between the east and west populations, just as at the southern base of Lake Michigan. In addition, seven unique haplotypes were found in the Beaver Archipelago, suggesting that genetic differentiation has, and is, occurring. Gartersnake colonization from the UP west of Beaver Archipelago would require crossing a greater expanse of water but is certainly possible (Schoener and Schoener 1984). Recently the Eastern Fox Snake (*Pantherophis vulpinus*) has been added to the list of species for the Beaver Archipelago, and the UP is apparently the only source population for this snake species (Leuck and Grassmick submitted, 2022). Such a dispersal could be enhanced by the prevailing westerly winds typical of areas such as this between 35° and 65° latitude (Ahrens and Henson 2017).

It is also important to note that not all the islands are the same age. Of the islands in the Beaver Archipelago that are the main focus of this chapter, Beaver Island and High Island are the oldest and are considered to have been connected to the LP via a land bridge in the last 8,000–9,000 years. However, they were separated from the UP by the Mackinac River, which may only have been 1.6 km wide in some areas (Hough 1958), and, as gartersnakes are good swimmers, colonization from the UP is likely. Garden Island likely became submerged 7,800–6,200 years ago and thus the populations there are younger than on Beaver Island and High Island (Hough 1958; Kapp *et al.* 1969).

Local Morphological Variation of Gartersnakes on Beaver Island

Our initial studies on gartersnakes focused on Beaver Island, which hosts species of snakes that occupy variable habitats (*e.g.*, field, forest, lake, marsh), and these often correspond with intraspecific differences in diet across a relatively small geographic range. Populations within species that are widely distributed across large geographic distances often show locally adaptive responses. For example, dietary differences are correlated with corresponding variation in behavioral and morphological measures (*e.g.*, body and head size), which has been reported in snakes separated by significant geographic distances (Aubret and Shine 2007; Bonnet *et al.* 2021; Burghardt and Schwartz 1999; Valencia-Flores *et al.* 2019). Geographic variation in the body size (mass and SVL) of snakes can be traced to different ecological sources, such as regional differences in size or type of prey (Forsman 1991a; Bronikowski and Arnold 1999), seasonality and climate (Ashton 2001), and population age structure (King 1989). The processes that account for geographic variation in body size in snakes can be traced to microevolutionary changes among populations (Bronikowski 2000), phenotypic plasticity (Bonnet *et al.* 2021; King 1989), or perhaps even genetic drift. In addition, there remains the understudied possibility that adaptive plasticity (*e.g.*, reaction norm evolution) plays a role in geographic variation in the behavior and morphology of snakes.

In addition to body size, prey size can also influence relative head size in snakes, which can be found in studies of intraspecific geographic variation where locally abundant prey differs in size (Forsman 1996; Grudzien *et al.* 1992; Queral-Regil and King 1998; see Schuett *et al.* 2005 for contrasting results). Forsman (1991b) reported phenotypic variation in relative head length between mainland and island populations of European adders (*Vipera berus*). Snakes from islands with large prey (voles, *Microtus agrestis*) had longer relative head sizes compared to adders from islands in which the average size of the same species of voles is smaller. Controlled laboratory testing confirmed that variation in relative head sizes between island and mainland populations of adders was attributable to phenotypic plasticity (Forsman 1996).

An interesting question regarding diet-induced variation in body and head size in snakes concerns whether it may occur at relatively close spatial scales, such as in groups separated by only a few kilometers. Beaver Island is an ideal location for answering this question because of its variable microhabitats and the widespread distribution of gartersnakes. Krause *et al.* (2003) examined dietary effects on body and head sizes of male and female gartersnakes with different natural diets from two nearby but ecologically dissimilar habitats on Beaver Island. Gartersnakes at one of the sites, McCafferty Farm (Figure 14.4), are highly abundant and are readily captured by hand beneath plywood boards distributed across a 6.0-ha field. The second site, Miller's Marsh (Figure 14.5), is located 10 km from the McCafferty farm and hosts a wider variety of prey, including several species of amphibians and earthworms (Dickinson 1979; J. C. Gillingham unpubl. data). McCafferty farm does not have a standing body of water in its vicinity. Stomach content analyses of gartersnakes

Figure 14.4 Gartersnake study site at McCafferty farm on Beaver Island.
Photo by J. C. Gillingham.

captured at McCafferty farm indicate that the snakes there primarily consume earthworms (Gillingham *et al.* 1990; Burghardt *et al.* 2000; Krause *et al.* 2003), while the following species were found in the stomachs of gartersnakes captured at Miller's Marsh: Red-backed Salamander (*Plethodon cinereus*), Green Frog (*Lithobates clamitans*), Gray Tree Frog (*Hyla versicolor*), American Toad (*Anaxyrus americanus*), Spotted Salamander (*Ambystoma maculatum*), Northern Short-tailed Shrew (*Blarina brevicauda*), a Snowshoe Hare (*Lepus americanus*), a Sora Rail (*Porzana carolina*) chick, an unidentified bird nestling, and earthworms (Lumbricidae).

Males and females from Miller's Marsh are larger than those from McCafferty farm (Figure 14.6a–d). Pregnant and nonpregnant females from Miller's Marsh had significantly longer SVLs than did females from McCafferty farm (Krause *et al.* 2003). Females from both sites were significantly longer than males at their respective sites, which is consistent with other reports of female-biased size dimorphism in *Thamnophis* (Manjarrez *et al.* 2014). There was an interaction between sex and site on SVL. Although males captured at Miller's Marsh were on average longer than males from McCafferty farm, the difference was not statistically significant.

Diet-induced plasticity of relative head size was also tested in adult gartersnakes captured at both sites. Female-biased size dimorphism in relative head size (with SVL as a covariate) was found for jaw and head lengths (but not widths; Krause *et al.* 2003). Similar to the analysis on SVL, males from both sites did not differ in measures of relative head size. Interestingly, and unexpected, was the finding that, controlling for SVL, females from McCafferty farm had relatively longer and wider heads, while females from Miller's Marsh had relatively longer jaws.

Figure 14.5 Gartersnake study site at Miller's Marsh on Beaver Island. Pond where both Common Gartersnakes (*Thamnophis sirtalis*) and Eastern Ribbonsnake (*Thamnophis saurita*) are found (above) and the adjacent North Road field where Common Gartersnakes are nearly ubiquitous (below).
Photos by G. M. Burghardt.

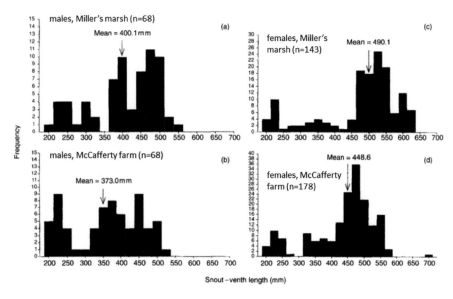

Figure 14.6 Size frequency histograms (SVL) of *Thamnophis sirtalis* for males from Miller's Marsh (a) and McCafferty farm (b) and females from Miller's Marsh (c) and McCafferty farm (d).

From Krause *et al.* (2003).

How did site variation in body and head size arise in adult snakes from the two sites? Possibly microgeographic differences between snakes at the two sites can be traced to population genetic differences, though the extent of any gene flow between the sites has not been quantified. Alternatively, the differences could be due to diet-induced phenotypic plasticity. Krause and Burghardt (2007) approached this question by measuring body and head sizes in neonates born to wild-caught mothers from both sites. Weight, SVL, and tail lengths were taken within 24 h of birth in 286 laboratory-born neonates from 22 litters (11 per site). The neonates did not differ in body mass or SVL between the sites, indicating that the postnatal basis for site differences in body size, at least in adult females, is due to phenotypic plasticity (although we cannot rule out the possibility of genetically based ontogenetic adaptive change). Head measurements (jaw length, head length, head width, and intraocular distance) were taken in 156 of the neonates. Interestingly, neonates born to mothers at Miller's Marsh had significantly longer jaws and wider interocular distances than did neonates from McCafferty farm (and sex did not interact with site). This finding warrants additional study because adult females from Miller's Marsh also had longer jaws (controlling for body size) than females from McCafferty farm. However, neonates born to mothers captured at McCafferty farm did not have longer and wider heads than those born to mothers captured at Miller's Marsh, which does not parallel the comparisons made between adults from the two sites.

Foraging and Feeding Behavioral Plasticity among Gartersnakes on Beaver Island

Phenotypic variation in physical measures of snakes, such as body and relative head size, interact with behavioral variation. We see this in snakes particularly with regard to interactions between various aspects of foraging (*e.g.*, prey searching, prey choice, capture, ingestion, diet quality), growth, and local ecology (Arnold 1981). Gillingham and colleagues (1990) studied gartersnake foraging in the field at a 3-ha site on the Jordan River (a small stream on Beaver Island). They constructed an elevated system to observe the plentiful animals quite unobtrusively (Figure 14.7). Typically, individual gartersnakes emerged from overnight refugia 0830–0930 hours daily. Foraging began following a variable basking period, and most animals had departed from the refugia and were foraging by 1100 h. Most foraging was recorded in the open grassy field, but some was observed in thicker vegetation along the stream. Snakes avoided areas to the north and west that were quite open and sandy. Snakes generally returned to the various refugia between 1530 and 1700 h, often entering the initial site of observation.

After visual location of a focal gartersnake was made from the scaffold walkway, snakes were followed until they could no longer be seen or until they exhibited a

Figure 14.7 Observational scaffold stretching across the gartersnake study site at Jordan River, Beaver Island.
Photo by J. C. Gillingham.

predatory event. During June 1984, foraging sequences of nine different marked gartersnakes were recorded, with episodes ranging from 21 to 50 min. Foraging gartersnakes exhibited the following activity patterns:

IMMOBILITY. Gartersnakes lie motionless with the body and tail outstretched while the head and neck is often raised 4–5 cm above the substrate. This could be described as an "alert" posture. In a few cases, the individual moved into a coiled posture.

MOBILITY. Snakes slowly move forward. The head and neck show no activity other than forward movement. Tongue-flicking is slow.

IMMOBILE PROBE. While the body is not moving forward, with its head, the gartersnake actively searches the details of its microhabitat (under leaves, twigs, grass tufts, in crevices and depressions, etc.) laterally back and forth. Tongue-flick rate is rapid.

MOBILE PROBE. As the snake slowly moves forward its head searches the details of its microhabitat as in immobile probe. Tongue-flick rate is comparable to immobile probe. This behavior preceded the strike at and capture of earthworms when exploring around earthworm burrows (Figure 14.8) especially when earthworm castings were present.

Because mobility, immobile probe, and mobile probe represent 90% of the foraging behaviors observed, *T. sirtalis* on Beaver Island can clearly be classified as widely foraging on diverse prey, but earthworms are the major prey (Gillingham *et al.* 1990). Significantly more time is spent actively searching than in a sit-and-wait stationary position. Earthworms are not diurnally active prey, so gartersnakes must forage widely to find them. Relatively high earthworm densities estimated through a quadrat survey at 39–177 per m^2 at this study site contribute to their feeding success (Gillingham *et al.* 1990). Although earthworms and sometimes American Toads are the main prey at this site, throughout their range *T. sirtalis* feeds on a wide variety of animals. This raises the issue of the plasticity and roles of learning and prior experience in their feeding behavior, and rearing experiments on Beaver Island gartersnakes were undertaken.

Neonatal gartersnakes born to adult females captured at McCafferty farm (as at the Jordan River site, earthworms were the almost exclusive diets) showed improvements in foraging behavior (*e.g.*, latency to approach and consume prey) over the course of their first 12 meals (Burghardt and Krause 1999) in a simplified laboratory setting. However, plasticity is constrained by the type of food consumed during these initial meals. For example, latency to approach prey declined if snakes were reared on a pure diet of either fish or worms, but not if fed a mixture of the two. The benefits of predatory experience were especially evident in snakes that fed exclusively on fish or worms, with mean total consumption times decreasing significantly after 12 meals. Snakes reared on a mixed diet showed a more peculiar pattern, with total consumption time declining significantly when fish were eaten, but mean consumption time

Figure 14.8 Common Gartersnake arching head and neck and exhibiting a thrusting attack on an earthworm in its burrow (above) and Common Gartersnake retrieving an earthworm from its burrow (below).
Photos by J. C. Gillingham.

for worms did not decline to a statistically significant degree. In addition, Burghardt and Krause (1999) found that snakes reared exclusively on worm, or on a mixed fish and worm diet, had greater mass and longer SVLs than did snakes reared only on fish, in spite of fish diet animals eating the highest percentage of prey offered, which were equalized by mass for each snake at 10–15% of body mass. This finding is consistent with plasticity of growth in wild-caught adult gartersnakes on Beaver Island (Krause *et al.* 2003).

Of additional interest is whether dietary effects on foraging behavior occur among wild-caught snakes and in neonates born to mothers collected at sites that differ in available prey. To test this, Krause and Burghardt (2001) studied the ontogeny of foraging behavior in neonates born to mothers from McCafferty farm ($n = 43$ snakes from nine litters) and Miller's Marsh ($n = 63$ snakes from seven litters). Similar to Burghardt and Krause (1999), latency to approach and consume prey declined over the course of the snakes' first several meals, regardless of their site of origin. However, Krause and Burghardt (2001) found no influence of site on the early development of foraging behavior, although it is possible that long-term experience with different types of prey would result in phenotypic variation in foraging wild-caught adult snakes. In a follow-up experiment, Krause and Burghardt (2001) captured male and female adult snakes from McCafferty farm and Miller's Marsh and, following a period of habituation to captivity, measured foraging efficiency on prey common to both sites (earthworms) and to more variable aquatic prey (frogs and fish) characteristic of Miller's Marsh. Adult gartersnakes from both sites were equally proficient at foraging for all three types of prey, indicating that early postnatal behavioral plasticity may not translate into phenotypic variation of adult foraging behavior, perhaps due to maturation of greater strength and mobility. It is possible that body and relative head size become more relevant predictors of foraging behavior in adult snakes and that morphology interacts with prey capture and ingestion. Experience has been shown to affect such behavior in ratsnakes (*Pantherophis* spp.) fed live versus dead mice (Almli and Burghardt 2006), and thus feeding on different prey types can have interference effects on predation efficiency. For example, feeding on prey that are more difficult to capture and ingest (fish) can have carryover effects on dealing with earthworms in neonates fed both prey types (Burghardt and Krause 1999). Among gartersnakes, earthworm specialist species that will eat fish in captivity take longer to capture them compared to both aquatic specialist and generalist species, and this seems to be in part due to morphological differences in head and body shape (Halloy and Burghardt 1990).

An intriguing study of the chemical cue prey preferences of snakes at McCafferty farm suggests additional avenues for future research on microevolution in general and Beaver Island in particular. As is well documented in many studies, pioneered by Stevan Arnold (*e.g.*, Arnold 1981), neonate gartersnakes show considerable heritability of prey chemical preference, and these remain even after subsequent numerous reports of widespread multiple paternity in natricine snakes (Schwartz *et al.* 1989; Wusterbarth *et al.* 2010). However, in a study of 17 litters of babies born to mothers from McCafferty farm, whose diet is almost exclusively earthworms, the heritability of the prey chemical cues was virtually zero (Burghardt *et al.* 2000)! In these studies, mosquito fish (*Gambusia affinis*), earthworm (*Lumbricus terrestris*), and deionized water controls were used for the chemical prey preference tests using the cotton swab method. Initially, the snakes responded fairly equally to both the fish and earthworm stimuli. However, after 12 feedings on fish, not only did the snakes show a significant increase in the chemical preference to fish (as well as overwhelmingly choosing fish in actual prey choice tests) but the snakes also now showed a significant heritability of response to both fish (0.323) and an even great heritability to worm (0.497)! Furthermore, the heritability of the worm/fish chemical cue preference went from a nonsignificant 0.074 to a highly significant 0.532, and the change in preference was itself significantly heritable. Furthermore, although all snakes were fed the exact same amount of prey over the 12 feedings, the snakes that grew the most in both SVL and mass had the greatest change in preference toward fish. This could reflect genetic differences in adapting to microbiotic changes in digestive efficiency. Would garter-snakes born to mothers from Miller's Marsh respond similarly? We do not know, but this is another indication of the plasticity potential of this generalist species and may underlie its widespread distribution and adaptability.

Geographic Variation across the Beaver Archipelago and Mainland Populations

The gartersnakes on Beaver Island differ across populations in sexual dimorphism and some morphological measures. There is little indication of major differences at birth, however. What about differences across the islands in the Beaver Archipelago and the mainland populations? John Placyk, Jr. carried out a major study of these populations in his dissertation (Placyk Jr. 2006), and we summarize here some of his published and unpublished findings on life history, morphological, and behavioral traits. He used animals from mainland populations from the UP and LP of Michigan as well as from Garden Island and High Island, and two populations from Beaver Island (Figure 14.9). One was from previously described Miller's Marsh and the other was from an old sawmill site, about 1 km from Miller's Marsh, but more similar to McCafferty farm in available prey (earthworms) and which also differs in predators. He studied body size, head morphology, litter measurements, antipredator behavior, growth, prey preference, and other behavior in both adult and neonate snakes. Due to difficulties in capturing mainland animals, the sample sizes for adults were much greater on the islands. Data were gathered from different years but not mentioned here unless differences were found for the traits discussed. Because of discontinued access to McCafferty farm, animals from that population were not studied and thus data from the sawmill site and McCafferty farm differ and were also collected in different years. Given the extensive data gathered, they are primarily summarized here, and the published papers and dissertation provide additional methodological and statistical details.

Figure 14.9 Gartersnake study site on Garden Island (above), where snakes were found in areas around long-abandoned buildings and fields. Gartersnake study site on High Island (below) where snakes were common along the beach and driftwood on Lake Maria.

Garden Island photograph by Photo by G. M. Burghardt; High Island photograph by Beth Leuck.

Predation

Potential predators on gartersnakes are far more diverse, and predation on garter-snakes more intense, on the mainland than on the Beaver Archipelago. Furthermore,

Beaver Island has far more potential predators than either High Island or Garden Island (Table 14.1). Several studies have evaluated the responses of Beaver Island gartersnakes to predatory events (Hampton and Gillingham 1989; Passek and Gillingham 1997; Schuett and Gillingham 1990). To assess predation effects on the various populations, captured snakes from all populations were surveyed for injuries, scoring tail stubs and body scars separately. The snakes on the mainland had significantly more injuries than the island populations, which did not differ (Placyk and Burghardt 2005). Thus, differences in predation pressure might underlie differences in life history traits. On the other hand, population densities on the islands are much greater than on the mainland (Carpenter 1952; Dickinson 1979) and thus intraspecific competition might be greater. This could be reflected in reproductive

Table 14.1 Potential predators of Common Gartersnakes, *Thamnophis sirtalis*, from the lower peninsula (LP), upper peninsula (UP), and Beaver Island (BI), Garden Island (GI), High Island (HI), Hog Island (Hgl), Squaw Island (SI), Trout Island (TI), and Whiskey Island (WI) of the Beaver Archipelago of Michigan, USA

Predator	LP	UP	BI	GI	HI	Hgl	SI	TI	WI
Large fish	X	X	X						
American Bullfrog (*Lithobates catesbeianus*)	X	X	X						
Snapping Turtle (*Chelydra serpentina*)	X	X	X	X	X				
Eastern Massasauga (*Sistrurus catenatus*)	X								
Blue Racer (*Coluber constrictor foxii*)	X								
Eastern Milksnake (*Lampropeltis triangulum*)	X		X	X	X				X
Common Raccoon (*Procyon lotor*)	X	X	X						
Red Fox (*Vulpes vulpes*)	X	X	X				X		X
Bobcat (*Lynx rufus*)	X	X	X						
Black Bear (*Ursus americanus*)	X	X							
Coyote (*Canis latrans*)	X	X	X	X		X			
Gray Wolf (*Canis lupus*)	X	X							
Domestic cat (*Felis catus*)	X	X	X						
Domestic dog (*Canis familiaris*	X	X	X						
Human (*Homo s. sapiens*)	X	X	X						

An **X** indicates the presence of a predator (Bowen and Gillingham 2002a, 2002b; Hatt *et al.* 1948; Harding 1996; Mifsud 2015; Phillips *et al.* 1965; Placyk and Gillingham 2002; Placyk and Burghardt 2005). Raptors and other large birds are found at all locations.

investment in terms of either number of offspring or neonate size, as well as in adult size measures.

Adult Body Size

For all adults captured, a condition index (CI) was calculated by taking the cube root of the mass and dividing by the SVL. As typical, adult females were longer and heavier than males, and pregnant females longer and heavier than both males and non-pregnant females. After accounting for sex and reproductive status, SVL did not differ across sites. CI and mass measures did, however. Using Tukey-Kramer multiple comparison tests, CIs for males did not differ across sites, except that Miller's Marsh animals had significantly lower CIs than the nearby sawmill males, although they did not differ significantly from any other site. Non-pregnant females showed a different pattern, with High Island and Miller's Marsh animals significantly lower in CI than Garden Island and sawmill females. Given the close proximity of Miller's Marsh and sawmill, the differences between them are surprising. The LP and UP animals did not differ from any other population, but the sample sizes were small. For pregnant females the pattern differed again, with Garden Island and LP animals significantly higher in CI than Miller's Marsh females, while all the other populations did not differ among themselves or the Garden Island, Miller's Marsh, and LP animals.

There were also some significant interactions of site and sex. For example, LP females were much larger than LP males, while UP males and females were more similar in size, mirroring the difference between McCafferty farm and Miller's Marsh sexual dimorphism differences. In fact, the sexual dimorphism in the LP snakes was higher than those of any of the Beaver Archipelago populations, and the UP snakes were lower than any of the Beaver Archipelago populations. Note that the sawmill and McCafferty farm animals (above) seemed to have different relationships in comparison with Miller's Marsh animals. Both sexual dimorphism and any adult sex ratios that differ across sites may be due to different predation pressures on males versus females at such sites or, less likely, differences in male–male aggression or mating competition.

Litter Traits

Although longer females, as is typical, had larger litters, after controlling for SVL there were also differences across sites in the number of offspring, with the LP females having larger litters than females from the island populations and the UP. Across the Beaver Archipelago, total litter mass was lower than in the LP. No significant litter or site differences were found for sex ratios. There were differences in offspring sizes, however, with the mass and SVL of the LP litters significantly lower than all Beaver Archipelago populations. Among the latter, Garden Island, High Island, and the sawmill site did not differ; Miller's Marsh was greater than Garden Island and High Island, but not significantly larger than SM. UP neonates were shorter than Beaver Archipelago animals but longer than LP neonates. Neonates from 2002 had lower CI values than those born in 2001. This could be due to the hotter, drier conditions in 2002 that might have limited prey availability for pregnant females. Overall, however, Beaver Archipelago snakes had fewer, larger offspring than LP females. This could be due to the more extreme climatic conditions on the islands across seasons than on the LP, especially, and represent greater female investment to deal with food availability, the earlier onset of cold weather, and other environmental differences, as well as reduced predation.

Head Morphology

Adult and neonate head measurements were taken across the sites. For adults, interocular distance (IOD) differed, with island adults having a larger IOD than mainland animals, but there were no sex or interaction effects. There was also a significant SVL effect, as also with eye diameter (ED). The latter measure uncovered a significant site difference in that sawmill animals had a significantly larger ED than all other populations overall, but there were more complex results if only females were compared. Sawmill females had significantly larger EDs than Garden Island, High Island, and LP animals, and LP females has significantly smaller EDs than Garden Island, High Island, sawmill, and Miller's Marsh females and did not differ from UP females.

Head measurements also showed site differences. In head length (HL), island females had longer heads than both mainland populations. Across sexes, sawmill snakes were the only ones significantly different, and larger, than the others. Jaw length (JL) and head width (HW) also differed. For HW, LP animals were larger than island animals, which did not differ from the UP. For JL, LP animals were significantly greater than all populations than sawmill. All these results are difficult to interpret, but LP animals having larger head sizes generally might reflect their greater diversity of prey. On the other hand, the similarity of Beaver Archipelago animals to UP animals might reflect the greater genetic influence of this source population.

After earthworms, American Toads (*Anaxyrus americanus*) are the second most abundant (11.4%) dietary item of Beaver Island Common Gartersnakes (Dickinson 1979). American toads on the Beaver Archipelago are significantly larger than those found on the mainland (Holman 2012, Gillingham 1988; Long 1982). Perhaps also of importance, a major predator on American Toads on the mainland, the Eastern Hog-nosed Snake (*Heterodon platirhinos*) is not found on Beaver Island (Harding and Mifsud 2017). The larger toads and absence of competition for them may be related to the difference between mainland and island animals in some of the morphological traits.

Multiple discriminant function models using all the head measurements were used to test whether animals could be assigned to their source population, either the six populations of the island versus mainland populations, or models incorporating geological history. All were significant (p < 0.05). The model using the six populations correctly classified 27.9% of the animals while the model comparing mainland and island populations correctly classified 75.8%. In testing for geological history, grouping island animals with LP adults correctly classified 77.2% of the animals, and grouping them with the UP adults correctly classified 76.9% of the animals.

Neonate head measurements gave different results in that there were no site differences, but litter, sex, and SVL differences were found when using individual measures with ANCOVAs. However, MANCOVA analysis using the five head measurements for neonates did reveal site, SVL, sex, and litter effects. Female neonates from High Island, Garden Island, Miller's Marsh, and the LP all had significantly greater JL than males, while UP and sawmill animals did not differ, though sample sizes were smaller. Neonate females had significantly larger values than males for all the other head measurements, but there were no site differences. Using the multiple discriminant function models, it was found that all the models showed significant classification of animals; grouping the island animals with the LP populations was somewhat more accurate (74.1%) than grouping with the UP animals (62.6%) though smaller numbers of UP than LP animals were measured.

Plasticity of head measurements was evaluated by raising 90 animals on either larger prey (pieces of nightcrawlers, *L. terrestris*) or small prey (whole or pieces of leafworms, *L. rubellus*) twice a week for 10 weeks. Animals were balanced across sex, litter, and site, but the main focus was on any changes. At both 25 and 35 weeks, all five measurements were significantly higher in the nightcrawler group, indicating that diet can influence adult head morphology. These results suggest that the greater differences in adults than in neonates is at least partly a consequence of plasticity.

Diet and Prey Chemical Preferences

Field studies of diet were not carried out, but Table 14.2 shows that available prey differs among the islands as well as on Beaver Island. Placyk carried out a series of neonate prey preference tests modeled after Burghardt (1969) using aqueous controls and water washes of nightcrawlers (*L. terrestris*), Green Frogs (*Lithobates clamitans*), American Toads (*A. americanus*), Fathead Minnows (*Pimephales promelus*), and Red-backed Salamanders (*Plethodon cinereus*) in 30 s cotton swab tests, with each stimulus tested twice when animals were 9 and 10 days old. Animals from 41 litters across the six sites, numbering 320, were tested. Results were analyzed using litter means. All prey stimuli led to higher tongue flick–attack scores (TFAS) than the controls, with earthworm eliciting the highest TFASs followed by toad, salamander, and then fish and frog stimuli. TFAS is a composite measure of chemical sensory response in squamate reptiles that measures the number of tongue flicks, attacks, and latency to attack (Cooper and Burghardt 1990). There were some site effects, although earthworms were responded to significantly more than all other prey and control stimuli.

Table 14.2 Prey available to Common Gartersnakes, *Thamnophis sirtalis*, from the lower peninsula (LP), upper peninsula (UP), and Beaver Island (BI), Garden Island (GI), High Island (HI), Hog Island (HgI), Squaw Island (SI), Trout Island (TI), and Whiskey Island (WI) of the Beaver Archipelago of Michigan, USA. Earthworms, American Toads (*Anaxyrus americanus*), and Eastern Red-backed Salamanders (*Plethodon cinereus*) occur in both the LP and UP, as well as on all/most of the islands.

Prey	LP	UP	BI	GI	HI	HgI	SI	TI	WI
Small pond fish (*e.g.*, Cyprinidae)	X	X	X		X				
Green Frog (*Lithobates clamitans*)	X	X	X	X	X				
Mink Frog (*Lithobates septentrionalis*)		X							
Northern Leopard Frog (*Lithobates pipiens*)	X	X	X						
American Bullfrog (*Lithobates catesbeianus*)	X	X	X						
Gray Treefrog (*Hyla versicolor*)	X	X	X					X	
Spring Peeper (*Pseudacris crucifer*)	X	X	X					X	
Eastern Newt (*Notophthalmus viridescens*)	X	X	X	X	X				

An X indicates the presence of a prey item (Hatt *et al.* 1948; Phillips *et al.* 1965; Harding 1996; Bowen and Gillingham 2002a, 2002b; Placyk and Gillingham 2002; Placyk *et al.* 2002; Placyk 2006).

Thus, there seems to be no initial relationship at birth between prey preference, head morphology, sex, or site. The generally greater response to earthworms could reflect the reliance on earthworm diets generally in young snakes.

Variation in Antipredator Behavior and Effects of Ontogeny

As noted earlier, there are indications of site differences in predation on the sites. The Eastern Milksnake (*Lampropeltis triangulum*) is a predator on snakes and is found throughout the LP and all three islands but is rare in the UP and only found in two counties (Holman 2012). Placyk and Burghardt (2011) asked whether adult snakes from the LP, High Island, Garden Island, Miller's Marsh, the sawmill site, and UP differ in their response to chemicals from milksnakes and the sympatric small ring-neck snake that does not seem to pose a threat to gartersnakes. Using a balanced de-sign where snakes were placed into either clean control cages or those previously housing a milksnake or ringneck snake, duration of moving and number of tongue-flicks were measured for 2 min after a 30 s acclimation period. It was found that both frequency of tongue-flicks and duration of mobility were significantly higher to milk-snake than to either ringneck or control stimuli across all animals. However, there was a clear site effect in that UP animals did not differ in number of tongue-flicks across the three conditions. LP animals did not differ in tongue-flicking to the ring-neck and control. The LP adults emitted significantly more tongue-flicks than did the UP adults. In terms of mobility, the results were similar, although there were also site differences in both measures. The UP adults moved less than those from all the other sites, although not significantly so.

What about neonate animals? Would they respond in a similar way to the adults? Testing 117 neonates from the six sites resulted in all populations showing more tongue-flicks to milksnake chemicals than to ringneck body chemicals or controls, and this was true for mobility duration as well. Without going into details, the main result is that recognition and discrimination of ophiophagous milksnakes is present at birth at all sites. The UP neonates did emit significantly fewer tongue-flicks than the LP neonates but did not differ from any of the Beaver Archipelago sites. The same was true of mobility in that the UP animals did not differ from any Beaver Archipelago sites but did show a significantly lower rate than the LP neonates. Site differences in adult behavioral responses may be a consequence of not encountering milksnakes or their odors, although a maturational effect could be present. The results suggest that the separation of the populations was recent enough that little predator recognition was lost, although both adults and neonates from the UP were overall less responsive to milksnake chemicals.

In another study (Placyk 2012), snakes were tested using a simulated predator encounter method based on Herzog and Burghardt (1986) and Mori et al. (1996), in which snakes were presented with a stationary or moving finger in front of their snout for 60 s each and then tapped on the body, front to back, for 60 s. Both adults and neonates from all six sites were studied. A series of behaviors were recorded including striking, biting, head hiding, tail wagging, body flattening, and musking (odiferous cloacal discharge). The results were pretty clear in that the adults from all populations except High Island and Garden Island readily fled during the tests, with all mean flees over 20 s, and they did not differ from each other, while there were no differences in the neonates across populations. Of the other behavior measures, High Island and Garden Island adults showed virtually none, and the Miller's Marsh and sawmill adults were intermediate between them and the mainland population in most measures. Neonates did not significantly differ across populations on any measures except musking and body flattening. LP neonates musked and body flattened less than other populations. Another experiment in Placyk (2012) involved gently handling snakes for 60 s weekly, not handling snakes except for weekly cage cleaning, and roughly handling snakes for 60 s weekly. Neonates from all six populations were tested shortly after birth and again after 6 weeks. The roughly handled snakes showed a significant increase in fleeing, whereas the control and gently handled snakes showed significant decrease in fleeing. In terms of strikes, both the gently and roughly handled snakes showed a significant increase from initial levels, but the roughly handled snakes much more so. Flattening also showed a significant increase, whereas other measures did not. These results replicated the results of Herzog (1990). Together, the results suggest that plasticity and experience, perhaps also maturation (Herzog *et al.* 1992), combine with innate factors in the antipredator behavior system, so that animals on the smaller islands with fewer predators (outside of milksnakes) are somewhat less responsive to threats than mainland and large island populations.

As a sidelight to these studies, Placyk also measured ventral scale counts because number of ventral scales is related to speed of escaping from predators (*e.g.,* Arnold and Bennett 1988). This number is fixed at birth. Comparing ventral scales in both adults and neonates can also indicate if selection is occurring in the population on such escape behavior. Both litter and site influenced the ventral scale counts. Burghardt and Schwartz (1999) report that snakes from Wisconsin had significantly higher scale counts than those from the LP, which may reflect genetic influences given the heritability of the measure. It is thus relevant that the UP and High Island (site closest to the UP) had the highest scale counts in both adults and juveniles, and the LP had the lowest, though only marginally lower than Beaver Island and Garden Island animals. The greater similarity of adults and neonates suggests that selection over the time span of separation has not led to significant changes in number of ventral scales in spite of different predator selection on the islands.

Conclusions, Limitations, and Future Prospects

The results here, briefly described though they are, show the promise of studying snake populations in recently formed archipelagos since differences were found in possible source populations, morphology, behavior, and ontogeny. Although there are fewer snake species in the archipelago than on the Michigan mainland, the population densities of the island snakes tend to be higher. The lack of human development, highways and roads, and decreased traffic in the Beaver Archipelago are certainly partially responsible for this difference. However, it may also be in part due to lower predation pressure. For example, Michigan's mainland supports more than 50 mammalian species compared to 22 found in the Beaver Archipelago, many of which may prey on snakes as part of their diet (Kurta 1995).

We have documented in this chapter differences in morphological, behavioral, and life history traits between a recently formed (less than 10,000 years old) island archipelago and nearby mainland populations of Common Gartersnakes (Table 14.3). As the islands are almost equidistant from two mainland source populations, genetic methods allowed us to show that animals came from both mainland populations and not just the LP, as was commonly thought. There are three main ways in which populations can diverge: selection, genetic drift, and phenotypic plasticity. Effects of selection can only be inferred; drift based on limited genetic diversity colonizing the various islands has not yet been tested and, given the swimming ability of the snakes, may be minimal. But we have established that plasticity in some measures, such as foraging, size, sexual dimorphism, and head morphology occurs, including in populations from nearby sites with different ecologies on the same island. On the other hand, head

Table 14.3 Summary of morphological and behavioral data collected for adult and neonate Common Gartersnakes (*Thamnophis sirtalis*).

Measure	Adults						Neonates					
	LP	UP	MM	SM	GI	HI	LP	UP	MM	SM	GI	HI
SVL (mm)	524.59	518.25	447.59	437.77	476.45	449.34	143.29	148.14	164.28	158.88	153.46	154.57
Mass (g)	89.59	69.06	39.07	42.99	58.80	39.25	1.37	1.61	2.01	1.90	1.79	2.00
CI	0.82	0.76	0.74	0.79	0.79	0.74	0.77	0.78	0.77	0.78	0.79	0.81
JL (mm)	20.29	19.02	17.81	18.30	18.67	17.50	9.44	9.19	9.78	9.63	9.59	9.43
HL (mm)	17.23	17.38	15.75	15.97	16.47	15.72	9.11	8.97	9.44	9.37	9.35	9.12
HW (mm)	12.10	11.30	9.59	9.14	10.17	9.36	5.11	5.17	5.34	5.33	5.18	5.37
IOD (mm)	7.56	7.33	7.07	7.07	7.39	7.14	4.17	4.03	4.35	4.48	4.32	4.38
ED (mm)	3.83	3.80	3.66	3.75	3.77	3.59	2.24	2.29	2.32	2.32	2.24	2.19
Flee	27.35	28.67	24.96	20.79	5.26	2.61	19.81	21.9	22.89	14.81	13.78	15.25
Bite	0.15	0.33	0	0.02	0	0	0.10	0.10	0.08	0.19	0.18	0.06
Strike	0.23	0.33	0.01	0.09	0	0	0.17	0.24	0.22	0.06	0.28	0.19
Tail-wag	0.12	0	0.12	0.14	0	0	0.54	0.57	0.51	0.56	0.58	0.75
Head-hide	0.12	0	0.11	0.09	0.01	0.02	0.15	0.10	0.11	0.13	0.08	0.06
Urinate	0.04	0.33	0.12	0.14	0.03	0	0.34	0.33	0.34	0.31	0.36	0.44
Defecate	0.08	0.33	0.07	0.12	0.01	0	0.0	0.05	0.03	0.0	0.06	0.06
Musk	0.08	0.67	0.15	0.05	0.03	0	0.51	0.67	0.72	0.69	0.56	0.75
Flatten	0.08	0	0	0	0	0	0.0	0.10	0.05	0.06	0.04	0.13

Means for SVL (mm), mass (g), condition index (CI), jaw length (JL) (cm), head length (HL) (cm), head width (HW) (cm), interocular distance (IOD) (cm), eye diameter (ED) (cm) and flee for both neonates and adults from the lower (LP) and upper (UP) peninsulas of Michigan, the Miller's Marsh (MM) and Sawmill (SM) sites from Beaver Island, Garden Island (GI) and High Island (HI). For bite, strike, tail-wag, head-hide, urinate, defecate, musk, and flatten, frequencies are given. Rows in bold print for a particular age class and measure indicate a significant site effect (from Placyk 2006).

measurements, innate prey chemical preferences, and foraging do not seem to differ greatly in neonates across populations.

The islands have generally reduced predation as compared to the mainland, particularly the LP, for which we have the most data. Island populations of gartersnakes are much denser than mainland ones, and, compared to the LP, for which we have more data than the UP, litter sizes are smaller and offspring size larger on the Beaver Archipelago. The greater predation on the mainland populations is shown in the significantly greater propensity to bite and strike in adult snakes from the mainland, although neonates show little population variation. Lessened predation pressure within the Beaver Archipelago may certainly be a factor in the observed higher insular gartersnake population density. It could well be that maturational changes are involved in the increase in antipredator behavior in some populations more than others. Both maturational and habituation can enhance or suppress antipredator behavior, as has been documented in this species (Herzog and Burghardt 1986; Herzog *et al.* 1989, 1992). Populations of common gartersnakes from more southerly populations in Michigan and Wisconsin differ in a number of traits at birth, including antipredator behavior and prey preferences, thus ruling out both maturation and experience as sole explanations for population differences in behavior (Burghardt and Schwartz 1999). Generally, however, our results suggest that neonates and young animals differ less across populations than do adults, which suggests that the selection pressures on animals change as animals grow and mature. The exact processes underlying these differences seem to be an important avenue for future research, which has implication for studies of the pace of evolutionary change and ecology in natural as compared to urban or more human-altered habitats.

The Beaver Archipelago is a valuable set of islands to study recent phenotypic evolution and behavior. It can complement the studies by King and colleagues on insular Lake Erie watersnakes, a threatened population now on the upswing due to active educational and mitigation efforts, along with an introduced food source (King *et al.* 2018). Most of the Beaver Archipelago other than Beaver Island is protected against development and habitat destruction, but threats remain. Mainland populations can be considered generally safe, although populations of even this most common species seem to be declining. Climate change and the effects of invasive species of earthworms on both native worms and the forest floor may be significant threats, although their effects may be mixed and unpredictable. Planned long-term studies such as on the McCafferty farm were aborted when the son of the owners took over the farm and both destroyed snakes and prohibited research access to the population that had been studied for decades. Many PIT-tagged snakes were lost to science after only a few recaptures. Although more than a third of Beaver Island is state-owned, which decreases the reliance on private property for snake studies, the island is developing as a tourist area, especially around St. James. Other areas are being developed as well and may attract owners less sympathetic to large numbers of snakes, even gartersnakes, than they are accustomed to on the mainland. Thus, while the Beaver Archipelago and mainland populations are most attractive populations for study, there are risks.

While there are two active biological research stations in the area (Central Michigan University Biological Station on Beaver Island and University of Michigan Biological Station in the upper LP), a new generation of herpetologists, ecologists, and ethologists need to continue studies on these animals.

Acknowledgments

The facilities, personnel, and logistical support provided by the Central Michigan University Biological Station on Beaver Island were essential to gartersnake field and laboratory studies dating back more than four decades. This chapter would not have been possible without that support. GMB thanks the many individuals and granting sources, including NSF, who supported and assisted in the studies described here.

MAK thanks Central Michigan University and the faculty and staff at the Beaver Island Biological Station for supporting his research efforts. Financial support for the research MAK carried out was provided by an Alumni Dissertation Support Grant, the Scholarly Activities Research Incentive Fund, and a Science Alliance in Psychology stipend upgrade from the University of Tennessee at Knoxville.

JSP thanks the many fellow students and colleagues who aided in his dissertation research, as well as the Beaver Island Biological Station staff and the NSF DDIG grant that helped support it (IBN 0309339). Due to unforeseen circumstances, John was unable to work on drafts of this chapter or review the final product, including the sections describing his research.

JCG wishes to thank his chapter coauthors, along with his CMU graduate students and Biology Department colleagues who contributed significantly to this body of gartersnake research, including but not limited to Ken Bowen, Howard Carbone, John Dickinson, Pedro do Amaral, Kerry Hansknecht, Jacque LeFreniere, Raymond Hampton, Charles Meyer, Kelly Passek, John Rowe, Gordon Schuett, and Mark Waters. JCG also thanks the Central Michigan University Faculty Research and Creative Endeavors Committee and the Michigan Department of Natural Resources for funding portions of his gartersnake research. An NSF grant to JCG provided funding to upgrade the research facilities at the CMU Biological Station.

References

Ahrens, C. D., and R. Henson. 2017. *Essentials of Meteorology: An Invitation to the Atmosphere.* 8th ed. Cengage Learning.

Almli, L. M., and G. M. Burghardt. 2006. Environmental enrichment alters the behavioral profile of ratsnakes (*Elaphe*). *Journal of Applied Animal Welfare Science* 9(2):85–109.

Arnold, S. J. 1981. The microevolution of feeding behavior. In A. Kamil and T. Sargent (eds.), *Foraging Behavior: Ecological, Ethological and Psychological Approaches.* Garland Press, pp. 409–453.

Arnold, S. J., and A. F. Bennett. 1988. Behavioural variation in natural populations. V. Morphological correlates of locomotion in the garter snake (*Thamnophis radix*). *Biological Journal of the Linnean Society* 34:175–190.

Ashton, K. G. 2001. Body size variation among mainland populations of the western rattlesnake (*Crotalus viridis*). *Evolution* 55:2523–2533.

Aubret, F., and R. Shine. 2007. Rapid prey-induced shift in body size in an isolated snake population (*Notechis scutatus*, Elapidae). *Austral Ecology* 32(8):889–899.

Bonnet, X., F. Brischoux, M. Briand, and R. Shine. 2021. Plasticity matches phenotype to local conditions despite genetic homogeneity across 13 snake populations. *Proceedings of the Royal Society B: Biological Sciences* 288:20202916.

Bowen, K. D., and J. C. Gillingham. 2002a. Geographic distribution. *Notophthalmus viridescens*. *Herpetological Review* 33:60.

Bowen, K. D., and J. C. Gillingham. 2002b. Geographic distribution. *Rana clamitans*. *Herpetological Review* 33:63.

Bronikowski, A. M. 2000. Experimental evidence for the adaptive evolution of growth rate in the garter snake *Thamnophis elegans*. *Evolution* 54:1760–1767.

Bronikowski, A. M., and S. J. Arnold. 1999. The evolutionary ecology of life history variation in the garter snake *Thamnophis elegans*. *Ecology* 80:2314–2325.Burghardt, G. M. 1969. Comparative prey-attack studies in newborn snakes of the genus Thamnophis. *Behaviour* 33:77–114.

Burghardt, G. M., and M. A. Krause. 1999. Plasticity of foraging behavior in garter snakes (*Thamnophis sirtalis*) reared on different diets. *Journal of Comparative Psychology* 113:277–285.

Burghardt, G. M., D. G. Layne, and L. Konigsberg. 2000. The genetics of dietary experience in a restricted natural population. *Psychological Science* 11:69–72.

Burghardt, G. M., and J. M. Schwartz. 1999. Geographic variations on methodological themes in comparative ethology: A natricine snake perspective. In S. A. Foster and J. A. Endler (eds.), *Geographic Variation in Behavior: Perspectives on Evolutionary Mechanisms*. Oxford University Press, pp. 69–94.

Carpenter, C. C. 1951. Young goldfinches eaten by garter snake. *Wilson Bulletin* 63:117–118.

Carpenter, C. C. 1952. Comparative ecology of the common garter snake (*Thamnophis s. sirtalis*), the ribbon snake (*Thamnophis s. sauritus*), and Butler's garter snake (*Thamnophis butleri*) in mixed populations. *Ecological Monographs* 22:235–258.

Casper, G. S., J. B. Leclere, and J. C. Gillingham. 2015. *Thamnophis sirtalis* (common garter-snake). Diet/scavenging. *Herpetological Review* 46:653–654.

Cooper, W. E., Jr., and G. M. Burghardt. 1990. A comparative analysis of scoring methods for chemical discrimination of prey by squamate reptiles. *Journal of Chemical Ecology* 16:45–65.

Dickinson, J. A. 1979. The Effects of Artificial Cover Availability on the Ecology and Movements of a Population of the Garter Snakes, *Thamnophis s. sirtalis*. Unpublished master's thesis, Central Michigan University.

Dietrich, R. V. 1988. The geological history of Beaver Island. *Journal of Beaver Island History* 3:59–77

do Amaral, J. P. S. 1999. Lip-curling in redbelly snakes (*Storeria occipitomaculata*): Functional morphology and ecological significance. *Journal of Zoology* 248:289–293.

Forsman, A. 1991a. Variation in sexual size dimorphism and maximum body size among adder populations: Effects of prey size. *Journal of Animal Ecology* 60:253–267.

Forsman, A. 1991b. Adaptive variation in head size in *Vipera berus* L. populations. *Biological Journal of the Linnean Society* 43:281–296.

Forsman, A. 1996. An experimental test for food effects on head size allometry in juvenile snakes. *Evolution* 50:2536–2542.

Gillingham, J. C. 1988. The amphibians and reptiles of Beaver Island. *Journal of Beaver Island History* 3:87–115.

Gillingham, J. C., J. Rowe, and M. A. Weins. 1990. Chemosensory orientation and earthworm location by foraging eastern garter snakes (*Thamnophis s. sirtalis*). In D. W. MacDonald, M. Muller-Schwarze, and S. E. Natynczuk (eds.), *Chemical Signals in Vertebrates* (vol. 5). Oxford University Press, pp. 522–532.

Grudzien, T. A, B. J. Huebner, A. Cvetkovic, and G. R. Joswiak. 1992. Multivariate analysis of head shape in *Thamnophis s. sirtalis* (Serpentes: Colubridae) among island and mainland populations from northeastern Lake Michigan. *American Midland Naturalist* 127:339–347.

Halloy, M., and G. M. Burghardt. 1990. Ontogeny of fish capture and ingestion in four species of garter snakes (*Thamnophis*), *Behaviour* 112(3–4):299–317.

Hampton, R., and J. C. Gillingham. 1989. Habituation of the alarm reaction in neonatal garter snakes, *Thamnophis sirtalis*. *Journal of Herpetology* 23:433–434.

Harding, J. H. 1996. *Amphibians and Reptiles of the Great Lakes Region.* University of Michigan Press.

Harding, J. H., and D. A. Mifsud. 2017. *Amphibians and Reptiles of the Great Lakes Region.* University of Michigan Press.

Hatt, R. T., J. V. Tyne, L. C. Stuart, C. H. Pope, and A. B. Grobman. 1948. *Island Life: A Study of the Land Vertebrates of the Islands of Eastern Lake Michigan.* Cranbrook Press.

Herzog, H. A. 1990. Experiential modification of defensive behaviors in garter snakes (*Thamnophis sirtalis*). *Journal of Comparative Psychology* 104:334–339.

Herzog Jr, H. A., B. B. Bowers, and G. M. Burghardt. 1989. Development of antipredator responses in snakes: IV. Interspecific and intraspecific differences in habituation of defensive behavior. *Developmental Psychobiology* 22:489–508.

Herzog Jr, H. A., B. B. Bowers, and G. M. Burghardt. 1992. Development of antipredator responses in snakes: V. Species differences in ontogenetic trajectories. *Developmental Psychobiology* 25(3):199–211.

Herzog, H. A., and G. M. Burghardt. 1986. Development of antipredator responses in snakes: I. Defensive and open-field behaviors in newborns and adults of three species of garter snakes (*Thamnophis melanogaster, T. sirtalis, T. butleri*). *Journal of Comparative Psychology* 100(4):372–379.

Holman, J. A. 2000. *Fossil Snakes of North America: Origin, Evolution, Distribution, Paleoecology.* Indiana University Press.

Holman, J. A. 2012. *The Amphibians and Reptiles of Michigan.* Wayne State University Press.

Holman, J. A., and J. H. Harding. 2006. *Michigan Snakes.* Michigan State University Extension.

Hough, J. L. 1958. *Geology of the Great Lakes.* University of Illinois Press.

Kapp, R. O., S. Bushouse, and B. Foster. 1969. A contribution to the geology and forest history of Beaver Island, Michigan. *Proceedings of the 12th Conference on Great Lakes Research*, 225–236.

King, R. B. 1989. Body size variation among island and mainland snake populations. *Herpetologica* 45:84–88.

King, R. B., K. M. Stanford, and P. C. Jones. 2018. Sunning themselves in heaps, knots, and snarls: The extraordinary abundance and demography of island watersnakes. *Ecology and Evolution* 8:7500–7521.

Krause, M. A., and G. M. Burghardt. 2001. Neonatal plasticity and adult foraging behavior in garter snakes (*Thamnophis sirtalis*) from two nearby, but ecologically dissimilar, habitats. *Herpetological Monographs* 15:100–123.

Krause, M. A., and G. M. Burghardt. 2007. Sexual dimorphism of body and relative head sizes in neonatal common garter snakes. *Journal of Zoology* 272:156–164.

Krause, M. A., G. M. Burghardt, and J. C. Gillingham. 2003. Body size plasticity and local variation of relative head and body size sexual dimorphism of garter snakes (*Thamnophis sirtalis*). *Journal of Zoology* 261:399–407.

Kurta, A. 1995. *Mammals of the Great Lakes Region.* University of Michigan Press.

Leuck, B. E., and P. M. Grassmick. 2022. Geographic distribution. *Pantherophis vulpinus.* *Herpetological Review* (submitted).

Lillywhite, H., and M. Martins, eds. 2019. *Islands and Snakes: Isolation and Adaptive Evolution.* Oxford University Press.

Long, C. A. 1982. Rare gigantic toads, *Bufo americanus*, from Lake Michigan Isles. *Reports of the Museum of Natural History, University of Wisconsin – Stevens Point* 18:15–19.

Manjarrez, J., J. Contreras-Garduño, and M. K. Janczur. 2014. Sexual size dimorphism, diet, and reproduction in the Mexican garter snake *Thamnophis eques*. *Herpetological Conservation and Biology* 9:163–169.

Mifsud, D. 2015. Reptile and amphibian community assessment and evaluation for the Beaver Island Archipelago. *Herpetological Resource Management.* Unpublished report, Chelsea, Michigan. 170 pp.

Mori, A., D. Layne, and G. M. Burghardt. 1996. Description and preliminary analysis of antipredator behavior of *Rhabdophis tigrinus tigrinus*, a colubrid snake with nuchal glands. *Japanese Journal of Herpetology* 16:94–107.

Passek, K., and J. C. Gillingham. 1997. Thermal influence on defensive behaviours of the Eastern garter snake, *Thamnophis sirtalis*. *Animal Behaviour* 54:629–633.

Phillips, C. J., J. T. Osoga, and L. C. Drew. 1965. The land vertebrates of Garden Island, Michigan. *Jack Pine Warbler* 43:20–25.

Phillips, J. G. 2016. Updated geographic distributions of Michigan herpetofauna: A synthesis of old and new sources. *Journal of North American Herpetology* 2016:45–69.

Placyk, Jr., J. S. 2006. Historical Processes, Evolutionary Change, and Phenotypic Plasticity: Geographic Variation in Behavior, Morphology, and Life-History Traits of Common Gartersnake, *Thamnophis sirtalis*, Populations. Unpublished PhD dissertation, University of Tennessee.

Placyk, Jr, J. S. 2012. The role of innate and environmental influences in shaping antipredator behavior of mainland and insular gartersnakes (*Thamnophis sirtalis*). *Journal of Ethology* 30:101–108.

Placyk, Jr., J. S., and G. M. Burghardt. 2005. Geographic variation in the frequency of scarring and tail stubs in eastern gartersnakes (*Thamnophis s. sirtalis*) from Michigan, USA. *Amphibia-reptilia* 26:353–358.

Placyk, Jr., J. S., and G. M. Burghardt. 2011. Evolutionary persistence of chemically elicited ophiophagous antipredator responses in gartersnakes, *Thamnophis sirtalis*. *Journal of Comparative Psychology* 125:134–142.

Placyk, Jr., J. S., G. M. Burghardt, R. L. Small, R. B. King, G. S. Casper, and J. W. Robinson. 2007. Post-glacial recolonization of the Great Lakes region by the common gartersnake (*Thamnophis sirtalis*) inferred from mtDNA sequences. *Molecular Phylogenetics and Evolution* 43:452–467.

Placyk, Jr., J. S., and J. C. Gillingham. 2002. Biogeography of the herpetofauna of the Beaver Archipelago: A synthesis and reevaluation. *Bulletin of the Chicago Herpetological Society* 37:210–215.

Placyk, Jr., J. S., M. J. Seider, and J. C. Gillingham. 2002. New herpetological records for High and Hog Islands of the Beaver Archipelago. *Herpetological Review* 33(3):230.

Queral-Regil, A., and R. B. King. 1998. Evidence for phenotypic plasticity in snake body size and relative head dimensions in response to amount and size of prey. *Copeia* 1998:423–429.

Rossman, D. A., N. B. Ford, and R. A. Seigel. 1996. *The Garter Snakes: Evolution and Ecology.* University of Oklahoma Press.

Schoener, A., and T. W. Schoener. 1984. Experiments on dispersal: Short-term floatation of insular anoles, with a review of similar abilities in other terrestrial animals. *Oecologia* 63(3):289–294.

Schuett, G. W., and J. C. Gillingham. 1990. The function of scream calling in nonsocial vertebrates: Testing the predator attraction hypothesis. *Bulletin of the Chicago Herpetological Society* 25:137–142.

Schuett, G., D. Hardy, R. Earley, and H. Greene. 2005. Does prey size induce head skeleton phenotypic plasticity during early ontogeny in the snake *Boa constrictor*? *Journal of Zoology* 267:363–369.

Schwartz, J. M., G. F. McCracken, and G. M. Burghardt. 1989. Multiple paternity in wild populations of the garter snake, *Thamnophis sirtalis*. *Behavioral Ecology and Sociobiology* 25(4):269–273.

Valencia-Flores E., C. S. Venegas-Barrera, V. Fajardo, and J. Manjarrez. 2019. Microgeographic variation in body condition of three Mexican garter snakes in central Mexico. *PeerJ* 7: e6601

Wusterbarth, T. L., R. B. King, M. R. Duvall, W. S. Grayburn, and G. M. Burghardt. 2010. Phylogenetically widespread multiple paternity in New World natricine snakes. *Herpetological Conservation and Biology* 5:86–93.

15

Prospects and Overview for Conservation of Snakes on Islands

Harvey B. Lillywhite and Marcio Martins

Introduction: Imperative for Conservation Action

There are more than 290,000 islands in oceans of the world ranging from large (>1 km^2) to small (<1 km^2 to ~0.0036 km^2) (Sayre *et al.* 2018). Therefore, we are talking about a great deal of total area potentially habitable for snakes. A database of snakes on islands (only marine landmasses, *i.e.*, excluding inland islands; Martins, unpublished; see further details on this database in Martins and Lillywhite 2019) shows that at least 1,300 species of snakes inhabit more than 1,080 islands worldwide. Snakes generally have high presence on world islands wherever the climate and food base are appropriate.

Numerous islands throughout the world have been influenced by cycles of changes in sea level that promoted the opening and closing of marine bridges and corridors as well as actual appearance and disappearance of islands themselves (*e.g.*, Sathiamurthy and Voris 2006; Voris 2000). Other factors, such as changes in salinity and the appearance and disappearance of mud flats and mangroves (Hanebuth *et al.* 2011), also provided scenarios favoring speciation and evolutionary changes of biota, including snakes (Heatwole *et al.* 2016, 2017). Insular snakes include terrestrial species and amphibious species such as sea kraits, a lineage of sea snakes that utilize terrestrial habitats on islands for oviposition, ecdysis, resting, and other activities (Heatwole 1999; Bonnet and Brischoux 2019). Hence, snakes and other biota are present on islands attributable to residence and speciation *in situ*, invasion from the sea, and variable immigration events including dispersal and human introductions (Paulay 1994).

Snakes are often important and integral parts of insular food webs and community structure, characteristically being mid-level or top predators in insular systems (*e.g.*, Wooten 2020). Importantly, the abundance and characteristics of snakes living on islands have provided numerous examples of spectacular natural phenomena of scientific as well as public interest related to numbers, behavior, and appearance of snakes (*e.g.*, Li 1995; Lillywhite and Sheehy 2019; Martins *et al.* 2019), as well as the evolution of unique traits (Arnaud and Martins 2019; Lillywhite and Sheehy

Harvey B. Lillywhite and Marcio Martins, *Prospects and Overview for Conservation of Snakes on Islands* In: *Islands and Snakes*. Edited by: Harvey B. Lillywhite and Marcio Martins, Oxford University Press. © Oxford University Press 2023. DOI: 10.1093/oso/9780197641521.003.0015

Figure 15.1 Some spectacular features of insular snakes. Upper left: Shedao Island Pitvipers (*Gloydius shedaoensis*) are exceptionally abundant and easily seen resting in trees and bushes when migratory birds utilize the island. Upper right: The evolutionary loss of a rattle is a remarkable character of the Santa Catalina Rattlesnake (*Crotalus catalinensis*), which evolved in the relative absence of predators on Isla Santa Catalina compared with populations of related rattlesnakes on the adjacent mainland. Lower left: Florida Cottonmouths (*Agkistrodon conanti*) on Seahorse Key feed on fish carrion that is abundant in the form of dropped or regurgitated fishes by nesting water birds that, prior to 2015, formed large nesting rookeries on the island. Lower right: Sea kraits have retained oviparity and are amphibious, utilizing islands for resting, ecdysis, and oviposition while foraging widely for fishes in adjacent marine waters. Shown here is a Yellow Sea Krait (*Laticauda saintgironsi*) moving to the sea following a period on land at New Caledonia.

Photographs by Xavier Bonnet (*G. shedaoensis, L. saintgironsi*), Gustavo Arnaud (*C. catalinensis*), and H. Lillywhite (*A. conanti*).

2019) (Figure 15.1). Hence, various insular populations of snakes have become the foci of efforts at conservation throughout the world (see examples in Lillywhite and Martins 2019). Because of increasing intensity of storms, climatic instability and changes, sea level rise, human disturbances, and other factors, we believe it is imperative that increasing attention be given to conservation of insular snakes and the islands on which they live (see also Fernández-Palacios *et al.* 2021).

Overview of Insular Biogeography

The classical theory of island biogeography proposed by MacArthur and Wilson (1963, 1967) states that species richness on islands results from a balance between immigration and extinction and that the size of the island and its distance from the source of potential immigrants (in general, a continent) affect species richness and composition. Some general patterns are evident when interpreting island biodiversity regarding immigration. For example, large islands are larger targets for potential immigrants and may thus show a higher richness than smaller islands. On the other hand, islands closer to the source of potential immigrants are more easily reached by them and thus tend to show higher richness as well. The combination of these features generally results in smaller and/or remote islands showing lower richness than larger and/or closer islands.

However, the origin of island biotas is not always the result of a series of successful immigrations because it depends largely on the origin of the island itself (Gillespie and Baldwin 2010; Meiri 2017; Martins and Lillywhite 2019). The biota of land-bridge islands (*e.g.*, Britain) is largely the result of a series of extinctions (a process known as *faunal relaxation*) and the persistence of some species after these islands were isolated from the continent during the sea-level rise at the end of Pleistocene glaciations. The biota of non-land-bridge islands (herein "oceanic" islands; see Ali 2017), on the other hand, is the result of successful colonization counterbalanced by extinctions, as predicted by MacArthur and Wilson (1963, 1967). Thus, communities from land-bridge islands are called *extinction-driven systems*, while those of "oceanic" islands are called *colonization-driven systems* (Patterson and Atmar 1986).

Snakes on Islands

As mentioned above, more than 1,300 species of snakes inhabit almost 1,080 marine islands around the world. Islands inhabited by snakes occur on all five continents, in all oceans, and from the equator to latitudes of 63 degrees north (Hitra Island, Norway) and 42 degrees south (Tasmania Island, Australia). These islands vary in size from 0.001 km^2 in some Japanese islands in the South China Sea to about 786,000 km^2 (New Guinea Island), with elevations ranging from less than 1 m (Huevos Island, Trinidad and Tobago) to almost 5,000 m above sea level (New Guinea Island). Snakes are also present on numerous islands in bodies of water on mainland habitat. An important generalization is that snakes have a successful and pervasive presence on islands, and, in many cases, the residence of snakes on islands is robust indeed. The various factors explaining the success of insular snakes are discussed below and elsewhere in this book.

Diversity and Endemism

Snake richness on islands varies from one species on many smaller islands to more than 130 in Borneo and Sumatra. Richness reaches higher values in land-bridge than in oceanic islands, and island area explains about one-third of the variation in snake richness in both island types (Martins and Lillywhite 2019; see Pyron and Burbrink 2014, and Fattorinni 2010, for additional examples). The richness of snakes on islands is also affected by the diversity of habitats that are available (Henderson 2004; Pyron and Burbrink 2014; Martins and Lillywhite 2019). Snake endemism tends to be high on islands, and approximately one-third of insular snake species are endemic, with the proportion of endemic species being greater on oceanic islands than on land-bridge islands (31.9% vs. 21.7%; Martins and Lillywhite 2019). Generally, islands have been identified as global centers of endemic richness, a pattern that is consistent across plants as well as vertebrates (although least pronounced for amphibians) (Kier *et al.* 2009).

Examples of islands with particularly high snake endemicity are Madagascar (71 endemics), Sri Lanka (46), and Borneo (31). The number of snake species that occur on land-bridge islands (about 720) is slightly smaller than on oceanic islands (about 780), with about half of the species occurring on just a few islands, while some occur on up to 60 islands. Although the composition of snake lineages tends to be a random sample of the mainland areas, some lineages are particularly well represented on islands (*e.g.*, boids, calamariines, pythonids, tropidopiids) while other lineages rarely occur on islands (*e.g.*, dipsadines, leptotyphlopids; Martins and Lillywhite 2019). In general, *in situ* diversification is less important than colonization from the mainland in explaining the phylogenetic composition of insular snake faunas (Pyron and Burbrink 2014).

Global Hotspots

The Eastern Hemisphere harbors most groups of islands inhabited by snakes, with the Malay Archipelago being the main global hotspot of insular snake diversity (O'Shea 2021). Indeed, 5 of the 10 islands with the highest diversity of snakes in the world are encompassed by the Malay Archipelago: Borneo (141 species), Sumatra (133), New Guinea (98), Java (96), and Sulawesi (63). Additional islands from the Eastern Hemisphere that are hotspots of snake diversity are Madagascar (90 species), Sri Lanka (89), Singapore (61), Hainan (57), and Hong Kong (50). In the Western Hemisphere, the Caribbean islands are by far the principal hotspot for insular snake diversity, with a total of 227 species on 171 islands. The five islands with higher snake diversity in the Caribbean region are Trinidad (47 species), Cuba (41), Hispaniola (33), Margarita (20), and Tobago (17). Finally, three islands on the coast of Brazil also have a high diversity of snakes: Grande (27 species), Santa Catarina (21), and São Sebastião (21).

Population Size and Genetic Diversity

Because small, isolated populations tend to be more prone to extinction (*e.g.*, Paulay 1994; Todd *et al.* 2010) and to have lower genetic variability (*e.g.*, Frankham 1997, 2005), studies on island organisms focusing on population size and genetic diversity are important for conservation. However, there are only a few studies focusing on these aspects for island snakes. Insular snake populations that have been studied exhibit highly variable densities. Some island populations of snakes are so small that their viability may be very low. Boback (2005) obtained very low estimates of population size for *Boa imperator* in five small islands on the coast of Belize, whereas Reed *et al.* (2007) estimated a population size of around 600 individuals for the same species in Cayo Cochino Pequeño (Honduras). Williams *et al.* (2016) estimated a population size of fewer than 50 individuals for the Saint Lucia Racer, *Erythrolamprus ornatus*. Densities of the Milos Viper on the island of Milos in the western Cyclades Islands are estimated to be only 0.5 snakes/ha in "optimal habitat" (Nilson 2019). Schwaner and Sarre (1988) estimated the population size of *Notechis ater* at Chappell Island (Australia) at about 1,100–1,400 snakes, representing densities of 4–13 snakes/ha. On the other hand, Carnac Island off Western Australia supports 300–400 adult tiger snakes in an area of approximately 19 ha, which estimates densities of 19–25 snakes/ha (Bonnet *et al.* 2002). The best estimates of the population size of the Golden Lancehead, *Bothrops insularis*, at Queimada Grande Island (Brazil) vary from 2,000 to 3,000 individuals, estimated to represent as many as 57–85 snakes/ha of forested habitat (Martins *et al.* 2008, 2019; Abrahão *et al.* 2021). Shedao Island pitvipers endemic to Shedao Island, China, have estimated densities of about 200 snakes/ha (Shine *et al.* 2002). Exceptionally large numbers of watersnakes have been reported from land-islands surrounded by fresh water and mainland areas (*e.g.*, "thousands of snakes" at Golem Grad, Chapter 12 of this volume; up to ~12,000 snakes distributed on US islands of western Lake Erie, comprising ~65 km of shoreline [where most snake activity occurs] and 2,440 ha of inland habitat, King and Stanford 2019).

Buckley and Jetz (2007) reported that, across 643 populations of lizards around the world, local abundances were an order of magnitude greater on islands than on mainlands, and this relationship held even when controlled for resource availability. This was attributed to sharp reductions of predator and competitor richness as dominant drivers of lizard abundance. We assume this relationship likely exists for snakes.

Regarding genetic diversity, Rodriguez-Robles *et al.* (2015) reported a lack of genetic diversity in the Caribbean boa, *Chilabothrus monensis*, in Mona island (Puerto Rico), suggesting that this pattern may be due to a founder effect or a low effective population size, among other hypotheses. Wen *et al.* (2021) found a low genetic diversity in the Shedao Island Pitviper, *Gloydius shedaoensis*, which population size was estimated at approximately 20,000 individuals (Li *et al.* 2007). Wen *et al.* (2021) suggest that the population is almost panmictic, with gene flow occurring homogeneously throughout Shedao Island. Salles-Oliveira *et al.* (2020) found low to medium genetic diversity in the Golden Lancehead, *Bothrops insularis*, at Queimada Grande Island (Brazil), with evidence of gene flow throughout the island.

Wharton (1969) estimated variable densities of Florida Cottonmouths on the island of Seahorse Key (Florida, USA), ranging from 4.6 snakes/ha on many parts of the island but 55 snakes/ha at bird rookeries where these snakes scavenge fish and carrion that are dropped by nesting birds. He reported seeing as many as 9 snakes beneath a single tree with nesting birds, and Lillywhite and Sheehy (2019) have also observed 5–8 snakes beneath rookery trees within an area of less than 4 m^2. Wharton (1969) further suggested there were approximately 600 snakes present on the entire 67-ha island, which gives an average density of 9 snakes/ha. Genetic and venom studies indicate the insular population of cottonmouths at Seahorse Key is genetically distinct and likely inbred compared with nearby mainland populations (McCleary 2009). See also Chapters 5 and 14 for further comments on the genetics of insular snake populations.

Immigration and Colonization of Islands by Snakes

Snakes are relatively successful inhabitants of islands, reflected in the fact that one-third of all snake species occur on islands (considering Martins unpublished island snakes database and the Reptile Database 2022). Previously, we have discussed the assembly of snake faunas on islands and the degree of insular endemism among snakes (Martins and Lillywhite 2019; see also Diversity and Endemism above). As pointed out above, land-bridge islands show extinction-driven snake assemblages, whereas oceanic islands show colonization-driven systems (*cf.* Patterson and Atmar 1986). On land-bridge islands, snakes remain in their habitats when the islands become isolated, although these habitats may change as a consequence of isolation (*e.g.*, the availability of some resources may change). On the other hand, oceanic islands may be colonized by individuals from a source population, whether mainland or other islands. A successful colonization depends on the dispersal ability and capacity to reach the island and to establish a population and a capacity to find the resources necessary to persist in the new habitat.

We suggest that an underappreciated aspect of the success of snakes in colonizing and adapting to islands is the evolutionary tendency for transitions to marine habits and habitat where conditions are appropriate (Murphy 2007, 2012; Lillywhite *et al.* 2008; Heatwole 2019).

Dispersal Abilities and Attributes Favoring Successful Colonization of Islands

Snakes are relatively successful at "island living," not only in terms of persistence but also in their ability to successfully invade and establish thriving populations. Perhaps foremost is the relatively low requirement for energy attributable to ec- tothermy and intermittent feeding (Lillywhite 2014). Many snake species con- sume bulky or relatively large prey and characteristically survive for weeks or months without feeding. Some species can survive long periods of starvation (1– 2 years: Wang *et al.* 2006; McCue *et al.* 2012). Such low energy requirements confer the ability of snakes to survive long overwater dispersal times as well as persistence on islands with limited, patchy, or ephemeral food resources (see further discus- sion in Martins and Lillywhite 2019). Plasticity of diet, including scavenging, has also favored survival of snakes on islands (Lillywhite *et al.* 2015). In more inte- grative context, low vagility, ectothermy, and seasonal inactivity can reduce energy expenditure and facilitate survival in habitats where food is available only during brief ephemeral periods (Shine *et al.* 2003).

Other attributes include life history features favoring comparatively rapid pop- ulation growth, at least for some species. Known examples with some documenta- tion and discussion include insular tiger snakes (Aubret 2015), invasive Brown Tree Snakes (Rodda *et al.* 1992), and endemic Shedao Island Pitvipers (Shine *et al.* 2002). Successful reproduction and population growth of snakes on islands is also evident from observations of the numbers and successes of insular snakes (*e.g.*, Lillywhite and Sheehy 2019; Shine *et al.* 2002). A special case involves the Brahminy blind snake (*Indotyphlops braminus*), which reproduces by obligate parthenogenesis. All popula- tions are unisexual and comprised of females. This species is small (typically <10 cm), fossorial, and often transported by humans to the extent that it now occurs on many oceanic islands and all continents except Antarctica (Wallach 2009; see Chapter 13 of this volume).

Blind snakes are easily transported to islands and have been called "Flowerpot Snakes" because of this means of introductions. Although most insular snakes are larger than this species, most are within a size range that is favorable for inhabiting even very small islands. Snakes are easily rendered cryptic by their behaviors, in- cluding secretive or fossorial habits, coloration, quiescence, and scansorial/arboreal habits in various species.

In addition to plasticity of diet, behavioral and physiological plasticity related to temperature (Aubret and Shine 2010; Yan *et al.* 2022) and water exchange (Moen *et al.* 2005; Miller and Lutterschmidt 2014; Lillywhite *et al.* 2009; Lillywhite and Sandfoss 2023) has been demonstrated in snakes and could likely be important with respect to future adjustments to climate change. Matching of phenotype to environmental conditions is complex, with many not well-understood facets. Evolving pheno- types that "match" a particular environment are considered adaptive, and mismatch of phenotypes to environmental conditions can result in costs to fitness. However, studies also show that mismatched phenotypes can have unexpected consequences over the long term if populations persist and may actually maximize fitness over a life course (Petrullo *et al.* 2023). These characters are in need of much further research.

Trophic Impacts

Snakes that occur on land-bridge islands likely are part of natural communities with likeness to mainland ecology where the biotic conditions are similar. In other cases, and particularly on oceanic islands, conditions might present trophic limitations, and snakes that invade or evolve adaptively and opportunistically might impact islands differently. More dramatic evolutionary changes in snakes attributed to insular conditions include changes of body size, which, relative to conspecifics on mainland areas, tend to be bimodal and are consistent with a "diet alteration hypothesis" involving insular prey that are larger or smaller than what is encountered on the mainland (Boback 2003; Martins and Lillywhite 2019; see also Chapters 4, 6, 8, and 14 of this volume). Thus, the driving influence for the evolution of body size in insular snakes appears to be related, generally, to the insular prey resources available to snakes, although the direction of change appears to have some dependency on the body size of the source population (Boback and Guyer 2003; Martins and Lillywhite 2019). Divergence of body size in insular tiger snakes (*Notechis*) inhabiting islands in southeastern Australia has involved repeated and rapid evolution of dwarf and gigantic populations that are geographically proximate to mainland tiger snakes (Schwaner and Sarre 1988; Keogh *et al.* 2005; Aubret 2015, 2019). Divergence of body size was shown to be related to prey resources and the isolation times of the populations, with rapid shifts attributable to phenotypic plasticity and more gradual divergence by genetic differentiation (Aubret and Shine 2007).

Snakes that are invasive to islands clearly can have negative impacts on the insular fauna and biotic communities. A well-known example that has received much research attention and discussion is the Brown Tree Snake (*Boiga irregularis*) that invaded Guam and, more recently, other islands (see Chapter 13, this volume). In the case of Guam, in spite of a generalist diet and consumption of several vertebrate prey, the impact of Brown Tree Snakes on bird populations rapidly caused the extinction of several endemic avian species (Savidge 1987) and continues to depress populations of some species of lizards (Campbell *et al.* 2012). Other examples of negative impacts on prey resources include Horseshoe Whip Snakes (*Hemorrhois hippocrepis*) on wall lizards (*Podarcis pityusensis*) on the island of Ibiza (Montes *et al.* 2022); California King Snakes (*Lampropeltis californiae*) on all the endemic herpetofauna of Gran Canaria (Piquet and López-Darias 2021); boas (*Boa constrictor*) on endemic fauna of Cozumel Island, Mexico (Martínez-Morales and Cuarón 1999), and Burmese Pythons (*Python bivittatus*) on indigenous mammals in Florida, including some of the Keys (Dorcas *et al.* 2012).

Predation by snakes on insular prey also can be an agent for evolutionary changes in the prey. Predation by egg-eating snakes has led to the evolution of parental care in a terrestrial lizard. Maternal care by Long-tailed Sun Skinks (*Eutropis longicaudata*) was shown to be genetically determined and to provide a significant increase in survival of eggs on Orchid Island (Taiwan) by reducing predation from egg-eating snakes (*Oligodon formosanus*) (Pike *et al.* 2016). The snakes are not abundant among other populations of the lizard, and those populations do not display parental care.

Body temperature and physiological processes of ectotherms, including behavioral performance, are dependent on immediate thermal environments. Survival of prey thus depends, in part, on access to temperatures that enable ectothermic prey to evade predation by means of rapid movements that outpace or outmaneuver that of their predators. Recent investigations by Yuan *et al.* (2021) have demonstrated that, on the Izu Islands of Japan, body temperatures of foraging endemic skinks (*Plestiodon latiscutatus*) have increased 1.3 °C from 1981 to 2019, yet are 2.9 °C warmer on islands where predatory rat snakes (*Elaphe quadrivirgata*) are present relative to islands that are free of snakes. This example illustrates the complexity of responses to climate change in consideration of the temporal and spatial occupancy of thermal environments by prey and predators that interact and may or may not respond to climate change. The responses of prey to climatic warming (or other climatic factors such as drought) might well be different depending on the presence or absence of snake predators.

Thus, snakes have potentially important impacts and functional roles on islands where they can be drivers of demographic dynamics or evolutionary responses of prey populations, and contrariwise, respond negatively or positively to fluctuations of prey densities attributable to other causes. Hence, insular snakes can have key roles in the dynamics and stability of trophic networks (Wooten 2020). See also Kemp (2023) for a discussion of how changes in environments and biodiversity can affect functional diversity and ecosystem services of reptiles on islands.

Evolution in Insular Populations

Although *in situ* diversification of snake lineages on islands is relatively rare, both in land-bridge and oceanic islands (especially the former; see review in Martins and Lillywhite 2019), there is plenty of evidence for evolution in island populations, probably because of their isolation from the source population and of differences in habitat features (*e.g.*, Boback 2003; Wüster *et al.* 2005; McCleary 2009; Arnaud and Martins 2019; Aubret 2019; Ammresh *et al.* 2023; see also Chapters 4, 6, 11, and 14 of this volume). For example, in a review on body size evolution in island snakes, Boback (2003) provided several examples of marked changes in body size in island populations compared to source populations and suggested that these changes reflect the consumption of prey of different sizes on islands, the so-called diet alteration hypothesis (see also Shine 1987). According to this hypothesis, giant and dwarf island snakes feed on prey that are larger and smaller, respectively, than those on the mainland. Aubret (2019) described a high amount of variation in the body size of island populations of a single species, the Mainland Tiger Snake, *Notechis scutatus*. Wüster *et al.* (2005) compared the morphology of two island pitvipers, *Bothrops insularis* and *Gloydius shedaoensis*, with their closest mainland relatives and found that both species evolved a larger head; *B. insularis* also exhibits a more anterior heart and a longer tail. The endemic rattlesnake, *Crotalus catalinensis*, of the Santa Catalina Island, Mexico, lacks the typical rattle of rattlesnakes (Figure 15.1), supposedly as a result of the absence of natural predators or as an adaptation for hunting (see review in Arnaud and Martins 2019; see also additional examples of rattle reduction in Rabatsky 2006). Other bizarre adaptations are known from islands, such as the jointed upper jaws of boid snakes on the Mascarenes (Frazetta 1970) and longer gastrointestinal tracts with cecal valves in insular lizards (Sagonas *et al.* 2015).

Equilibrium Versus Non-equilibrium States

The theoretical ideas of MacArthur and Wilson (1967) stimulated many studies of insular biotas and attempts to test various aspects of island biogeography (see Whittaker 1998; Morrison 2013). Although numerous investigators have published studies of insular snake populations, we think there are few data that enable robust judgment concerning the concepts of equilibrium or non-equilibrium of insular ophidian faunas. MacArthur and Wilson (1967) focused on numbers of species, while other ensuing investigations of insular biotas have also examined population numbers. If an island is in "stasis" (or "static equilibrium" *sensu* Whittaker 1998), it experiences neither immigration of new species nor extinctions, and the composition of species remains unchanged over time. However, equilibrium and non-equilibrium states were never explored for island snake communities. Fattorini (2010) reported, however, that in the Tuscan Archipelago of the Mediterranean, colonization of islands by reptiles via land bridges occurred during Pleistocene connections to the mainland, and, following disconnection, the reptilian fauna (including snakes) equilibrated according to island area during a long period of relaxation. A similar process was suggested for land-bridge islands from southeastern Brazil (Martins and Lillywhite 2019).

Regarding population equilibrium, there seems little doubt that snake populations on islands change over time (*e.g.*, King and Stanford 2019; Sandfoss *et al.* 2018; Wen *et al.* 2022; see also Chapters 10 and 12 of this volume), with immigrations and extinctions of species being known from recent examples (*e.g.*, see Chapter 13 of this volume). Many insular snake populations likely shift equilibrium over time (*e.g.*, Chapter 10 of this volume), although some populations appear to have been relatively stable for long periods as well (Guimarães *et al.* 2014). The stability of insular populations of snakes is largely related to (1) relative isolation, (2) abundant food resources, and (3) relative absence of predators compared with mainland populations.

We suggest that population equilibrium analyses have not been published for snakes because of issues related to sampling intensity and temporal scale. Nonetheless, it seems certain going forward that non-equilibrium dynamics will increase among insular snake populations owing to the degradation of islands attributable to climate change coupled with increasing human impacts, including anthropogenic introductions of invasive species (Paulay 1994; Chapter 13 of this volume). We propose that future research of insular snakes could significantly benefit conservation efforts by providing some indicators of the equilibrial status of populations as well as improved understanding of negative impacts and priorities for conservation actions.

Threats to Insular Snake Faunas

Catastrophic Events

There is relatively little information concerning how catastrophic events such as cyclones, earthquakes, fire, or volcanic eruptions affect snakes living on islands. The direct impacts of fire, moving lava, crashing waves, etc. might kill snakes directly, and these and other factors such as intense storms will likely induce secondary mortality as a result of various causes (*e.g.*, predation on snakes in opened habitat, reductions of prey populations, flooding with seawater). The impact of storms, especially tropical cyclones, will vary among insular snake populations depending on the habitats and habits of species. Vulnerability to wind and waves should be quite different comparing, for examples, species that are arboreal with those that are terrestrial/fossorial, as well as location, whether interior or near shorelines. In cases of severe impacts on snake populations, recovery and the time course of non-equilibrial states will depend on the dispersal abilities and life histories of a species relative to the frequency of such storms (see Morrison 2013 for further discussion of these principles).

On the island of Seahorse Key (Florida, USA), an intense, no-name storm pushed seawater over much of the low-lying habitat of the island in 1993, and numerous dead Florida Cottonmouths (*Agkistrodon conanti*) were observed on the landscape after the storm passed (Allen Dinsmore, unpublished observations). Typhoon Morakot ravaged the island of Lanyu (Orchid Island), Taiwan, during early August of 2009. Three species of sea kraits (*Laticauda* spp.) inhabit the coastal region of the island and forage in surrounding marine habitats. These snakes are normally abundant in littoral habitats surrounding the island, but disappeared coincident with falling barometric pressure prior to arrival of the typhoon (Liu *et al.* 2010). The typhoon impacted the island severely, destroying some coastal buildings and sections of the ring coastal highway when large rocks and other debris were moved by wind and water and impacted the coastal perimeter of the island (Figure 15.2). Sea kraits were again seen at pre-storm abundance levels in the coastal waters and intertidal habitat within days after the storm passed. Liu *et al.* (2010) hypothesize that snakes anticipated the cyclone by sensing changes of pressure and secluded

Figure 15.2 Coastal damages from typhoon Morakot interacting with Orchid Island, Taiwan, August 7–9, 2009. Wind and rough surf created coastal damage, as shown by downed trees and displaced rocks and boulders on the coastal road. Coastal sea kraits disappeared during the storm but reappeared following its passage.
Photos by H. Lillywhite.

themselves in places that offered protection from the greater impacts of the storm. The authors found no evidence of snakes being killed directly, in spite of the extensive coastal damage that was witnessed. Sea snakes have the option of accessing deeper recesses in corals, rocks, or sea caves during passage of a storm, but such places are not accessible to terrestrial species on the island. As example, the arboreal Chinese Green Tree Viper (*Trimeresurus stejnegeri*) might have been negatively impacted by the extremes of wind and sea spray that covered the island during the cyclone, but confirming data are not available.

Impacts of Climate Change

Globally, islands and coastal environments are a major concern regarding impacts of anthropogenic climate change (Harley *et al.* 2006). As consequences of ice melting and warming seawater, the world's oceans are expanding and affecting coastlines with rising sea levels, upwelling, currents, storm frequency, and wind fields. Increasing atmospheric carbon dioxide (CO_2) is decreasing the pH of seawater and potentially depleting the atmospheric ozone layer and thereby enhancing levels of ultraviolet radiation at the Earth's surface. The nature and magnitude of climatically driven changes in the physical environment will strongly influence biotic responses, especially the distribution and abundance of species, which will shift according to physiological tolerances and abilities to adjust genetic expression, move, or adapt. Changes in productivity are also an expected emergent response.

The impact of climate change on island snakes is complex for two reasons. First, terrestrial insular snakes are relatively isolated, with varying degrees of dispersal capabilities. Second, sea kraits occupy numerous smaller islands as well as coastal areas of larger land masses (Heatwole 2019), and they also enter the sea, sometimes moving long distances for feeding (Brischoux *et al.* 2007). Hence, these snakes are influenced by oceanic processes and changes as well as terrestrial impacts. Similarly, semi-marine, amphibious, or marine species associated with aquatic habitats (*e.g.*, acrochordids, homalopsids, some colubrids) will be affected by ecological changes in coastal marine communities. Because terrestrial snakes inhabiting islands occupy discrete units of habitat, climatic corridors with connectivity of protected networks are either not possible or challenging at best.

Flooding and seawater incursion resulting from rise of sea level (see Wetzel *et al.* 2013; Bellard *et al.* 2014) are perhaps the more protracted threat to insular snakes. Many smaller islands have maximum elevations of just a few meters and mean elevations of 1–2 meters. The global mean water level in the ocean rose by 3.6 mm per year from 2006 to 2015, which was 2.5 times the average rate per year in the previous century (NOAA: Climate.gov). By the end of the century, global mean sea level is likely to rise at least 0.3 m above 2000 levels. The 2020 National Climate Assessment concluded that sea levels will rise from 0.2 to 2.0 meters by 2100, and sea levels in the western Pacific Ocean are increasing at 2–3 times the global average. High surf events might be expected to inundate low-lying islands and threaten entire insular as well as coastal habitats.

Tropical cyclones are severely impacting small islands at a global scale, and, according to the Intergovernmental Panel on Climate Change Working Group II (IPCC WGII) Sixth Assessment Report, 2021, both the intensity and intensification rates have increased in the past 40 years (see also Kossin *et al.* 2020). Significant levels of terrestrial species diversity and endemism, generally, occur on small islands where snakes are often an associated part of the fauna. Almost 50% of terrestrial species that are considered to be at risk of extinction also occur on islands where they are subjected to elevation of sea levels, heavy precipitation events, tropical cyclones, and storm surges. Furthermore, small islands are predicted to experience severe coral bleaching, aridity, and freshwater stress. Marine biodiversity hotspots appear to be particularly sensitive to global warming; they have warmed more than non-hotspots and are projected to experience higher levels of future warming (Kocsis *et al.* 2021). The rate of sea level rise is accelerating and has more than doubled from the beginning of the 20th century to the early 21st century.

Climatic impacts on islands will escalate disruptions of ecological communities and eliminate or fragment habitat, and these processes will interact with other anthropogenic drivers that threaten species survival, such as overexploitation or alterations of landscape. Warming temperatures also favor pathogens relative to hosts and thus the negative impacts of disease are expected to become more severe as temperature warms (Harvell *et al.* 2002). Future protection of insular snake populations, especially endemic species, will need to address risks that are attributable to climate change.

Predation and Invasive Alien Species

Insular snake populations are vulnerable to top-down predators, especially raptors and other large birds, as well as introduced mammals such as cats, mongooses, pigs, rats, and other feral species that forage for their eggs (see, *e.g.*, Dueñas *et al.* 2021). Biological invasions are currently one of the principal causes of global ecological change (Vitousek *et al.* 1997; Mack *et al.* 2000), and mammals are one of the more important taxa causing losses of biodiversity following biological invasions on islands (Courchamp *et al.* 2003). Examples of snakes being preyed upon and sometimes decimated are largely known from gray literature (but see Seaman and Randall 1962; Pregil *et al.* 1988; Case and Bolger 1991; Márquez *et al.* 2017; Aubret 2019; Henderson 2019; Martins *et al.* 2019). Sea snakes (including amphibious *Laticauda* spp.) are also preyed on by sharks as well as large birds when surfacing to air breathe (Heatwole 1975; Guinea 1986; Bhaskar 1996; Heatwole *et al.* 2005; Masunaga *et al.* 2007; Sheehy *et al.* 2011; Bonnet and Brischoux 2019; Somaweera *et al.* 2021). We are not aware of robust studies that quantify predation on insular snakes by either indigenous or introduced predators.

On the other hand, numerous islands provide refugia for snake populations owing to the absence of large predators that might occur on adjacent mainland areas but not offshore islands. This likely explains, at least in part, the dense and relatively stable populations of snakes that have characterized some islands.

Human Impacts

Human activities have caused dramatic insult to snakes and other biota living on islands. Indeed, human activity is responsible for tremendous environmental degradation and losses of insular biodiversity worldwide (Olson 1989; WWF 2022), and the overriding impact now and going forward is the cascade of consequences related to human-induced climate change (Lillywhite 2013). The main negative impacts of humans to snake populations living on islands include the destruction, fragmentation, and degradation of natural ecosystems causally related to agriculture, residential housing, and harvesting of resources, including mining (*e.g.*, Nilson 2019), timber (*e.g.*, Henderson 2019), and snakes themselves (Reed *et al.* 2007; Auliya *et al.* 2016; Martins *et al.* 2019; other examples are cited in Martins and Lillywhite 2019). Development can also bring hazards to snakes associated with mortality from crossing roads and highways (Ashley and Robinson 1996; Nilson 2019). Other downstream effects from human development on islands include pollution (*e.g.*, related to agriculture), direct killing (especially venomous species), disturbance from domestic animals, alteration or elimination of water sources, disturbance to or elimination of prey, and introduction of alien predators. Species that are amphibious (sea kraits, watersnakes, homalopsids) can also be subject to incidental deaths when snakes become lethal bycatches in fishing nets. One of the more bizarre impacts of humans on snakes occurred in ancient times when vipers were thrown as weapons to attack vessels coming from Africa and Sardinia to pirate Etruscan towns (Barbanera *et al.* 2009). Vipers were also carried by Greek militia during the 8th to 3rd centuries BC, and this could explain the origins of some insular viperid populations (Barbanera *et al.* 2009).

Tourism and related development can also pose threats to insular snake populations, although on some Pacific islands tourism has incentivized efforts to protect local snake populations (Taiwan, Tu, personal communication; Fiji, Lillywhite, personal observations; see also Lillywhite *et al.* 2017). Nonetheless, mass tourism can lead to rapid degradation of fragile natural habitats that become attractive to tourists (Fattorini 2010). The influence of humans on insular snakes can be very complex, historically and as a trajectory of influences both now and in the future.

Strategies for Conservation and Management

A diversity of measures can be used to conserve island snakes, including (i) increasing knowledge about species distribution, life history, etc.; (ii) assessing the effectiveness of protected areas; (iii) selecting priority locations to set new protected areas; (iv) predicting the vulnerability of species to future threats; (v) preparing conservation action plans; and (vi) implementing conservation actions (*e.g.*, population monitoring and management, translocation, control of invasive species, maintenance of *ex situ* populations, education). Measures focused on small population size include population modeling to predict future trends in population size, population monitoring to track variation in time, translocation to increase local population size, and maintenance of *ex situ* populations to allow reintroductions. On the other hand, measures to tackle overcollection and excess harvesting include creating new legislation or enhancing existing protections, improving enforcement, and developing effective education programs.

Probably the most important aspect of protecting vulnerable island populations of snakes is to protect habitat and ensure that sufficient area and heterogeneity of spatial and biotic resources remain intact and viable (see Chapter 12 of this volume). Depending on the island and species, this requirement may not be simple to achieve. To illustrate the complexity and challenges involved, we highlight the terrestrial sites that are utilized by sea kraits (*Laticauda* spp.). In New Caledonia, the distribution and abundance of sea kraits depends on a range of conditions and microhabitats, and the large-scale functioning of metapopulations is influenced by interdependence among different colonies (Bonnet and Brischoux 2019). These snakes depend on variable locations for feeding, mating, oviposition, and other activities involving a diversity of sites that are partly connected. Hence, preserving connectivity among diverse sites is essential for maintaining the global functioning of multiple colonies.

Existing Programs and Successes

There are successful conservation programs for many threatened island snakes. For example, the critically endangered Antiguan Racer, *Alsophis antiguae*, used to be abundant and widespread in Antigua, Barbuda, and other nearby islands (Sajdak and Henderson 1991). However, no individuals were seen in Antigua since the 1940s, probably due to the introduction of mongoose on the island (Henderson *et al.* 1996). It became extinct also in Barbuda, for unknown reasons (Daltry and Mayer 2016). In 1995, the only remaining population, at Great Bird Island, was estimated to be about 50 individuals (Daltry 2006; Daltry *et al.* 2017). By then, the Black Rat (*Rattus rattus*) was identified as a serious threat for the snake, and a rat eradication program was successfully implemented. As a result of rat eradication, the racer population increased by 300% from 1995 to 2004 (Daltry 2006). Furthermore, a reintroduction plan began in 1999, with the successful translocation of individuals from Great Bird Island to three other islands: Rabbit, Green, and York. In 2015, the total population on these four islands was estimated at more than 1,100 individuals (Daltry and Mayer 2016; Daltry *et al.* 2017). The successful reintroduction of the Antiguan Racer was made possible by field research, fundraising, and an education campaign (Daltry *et al.*

2017). Although it is still a threatened species, *A. antiguae* would probably be completely extinct if these conservation measures had not been taken.

Another successful conservation program recovered the vulnerable Round Island Ground Boa, *Casarea dussumieri*, from the Mauritius Islands. Once widespread in the Mauritian archipelago, in the mid-1800s, this species became restricted to the small Round Island (2.15 km²) following the arrival of invasive mammalian predators on the other islands (see Cole *et al.* 2018). Successful conservation actions started in the 1980s (*e.g.*, eradication of invasive species, habitat restoration, reintroduction in Gunner's Quoin), and, by 2018, its total population was estimated to be around 2,000 individuals (Cole *et al.* 2018).

The critically endangered Golden Lancehead, *Bothrops insularis*, from the Queimada Grande Island of southeastern Brazil has been also the focus of several conservation actions. Since its discovery in the early 1900s, this species was very common in the island, with estimates of about 3,000–4,000 individuals (Duarte *et al.* 1995). Indeed, even with the collection of more than 1,000 snakes from early 1900s to 1993 (Duarte *et al.* 1995), the snake was still very common in the 1990s (Marques *et al.* 2002). In the 2000s, Martins *et al.* (2008) provided the first estimate of the population size of the Golden Lancehead at about 2,300 individuals. However, from the 1990s to the 2000s, the encounter rate of snakes in the only trail that crosses the island dropped to half and various signs of illegal collection were detected (Martins *et al.* 2008; Guimarães *et al.* 2014). Based on these observations, it is possible to argue that the population size in the 1990s was around 4,000 individuals. However, since the 2000s, an increased enforcement of visits to the island resulted in a significant decrease in the signs of illegal collection, and a recent study estimated the population size at about 2,900 individuals (Abrahão *et al.* 2021). Besides the estimates of population size and the enforcement of visits to the island, other conservation actions took place over the past decades: the establishment of an *ex situ* population, further studies on its biology (*e.g.*, habitat use, reproduction, diet), and education focused on children from the region of the island (see Martins *et al.* 2019).

For other success stories related to insular snake conservation see Nilson (2019), Arnaud and Martins (2019), King and Stanford (2019), and Chapters 7, 9, and 12 of this volume.

Thoughts about the Future

Numerous threats will continue to impact insular populations of snakes and most will affect insular physiographic features and biological communities, including other important elements of biodiversity. The long-term impacts of habitat disturbance/loss and climate change are perhaps the most serious and ongoing threats to insular snakes. In the absence of management and active conservation schemes, snake populations will need to adjust to possible changes in prey base and other biological interactions, as well as physical features of the environment—especially temperature, water, and bioclimatic features of microenvironment such as shade. Long-term studies with suitable datasets for endotherms suggest that many populations harbor biologically meaningful levels of additive genetic variance in fitness enabling rapid adaptive evolution, hence natural selection has the potential to mitigate the effects of environmental change (Bonnet *et al.* 2022). Insular snake populations are relatively small in many cases, however, and genetic variance is either small or unknown. Evidence from several studies of insular snakes suggests that adjustments to change will involve both plastic responses as well as adaptive evolution (Inger and Voris 2001; Palkovacs 2003; Keogh *et al.* 2005; Pyron and Burbrink 2014; Martins and Lillywhite 2019).

Ambitious efforts in conservation will be needed to mitigate the growing crisis in global losses of biodiversity, especially on islands (Fernández-Palacios *et al.* 2021). Generally, some minimum land area is important to preserve biodiversity, ecologically intact areas, optimal locations for representative species ranges, and ecoregions. Spatially explicit land-use scenarios and adaptive, sustainable management may be necessary to safeguard biodiversity. Islands are problematic in these contexts because land areas can be small, populations are to varying degrees isolated, and connectivity is absent to enable movement and compensate for local extirpations of populations. On the other hand, in many cases, islands offer refugia from predators, human impacts, and other elements of disturbance in scenarios of environmental stability (*e.g.*, Chapters 8 and 12 of this volume). The optimal situation for protecting snake populations on islands is for a protected area to include an island or archipelago that is wild and without human habitation and characterized by sufficient area and height to be buffered from rising sea levels and provide effective refugia from storms and other impacts of climatic changes. Given the unprecedented rapidity of climatic changes today, these aspects of conservation will be critical for maintaining populations of insular snakes, especially in low-lying areas. Because islands are characteristically isolated by surrounding water without terrestrial connectivity to other areas, the vagility and dispersal capabilities of snakes must be given consideration. And for amphibious species with off-island foraging sites, schemes for conservation of species have added complexity.

Considering the variable sizes, number, isolation, and wealth of snake fauna associated with the world's islands, we believe the following list of maxims can provide a helpful guide to the conservation of insular snakes. See also the urgency statements in Fernández-Palacios *et al.* (2021).

1. It is imperative that national governments or associated agencies take immediate actions worldwide to mitigate global warming and climate change. All globally distributed ecosystems are at risk if immediate actions are not implemented.
2. Herpetologists should continue working to identify and, where possible, prioritize insular snake faunas that include endemic, endangered, and at-risk populations requiring protection. New and interdisciplinary tools will be helpful with respect to research and monitoring the health and diversity of insular snake

populations, including, for example, the use of environmental DNA (eDNA; Ogden 2022) and employing tools for taxonomic revisions or discoveries that include insular species (Guedes *et al.* 2020).

3. We consider it important to preserve for future generations insular species of snakes that represent endemism, small distributions, rarity, unusual biological properties, and populations considered to be "evolutionarily significant units" (*e.g.*, Grismer 1999, 2001).

4. Efforts should be taken to keep uninhabited islands with robust snake populations wild and with minimal human contact. We believe the best conservation policy is for islands to remain wild and undisturbed where feasible and without intensive management, even though such action might be well-intended. Although single components of an ecosystem might be targeted (snake species), the entire system must be taken into consideration.

5. Protected areas or reserves are not static systems, and variable rates of change will need to be addressed with ongoing effort. Climate change, as well as other factors, will perturb seemingly stable systems, and conservation programs will need to adaptively manage islands as non-equilibrial systems with alternative stable states. Long-term ecological research should continue to monitor ecosystem responses to climate change (Jones and Driscoll 2022). Conservation of insular biodiversity and resources under conditions of expanding uncertainty and transforming ecosystems will likely require novel approaches of management (including resist-accept-direct [RAD] adaptive management: Lynch *et al.* 2022) and integration of evolving paradigms of ecological and conservation science into management and policy (Wallington *et al.* 2005). Methodological advances for studying and understanding local adaptation—especially important for islands—should enable development of strategies for improving the likelihood of population persistence relative to risks of decline attributable to climate change (Meek *et al.* 2023).

6. Societal values will need to be considered in decisions related to conservation and management of biodiversity, and the importance of involving the public must be considered. It is essential that scientifically correct and up-to-date information be published and provided to the public concerning the wider community and adverse human impacts (*e.g.*, deforestation, fragmentation of habitat, invasive species, overexploitation in fisheries or agriculture, etc.). "Conservation efforts that attempt to wall off nature and safeguard it from humans will ultimately fail" (Wallington *et al.* 2005).

7. In cases where recovery programs are required to restore populations of insular snakes that might have experienced serious decline in contact with humans, outreach programs should be aimed at residents and visitors, if possible coupled with research efforts to restore and/or monitor the snake population. An example of successful conservation of insular snakes involving the public was reported by King and Stanford (2019) (Figure 15.3).

Figure 15.3 Left: Park rangers releasing sea kraits (*Laticauda saintgironsi*) that are part of a mark-and-recapture study on New Caledonia. These snakes utilize terrestrial habitat and also swim long distances to forage at sea, thus complicating areas that require protection for connecting corridors of critical habitat. Right: Kristin Sanford leads a field trip with secondary school students and holds a Fox Snake to acquaint the students with a gentle "first snake" at an island state park near Lake Erie, USA. Photographs by Xavier Bonnet (left) and V. Mettler (right).

8. In the case of human-inhabited or developed islands, research, monitoring, public education, and civic involvement will be helpful, or required, to mitigate the loss or fragmentation of habitat, usually related to deforestation or expansion of agriculture, pasturelands, and urbanization (Martins and Lillywhite 2019). In cases of snake populations that are amphibious, protection of sites away from the island might be required to provide corridors of important habitat and connections of ecosystems vital to the survival of a species.

9. Herpetologists and conservation scientists will need to establish and implement priorities for education related to mitigating overexploitation of insular snake populations. Anthropogenic depletion of snake populations can be related to food or medicinal uses, pet trade, scientific research, and wanton killing out of ignorance of the importance of all elements of biodiversity.

10. Conservation of insular snakes is important for aesthetic, cultural, and scientific values in addition to other reasons discussed in this chapter and volume.

This chapter and others in this volume have been contributed to assess and promote understanding of the diversity and conservation of insular snakes and to focus attention on what we consider to be one of the world's treasures of natural phenomena. Human responses to habitat disturbance/loss and climate change, as well as other threats to insular snakes, are varied and hopefully will multiply and intensify as challenges for conservation accelerate. Knowledge and interest related to snake populations living on islands attest to the need for and value of long-term research efforts to understand, mitigate, and adapt to changing insular ecosystems that threaten the diversity of their unique snake fauna.

Acknowledgments

We are grateful to Jeremy Lewis, Oxford University Press, for encouraging us to compile chapters for a second volume of the book we first edited and published in 2019, and for editorial guidance throughout this project. MM thanks Fundação de Amparo à Pesquisa do Estado de São Paulo (FAPESP) for a grant (#2020/12658-4) and Conselho Nacional de Desenvolvimento Científico e Tecnológico (CNPq) for a research fellowship (#309772/2021-4).

References

Abrahão, C. R., Amorim, L. G., Magalhães, A. M., Azevedo, C. R., Grisi-Filho, J. H. H., and Dias, R. A. 2021. Extinction risk evaluation and population size estimation of Bothrops insularis (Serpentes: Viperidae), a critically endangered insular pitviper species of Brazil. *South American Journal of Herpetology* 19:32–39.

Ali, J. R. 2017. Islands as biological substrates: Classification of the biological assemblage components and the physical island types. *Journal of Biogeography* 44:984–994.

Ammresh, E. Sherratt, V. A. Thomson, M. S. Y. Lee, N. Dunstan, L. Allen, J. Abraham, and A. Palci. 2023. Island tiger snakes (*Notechis scutatus*) gain a "head start" in life: How both phenotypic plasticity and evolution underlie skull shape differences. *Evolutionary Biology* https://doi.org/10.1007/s11692-022-09591-z.

Arnaud, G., and M. Martins. 2019. Living without a rattle. The biology and conservation of the rattlesnake, *Crotalus catalinensis*, from Santa Catalina Island, Mexico. In H. B. Lillywhite and M. Martins (eds.), *Islands and Snakes. Isolation and Adaptive Evolution.* Oxford University Press, pp. 241–257.

Ashley, P. E., and J. T. Robinson. 1996. Road mortality of amphibians, reptile and other wildlife on the Long Point Causeway, Lake Erie, Ontario. *Canadian Field Naturalist* 110:403–412.

Aubret, F. 2015. Island colonisation and the evolutionary rates of body size in insular neonate snakes. *Heredity* 115:349–356.

Aubret, F. 2019. Pleasure and pain. Insular tiger snakes and seabirds in Australia. In H. B. Lillywhite and M. Martins (eds.), *Islands and Snakes. Isolation and Adaptive Evolution.* Oxford University Press, pp. 138–155.

Aubret, F., and R. Shine. 2007. Rapid prey-induced shift in body size in an isolated snake population (*Notechis scutatus*, Elapidae). *Austral Ecology* 32:889–899.

Aubret, F., and R. Shine. 2010. Thermal plasticity in young snakes: How will climate change affect the thermoregulatory tactics of ectotherms? *Journal of Experimental Biology* 213:242–248.

Auliya, M., S. Altherr, D. Ariano-Sanchez, E. H. Baard, C. Brown, R. M. Brown, J. C. Cantu, et al. 2016. Trade in live reptiles, its impact on wild populations, and the role of the European market. *Biological Conservation* 204:103–119.

Barbanera, F., M. A. L. Zuffi, M. Guerrini, A. Gentil, S. Tofanelli, M. Fasola, and F. Dini. 2009. Molecular phylogeography of the asp viper *Vipera aspis* (Linnaeus, 1758) in Italy: Evidence for introgressive hybridization and mitochondrial DNA capture. *Molecular Phylogenetics and Evolution* 52:103–114.

Bellard, C., C. Leclerc, and F. Courchamp. 2014. Impact of sea level rise on the 10 insular biodiversity hotspots. *Global Ecology and Biogeography,* 23:203–212.

Bhaskar, S. 1996. Sea kraits on South Reef Island, Andaman Islands, India. *Hamadryad* 21:27–35.

Boback, S. M. 2003. Body size evolution in snakes: Evidence from island populations. *Copeia* 2003:81–94.

Boback, S. M. 2005. Natural history and conservation of island boas (*Boa constrictor*) in Belize. *Copeia* 2005:880-885.

Boback, S. M., and C. Guyer. 2003. Empirical evidence for an optimal body size in snakes. *Evolution* 57:345–351.

Bonnet, X., and F. Brischoux. 2019. Terrestrial habitats influence the spatial distribution and abundance of amphibious sea kraits. Implications for conservation. In H. B. Lillywhite and M. Martins (eds.), *Islands and Snakes. Isolation and Adaptive Evolution.* Oxford University Press, pp. 72–95.

Bonnet, X., D. Pearson, M. Ladyman, O. Lourdais, and D. Bradshaw. 2002. "Heaven" for serpents? A mark-recapture study of tiger snakes (*Notechis scutatus*) on Carnac Island, Western Australia. *Austral Ecology* 27:442–450.

Bonnet, T., M. B. Morrissey, P. de Villemereuil, S. C. Alberts, P. Arcese, L. D. Bailey, S. Boutin, *et al.* 2022. Genetic variance in fitness indicates rapid contemporary adaptive evolution in wild animals. *Science* 376:1012–1016.

Brischoux, F., X. Bonnet, and R. Shine. 2007. Foraging ecology of sea kraits (*Laticauda* spp.) in the Neo-Caledonian Lagoon. *Marine Ecology Progress Series* 350:145–151.

Buckley, L. B., and W. Jetz. 2007. Insularity and the determinants of lizard population density. *Ecology Letters* 10:481–489.

Campbell, E. W. III, A. A. Yackel Adams, S. J. Converse, T. H. Fritts, and G. H. Rodda. 2012. Do predators control prey species abundance? An experimental test with brown treesnakes on Guam. *Ecology* 93:1194–1203.

Case, T. J., and D. T. Bolger. 1991. The role of introduced species in shaping the distribution and abundance of island reptiles. *Evolutionary Ecology* 5:272–290.

Cole, N., A. Hector, P. Roopa, R. Mootoocurpen, and M. Goder. 2018. *Casarea dussumieri* (errata version published in 2019). The IUCN Red List of Threatened Species 2018:e. T3989A152276140. http://dx.doi.org/10.2305/IUCN.UK.2018-2.RLTS.T3989A152276 140.en.

Courchamp, F., J.-L. Chapuis, and M. Pascal. 2003. Mammal invaders on islands: Impact, control and control impact. *Biological Reviews* 78:347–383.

Daltry, J. C. 2006. The effect of black rat *Rattus rattus* control on the population of the Antiguan racer snake *Alsophis antiguae* on Great Bird Island, Antigua. *Conservation Evidence* 3:30–32.

Daltry, J. C., K. Lindsay, S. N. Lawrence, M. N. Morton, A. Otto, and A. Thibou. 2017. Successful reintroduction of the Critically Endangered Antiguan racer *Alsophis antiguae* to offshore islands in Antigua, West Indies. *International Zoo Yearbook* 51:97–106.

Daltry, J. C., and G. C. Mayer. 2016. *Alsophis antiguae*. The IUCN Red List of Threatened Species. 2016:e.T939A71739009. http://dx.doi.org/10.2305/IUCN.UK.2016-3.RLTS.T939A71739009.en.

Dorcas, M. E., J. D. Willson, R. N. Reed, R. W. Snow, M. R. Rochford, M. A. Miller, W. E. Meshaka Jr., *et al.* 2012. Severe mammal declines coincide with proliferation of the invasive Burmese pythons in Everglades National Park. *Proceedings of the National Academy of Sciences USA* 109:2418–2422.

Duarte, M. R., G. Puorto, and F. L. Franco. 1995. A biological survey of the pitviper *Bothrops insularis* Amaral (Serpentes, Viperidae): An endemic and threatened offshore island snake of southeastern Brazil. *Studies of Neotropical Fauna and Environment* 30:1–13.

Dueñas, M. A., D. J. Hemming, A. Roberts, and H. Diaz-Soltero. 2021. The threat of invasive species to IUCN-listed critically endangered species: A systematic review. *Global Ecology and Conservation* 26:e01476.

Fattorini, S. 2010. Influence of recent geography and paleogeography on the structure of reptile communities in a land-bridge archipelago. *Journal of Herpetology* 44:242–252.

Fernández-Palacios, J. M., H. Kreft, S. D. H. Irl, S. Norder, C. Ah-Peng, P. A. V. Borges, K. C. Burns, *et al.* 2021. Scientists' warning: The outstanding biodiversity of islands is in peril. *Global Ecology and Conservation* 31 e01847.

Frankham, R. 1997. Do island populations have less genetic variation than mainland populations? *Heredity* 78:311–327.

Frankham, R. 2005. Genetics and extinction. *Biological Conservation* 126:131–140.

Frazetta, T. H. 1970. From hopeful monsters to bolyerine snakes? *American Naturalist* 104:55–72.

Gillespie, R. G., and B. G. Baldwin. 2010. Island biogeography of remote archipelagoes: Interplay between ecological and evolutionary processes. In J. B. Losos and R. E. Ricklefs (eds.), *The Theory of Island Biogeography Revisited*. Princeton University Press, pp. 358–387.

Grismer L. L. 1999. An evolutionary classification of reptiles on islands in the Gulf of California, Mexico. *Herpetologica* 55:446–469.

Grismer L. L. 2001. An evolutionary classification and checklist of amphibians and reptiles on the Pacific islands of Baja California, Mexico. *Bulletin of the Southern California Academy of Sciences* 100:12–23.

Guedes, J. J. M., R. N. Feio, S. Meiri, and M. R. Moura. 2020. Identifying factors that boost species discoveries of global reptiles. *Zoological Journal of the Linnean Society* 190:1274–1284.

Guimarães, M., R. Munguía- Steyer, P. F. Doherty, Jr., M. Martins, and R. J. Sawaya. 2014. Population dynamics of the critically endangered golden lancehead pitviper, *Bothrops insularis*: Stability or decline? *PLoS One* 9:e95203.

Guinea, M. L. 1986. Aspects of the Biology of the common Fijian Sea Snake *Laticauda colubrina* (Schneider). M.Sc. thesis, University of the South Pacific, Suva, Fiji.

Hanebuth, T. J. J., H. K. Voris, Y. Yokoyama, Y. Saito, and J. Okuno. 2011. Formation and fate of sedimentary depocentres on Southeast Asia's Sunda Shelf over the past sea-level cycle and biogeographic implications. *Earth-Science Reviews* 104:92–110.

Harley, C. D. G., A. R. Hughes, K. M. Hultgren, B. G. Miner, C. J. B. Sorte, C. S. Thornber, L. F. Rodriguez, *et al.* 2006. The impacts of climate change in coastal marine systems. *Ecology Letters* 9:228–241.

Harvell, C. D., C. E. Mitchell, J. R. Ward, S. Altizer, A. P. Dobson, R. S. Ostfeld, *et al.* 2002. Climate warming and disease risks for terrestrial and marine biota. *Science* 296:2158–2162.

Heatwole, H. 1975. Predation on sea snakes. In W. A. Dunson (ed.), *The Biology of Sea Snakes*. University Park Press, pp. 233–249.

Heatwole, H. 1999. *Sea Snakes*. University of New South Wales Press.

Heatwole, H. 2019. Isolation, dispersal, and changing sea levels. How sea kraits spread to far-flung islands. In H. B. Lillywhite and M. Martins (eds.), *Islands and Snakes. Isolation and Adaptive Evolution*. Oxford University Press, pp. 45–71.

Heatwole, H., S. Busack, and H. Cogger. 2005. Geographic variation in sea kraits of the *Laticauda colubrina* complex (Serpentes: Elapidae: Hydrophiinae: Laticaudini). *Herpetological Monographs* 19:1–136.

Heatwole, H., H. Lillywhite, A. Grech, and H. Marsh. 2016. Physiological, ecological, and behavioral correlates of the geographic distributions of sea kraits (*Laticauda* spp.): A review and critique. *Journal of Sea Research* 115:18–25.

Heatwole, H., A. Grech, and H. Marsh. 2017. Paleoclimatology, paleogeography, and the evolution and distribution of sea kraits (Serpentes; Elapidae; *Laticauda*). *Herpetological Monographs* 31:1–17.

Henderson, R. W. 2004. Lesser Antillean snake faunas: Distribution, ecology, and conservation concerns. *Oryx* 38:311–320.

Henderson, R. W. 2019. The eyes have it. Watching treeboas on the Grenada Bank. In H. B. Lillywhite and M. Martins (eds.), *Islands and Snakes. Isolation and Adaptive Evolution.* Oxford University Press, pp. 156–180.

Henderson, R. W., R. Powell, J. C. Daltry, and M. L. Day. 1996. *Alsophis antiguae* Parker. *Catalogue of American Amphibians and Reptiles* 632:11–13.

Inger, R. F., and H. K. Voris. 2001. The biogeographical relations of the frogs and snakes of Sundaland. *Journal of Biogeography* 28:863–891.

Jones, J. A., and C. T. Driscoll. 2022. Long-term ecological research on ecosystem responses to climate change. *BioScience* 72:814–826.

Kemp, M. E. 2023. Defaunation and species introductions alter long-term functional trait diversity in insular reptiles. *Proceedings of the National Academy of Sciences USA* 120(7): E2201944119.

Keogh, J. S., I. A. W. Scott, and C. Hayes. 2005. Rapid and repeated origin of insular gigantism and dwarfism in Australian tiger snakes. *Evolution* 59:226–233.

Kier, G., H. Kreft, T. M. Lee, W. Jetz, P. L. Ibisch, C. Nowicki, J. Mutke, and W. Barthlott. 2009. A global assessment of endemism and species richness across island and mainland regions. *Proceedings of the National Academy of Sciences* (USA) 106:9322–9327.

King, R. B., and K. M. Stanford. 2019. Decline and recovery of the Lake Erie Watersnake. A story of success in conservation. In H. B. Lillywhite and M. Martins (eds.), *Islands and Snakes. Isolation and Adaptive Evolution.* Oxford University Press, pp. 258–287.

Kocsis, Á. T., Q. Zhao, M. J. Costello, and W. Kiessling. 2021. Not all biodiversity rich spots are climate refugia. *Biogeosciences* 18:6567–6578.

Kossin, J., K. Knapp, T. Olander, and C. Velden. 2020. Global increase in major tropical cyclone exceedance probability over the past four decades. *Proceedings of the National Academy of Sciences* 117:11975–11980.

Li, J.-L. 1995. *China Snake Island.* Dalian, China: Liaoning Science and Technology Press.

Li, J., L. Sun, X. Wang, H. Bi, L. Wang, C. Wu, Y. F. Yu, *et al.* 2007. Influence of population distribution pattern of Gloydius shedaoensis Zhao on predatory rate. *Journal of Snake* 19:12–16.

Lillywhite, H. B. 2013. Climatic change and reptiles. In K. Rohde (ed.), *The Balance of Nature and Human Impact.* Cambridge University Press, pp. 279–294.

Lillywhite, H. B. 2014. *How Snakes Work. Structure, Function and Behavior of the World's Snakes.* Oxford University Press.

Lillywhite, H. B., and M. Martins (eds.). 2019. *Islands and Snakes. Isolation and Adaptive Evolution.* Oxford University Press.

Lillywhite, H. B., and M. R. Sandfoss. 2023. Drinking behavior and water balance in insular cottonmouth snakes. In D. Penning (ed.), *Snakes: Morphology, Function and Ecology.* Nova Publishers, pp. 261–277.

Lillywhite, H. B., and C. M. Sheehy III. 2019. The unique insular population of cottonmouth snakes at Seahorse Key. In H. B. Lillywhite and M. Martins (eds.), *Islands and Snakes. Isolation and Adaptive Evolution.* Oxford University Press, pp. 201–240.

Lillywhite, H. B., C. M. Sheehy III, and F. Zaidan III. 2008. Pitviper scavenging at the intertidal zone: An evolutionary scenario for invasion of the sea. *BioScience* 58:947–955.

Lillywhite, H. B., J. B. Pfaller, and C. M. Sheehy III. 2015. Feeding preferences and responses to prey in insular neonatal Florida cottonmouth snakes. *Journal of Zoology* 297:156–163.

Lillywhite, H. B., J. G. Menon, G. K. Menon, C. M. Sheehy III, and M.-C. Tu. 2009. Water exchange and permeability properties of the skin in three species of amphibious sea snakes (*Laticauda* spp.). *Journal of Experimental Biology* 212:1921–1929.

Lillywhite, H. B., C. M. Sheehy III, H. Heatwole, F. Brischoux, and D. W. Steadman. 2017. Why are there no sea snakes in the Atlantic? *BioScience* 68:15–24.

Liu, Y.-L., H. B. Lillywhite, and M.-C. Tu. 2010. Sea snakes anticipate tropical cyclone. *Marine Biology* 157:2369–2373.

Lynch, A. J., L. M. Thompson, J. M. Morton, E. A. Beever, M. Clifford, D. Limpinsel, R. T. Magill, *et al.* 2022. RAD adaptive management for transforming ecosystems. *BioScience* 72:45–56.

MacArthur, R. H., and E. O. Wilson. 1963. An equilibrium theory of insular zoogeography. *Evolution* 17:373–387.

MacArthur, R. H., and E. O. Wilson. 1967. *The Theory of Island Biogeography.* Princeton University Press.

Mack, R. N., D. Simberloff, M. Lonsdale, H. Evans, M. Clout, and F. A. Bazzaz. 2000. Biotic invasions: Causes, epidemiology, global consequences and control. *Ecological Applications* 10:689–710.

Marques, O. A. V., M. Martins, and I. Sazima. 2002. A jararaca da Ilha da Queimada Grande. *Ciência Hoje* 31:56–59.

Márquez, C., D. F. Cisneros-Heredia, and M. Yánez-Muñoz. 2017. *Pseudalsophis biserialis.* IUCN Red List of Threatened Species. e.T190541A56253872. http://dx.doi.org/10.2305/ IUCN.UK.2017-2.RLTS.T190541A56253872.en.

Martínez-Morales, M. A., and A. D. Cuarón. 1999. *Boa constrictor,* an introduced predator threatening the endemic fauna on Cozumel Island, Mexico. *Biodiversity and Conservation* 8:957–963.

Martins, M., and H. B. Lillywhite. 2019. Ecology of snakes on islands. In H. B. Lillywhite and M. Martins (eds.), *Islands and Snakes. Isolation and Adaptive Evolution.* Oxford University Press, pp. 1–44.

Martins, M., R. J. Sawaya, S. Almeida-Santos, and O. A. V. Marques. 2019. The Queimada Grande Island and its biological treasure: The golden lancehead. In H. B. Lillywhite and M. Martins (eds.), *Islands and Snakes. Isolation and Adaptive Evolution.* Oxford University Press, pp. 117–137.

Martins, M., R. J. Sawaya, and O. A. V. Marques. 2008. A first estimate of the population size of the critically endangered lancehead, *Bothrops insularis. South American Journal of Herpetology* 3:168–174.

Masunaga, G., T. Kosuge, N. Asai, and H. Ota. 2007. Shark predation of sea snakes (Reptilia: Elapidae) in the shallow waters around the Yaeyama Islands of the southern Ryukyus, Japan. *Marine Biodiversity Records* 1(e96):1–4.

McCleary, R. J. R. 2009. Evolution of venom variation in the Florida cottonmouth, *Agkistrodon piscivorus conanti.* Unpublished doctoral dissertation, University of Florida.

McCue, M. D., H. B. Lillywhite, and S. J. Beaupre. 2012. Physiological responses to starvation in snakes: Low energy specialists. In M. D. McCue (ed.), *Comparative Physiology of Fasting, Starvation, and Food Limitation.* Springer-Verlag, pp. 103–131.

Meek, M. H., E. A. Beever, S. Barbosa, S. W. Fizpatrick, N. K. Fletcher, C. S. Mittan-Moreau, B. N. Reid, *et al.* 2023. Understanding local adaptation to prepare populations for climate change. *BioScience* 73:36–47.

Meiri, S. 2017. Oceanic island biogeography: Nomothetic science of the anecdotal. *Frontiers of Biogeography* 9:e32081.

Miller, M. A., and W. I. Lutterschmidt. 2014. Cutaneous water loss and epidermal lipids in two sympatric and congeneric pitvipers. *Journal of Herpetology* 48:577–583.

Moen, D. S., C. T. Winne, and R. N. Reed. 2005. Habitat-mediated shifts and plasticity in the evaporative water loss rates of two congeneric pit vipers (Squamata, Viperidae, *Agkistrodon). Evolutionary Ecology Research* 7:759–766.

Montes, E., F. Kraus, B. Chergui, and J. M. Pleguezuelos. 2022. Collapse of the endemic lizard *Podarcis pityusensis* on the island of Ibiza mediated by an invasive snake. *Current Zoology* 68:295–30.

Morrison, L. W. 2013. Island flora and fauna: Equilibrium and nonequilibrium. In K. Rohde (ed.), *The Balance of Nature and Human Impact.* Cambridge University Press, pp. 121–132.

Murphy, J. C. 2007. *Homalopsid Snakes: Evolution in the Mud.* Krieger.

Murphy, J. C. 2012. Marine invasions by non-sea snakes, with thoughts on terrestrial–aquatic–marine transitions. *Integrative and Comparative Biology* 52:217–226.

Nilson, G. 2019. The ecology and conservation of the Milos viper, *Macrovipera schweizeri*. In H. B. Lillywhite and M. Martins (eds.), *Islands and Snakes. Isolation and Adaptive Evolution.* Oxford University Press, pp. 181–200.

Ogden, L. E. 2022. The emergence of eDNA. *BioScience* 72:5–12.

Olson, S. L. 1989. Extinction on islands: Man as a catastrophe. In D. Western and M. C. Pearl (eds.), *Conservation for the Twenty-First Century.* Oxford University Press, pp. 50–53.

O'Shea, M. 2021. Foreword: Wallacea: A hotspot of snake diversity in Biodiversity. In Dmitry Telnov (exec. ed.), Maxwell V. L. Barclay, and Olivier S. G. Pauwels (eds.), *Biogeography and Nature Conservation in Wallacea and New Guinea Volume IV.* The Entomological Society of Latvia, pp. 7–17.

Palkovacs, E. P. 2003. Explaining adaptive shifts in body size on islands: A life history approach. *Oikos* 103:37–44.

Patterson, B. D., and W. Atmar. 1986. Nested subsets and the structure of insular mammalian faunas and archipelagos. *Biological Journal of the Linnean Society* 28:65–82.

Paulay, G. 1994. Biodiversity on oceanic islands: Its origin and extinction. *American Zoologist* 34:134–144.

Petrullo, L., S. Boutin, J. E. Lane, A. G. McAdam, and B. Dantzer. 2023. Phenotype-environment mismatch errors enhance lifetime fitness in wild red squirrels. *Science* 379:269–272.

Pike, D. A. R. W. Clark, A. Manica, H.-Y. Tseng, J.-Y. Hsu, and W.-S. Huang. 2016. Surf and turf: Predation by egg-eating snakes has led to the evolution of parental care in a terrestrial lizard. *Scientific Reports* 6:22207.

Piquet, J. C., and M. López-Daria. 2021. Invasive snake causes massive reduction of all endemic herpetofauna on Gran Canaria. *Proceedings of the Royal Society B* 288:20211939.

Pregill, G. K., D. W. Steadman, S. L. Olson, and F. V. Grady. 1988. Late Holocene fossil vertebrates from Burman Quarry, Antigua, Lesser Antilles. *Smithsonian Contributions to Zoology* 463:1–47.

Pyron, R. A., and F. T. Burbrink. 2014. Ecological and evolutionary determinants of species richness and phylogenetic diversity for island snakes. *Global Ecology and Biogeography* 23:848–856.

Rabatsky, A. 2006. Rattle reduction and loss in rattlesnakes endemic to islands in the Sea of Cortés. *Sonoran Herpetologist* 19:80–81.

Reed, R. N., S. M. Boback, C. E. Montgomery, S. Green, Z. Stevens, and D. Watson. 2007. Ecology and conservation of an exploited insular population of *Boa constrictor* (Squamata: Boidae) on the Cayos Cochinos, Honduras. In R. W. Henderson and R. Powell (eds.), *Biology of the Boas and Pythons.* Eagle Mountain Publishing, pp. 389–403.

Rodda, G. H., T. H. Fritts, and P. J. Conry. 1992. Origin and population growth of the brown tree snake, *Boiga irregularis,* on Guam. *Pacific Science* 46:46–57.

Rodríguez-Robles, J. A., Jezkova, T., Fujita, M. K., Tolson, P. J., and García, M. A. 2015. Genetic divergence and diversity in the Mona and Virgin Islands boas, *Chilabothrus monensis* (*Epicrates monensis*)(Serpentes: Boidae), West Indian snakes of special conservation concern. *Molecular Phylogenetics and Evolution* 88:144–153.

Sagonas, K., P. Panayiotis, and E. D. Valakos. 2015. Effects of insularity on digestion: Living on islands induces shifts in physiological and morphological traits in island reptiles. *Science of Nature* 102:1–7.

Sajdak, R. A., and R. W. Henderson. 1991. Status of West Indian racers in the Lesser Antilles. *Oryx* 25:33–38.

Salles-Oliveira, I., Machado, T., Banci, K. R. D. S., Almeida-Santos, S. M., and Silva, M. J. D. J. 2020. Genetic variability, management, and conservation implications of the critically endangered Brazilian pitviper *Bothrops insularis*. *Ecology and Evolution* 10:12870–12882.

Sandfoss, M. R., C. M. Sheehy III, and H. B. Lillywhite. 2018. Collapse of a unique insular bird-snake relationship. *Journal of Zoology* 304:276–283.

Sathiamurthy, E., and H. K. Voris. 2006. Maps of Holocene sea level transgression and submerged lakes on the Sunda Shelf. *Natural History Journal of Chulalongkorn University* suppl. 2:1–43.

Savidge, J. A. 1987. Extinction of an island forest avifauna by an introduced snake. *Ecology* 68:660–668.

Sayre, R., S. Noble, S. Hamann, R. Smith, D. Wright, S. Breyer, K. Butler, *et al.* 2018. A new 30 meter resolution global shoreline vector and associated global islands database for the development of standardized global ecological coastal units. *Journal of Operational Oceanography: A Special Blue Planet Edition.* doi:10.1080/1755876X.2018.1529714.

Schwaner, T. D., and S. D. Sarre. 1988. Body size of tiger snakes in southern Australia, with particular reference to *Notechis ater serventyi* (Elapidae) on Chappell Island. *Journal of Herpetology* 22:24–33.

Seaman, G. A., and J. E. Randall. 1962. The mongoose as a predator in the Virgin Islands *Journal of Mammalogy* 43:544–546.

Sheehy III, C. M., J. B. Pfaller, H. B. Lillywhite, and H. F. Heatwole. 2011. *Pelamis platurus.* Predation. *Herpetological Review* 42:3.

Shine, R. 1987. Ecological comparisons of island and mainland populations of Australian tiger-snakes (*Notechis*: Elapidae). *Herpetologica* 43:233–240.

Shine, R., L.-X. Sun, M. Fitzgerald, and M. Kearney. 2003. A radiotelemetric study of movements and thermal biology of insular Chinese pit-vipers (*Gloydius shedaoensis*, Viperidae). *Oikos* 100:342–352.

Shine, R., L.-X. Sun, E. Zhao, and X. Bonnet. 2002. A review of 30 years of ecological research on the Shedao pitviper, *Gloydius shedaoensis. Herpetological Natural History* 9:1–14.

Shine, R., L. X. Sun, M. Kearney, and M. Fitzgerald. 2002. Thermal correlates of foraging-site selection by Chinese pit-vipers (*Gloydius shedaoensis*, Viperidae). *Journal of Thermal Biology* 27:405–412.

Somaweera, R., V. Udyawer, M. L. Guinea, D. M. Ceccarelli, R. H. Clarke, M. Glover, M. Hourston, *et al.* 2021. Pinpointing drivers of extirpation in sea snakes: A synthesis of evidence From Ashmore Reef. *Frontiers in Marine Science* 8:658756.

Todd, B. D., J. D. Willson, and J. W. Gibbons. 2010. The global status of reptiles and causes of their decline. In Sparling, D. W., C. A. Bishop, and S. Krest (eds.), *Ecotoxicology of Amphibians and Reptiles*, 2nd ed. CRC Press, pp. 47–67.

Vitousek, P. M., C. M. Dantonio, L. L. Loope, M. Rejmanek, and R. Westbrooks. 1997. Introduced species: A significant component of human-caused global change. *New Zealand Journal of Ecology* 21:1–16.

Voris, H. K. 2000. Maps of Pleistocene sea levels in Southeast Asia: Shorelines, river systems and time durations. *Journal of Biogeography* 27:1153–1167.

Wallach, V. 2009. *Ramphotyphlops braminus* (Daudin): A synopsis of morphology, taxonomy, nomenclature and distribution (Serpentes: Typhlopidae). *Hamadryad* 34:34–61.

Wallington, T. J., R. J. Hobbs, and S. A. Moore. 2005. Implications of current ecological thinking for biodiversity conservation: A review of the salient issues. *Ecology and Society* 10:15.

Wang, T., C. C. Y. Hung, and D. J. Randall. 2006. The comparative physiology of food deprivation: From feast to famine. *Annual Review of Physiology* 68:223–251.

Wen, G., L. Jin, Y. Wu, X. Wang, J. Fu, and Y. Qi, 2022. Low diversity, little genetic structure but no inbreeding in a high-density island endemic pit-viper *Gloydius shedaoensis, Current Zoology* 68:526–534.

Wetzel, F. T., H. Beissmann, D. J. Penn, and W. Jetz. 2013. Vulnerability of terrestrial island vertebrates to projected sea-level rise. *Global Change Biology,* 19:2058–2070.

Wharton, C. H. 1969. The cottonmouth moccasin on Sea Horse Key, Florida. *Bulletin of the Florida State Museum, Biological Science* 14:227–272.

Whittaker, R. J. 1998. *Island Biogeography*. Oxford University Press.

Williams, R. J., T. Ross, M. Morton, J. Daltry, and L. Isidore. 2016. Update on the natural history and conservation status of the Saint Lucia racer, *Erythrolamprus ornatus* Garman, 1887 (Squamata, Dipsadidae). *Herpetology Notes* 9:157–162.

Wooten, D. A. 2020. Trophic ecology of Seahorse Key, Florida: A unique bird-snake interaction network analysis. *American Midland Naturalist* 184:177–187.

Wüster, W., M. R. Duarte, and M. G. Salomão. 2005. Morphological correlates of incipient arboreality and ornithophagy in island pitvipers, and the phylogenetic position of *Bothrops insularis*. *Journal of Zoology* 266:1–10.

World Wildlife Fund (WWF). 2022. *Living Planet Report 2022: Building a nature-positive society*. R. E. A. Almond, M. Grooten, D. Juffe Bignoli, and T. Petersen (eds.). WWF.

Yan C., W. Wu, W. Dong, B. Zhu, J. Chang, Y. Lv, S. Yang, and J.-T. Li. 2022. Temperature acclimation in hot-spring snakes and the convergence of cold response. *The Innovation* 3(5), 100295.

Yuan, F. L., S. Ito, T. P. N. Tsang, T. Kuriyama, K. Yamasaki, T. C. Bonebrake, and M. Hasegawa. 2021. Predator presence and recent climatic warming raise body temperatures of island lizards. *Ecology Letters* 24:533–542.

Taxonomic Index

For the benefit of digital users, indexed terms that span two pages (e.g., 52–53) may, on occasion, appear on only one of those pages.

Tables and figures are indicated by *t* and *f* following the page number

Subject Index

For the benefit of digital users, indexed terms that span two pages (e.g., 52–53) may, on occasion, appear on only one of those pages.

Tables and figures are indicated by *t* and *f* following the page number